Paralysed with Fear

The Story of Polio

Gareth Williams
Professor of Medicine, Faculty of Medicine & Dentistry,
The University of Bristol

Illustrations by Ray Loadman

palgrave
macmillan

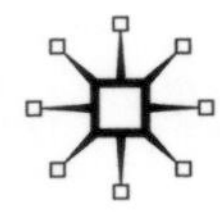

First published 2013 by
PALGRAVE MACMILLAN

Palgrave Macmillan in the UK is an imprint of Macmillan Publishers Limited, registered in England, company number 785998, of Houndmills, Basingstoke, Hampshire RG21 6XS.

Palgrave Macmillan in the US is a division of St Martin's Press LLC, 175 Fifth Avenue, New York, NY 10010.

Palgrave Macmillan is the global academic imprint of the above companies and has companies and representatives throughout the world.

Palgrave® and Macmillan® are registered trademarks in the United States, the United Kingdom, Europe and other countries

ISBN: 978–1–137–29975–8

This book is printed on paper suitable for recycling and made from fully managed and sustained forest sources. Logging, pulping and manufacturing processes are expected to conform to the environmental regulations of the country of origin.

A catalogue record for this book is available from the British Library.

A catalog record for this book is available from the Library of Congress.

Let me assert my firm belief that the only thing we have to fear is fear itself.

President Franklin D. Roosevelt (1882–1945),
First Inaugural Address to the American Nation,
4 March 1933

In memory of my parents, Joan and Alwyn

And to Caroline, Tim, Jo, Sally and Pippa

June and Geoff *

* All royalties from sales of this book will help to support the Edward Jenner Museum at Dr. Jenner's House, Berkelely, Gloucestershire (UK registered charity number 284085). Visit "http://www.jennermuseum.com" www.jennermuseum.com

Contents

List of Figures

Acknowledgements

The discerning reader may be troubled to see 'shades of grey' mentioned on page 17. To explain: this is an innocent play on words (*polio* being the Greek for 'grey'), which was written several months before a best-selling book with a similar name burst on to the market.

My long-suffering family will confirm that *Paralysed with Fear* also involved bondage and masochism – but then writing a book always does. Luckily, this one has been great fun too, at least for me. So my thank-you list begins with Caroline (my wife), Tim and Jo (too big now to be described as children), and Sally and Pippa (the dogs). As always, their tolerance and encouragement ('Aren't you nearly there yet?') have been much more than I deserved. I still don't know how they fight the temptation to turn me over to Social Services and the RSPCA for neglecting them so badly. As always, Caroline has been a soft target for early drafts. One of these affected her so profoundly that she overshot Bristol on the train and came back to reality somewhere in Wales. This was a helpful prompt to rewrite the section that had put her to sleep.

Next in line for thanks is a group of friendly critics who have read more versions of the manuscript than can be good for their health: Jenifer Roberts, Paul Beck, Bob and Tracy Spencer, Ray and Jeanne Loadman, Kathryn and Freyja Atkins, Cherry Lewis, Martine Fisbach and Thomas Berglindh. They are all avid readers, with additional skills that include writing, infectious diseases, science and running an independent bookshop. They can spot waffle and a misplaced apostrophe at 50 paces, and they have made a formidable Hanging Committee. I am indebted to them for pointing out, firmly and often humanely, all the things that have needed fixing.

Two members of the Hanging Committee deserve special mention for services beyond the call of duty. Four years ago, I was lucky to have tempted Ray Loadman away from his acrylics long enough to illustrate *Angel of Death*. I was delighted when he agreed to do the same for *Paralysed with Fear*. As well as being a superb artist, Ray has the magician's touch needed to transform a sow's ear (my rough sketches) into a silk purse, namely all the drawings which add elegance and style to this book. I have also been extremely fortunate to have had Freyja Atkins working with me as research assistant. Freyja's placid exterior hides an inner persona that combines all the best qualities of bloodhound and Rottweiler. She has worked wonders in chasing up elusive references, images and permissions. Without her help, I doubt that I would have made the deadline – or indeed survived it.

In the same vein, I'm deeply grateful to several experts: Stanley Plotkin, for helping me to navigate between fact and fiction; Barry North, for guidance on topics ranging from *This Is Your Life* to Sister Kenny; Helma Ruijs, for putting me right about football and much else in the Dutch Bible Belt; Preben Berthelsen, for insights into anaesthesia and Copenhagen, 1952; the Ulster collective of Sheila Patrick, Jimmy McCallister, Hugh O'Neil, Margaret Haire and Alan Trudgett, for memories of Belfast *ca* 1956; and Dominic Lopez-Real, for many helpful comments.

With backup like that, this book should be completely error free. If any mistakes have crept through, they can only be my fault – so if you spot anything wrong, please let me know and I shall put it right.

It is impossible, especially now, to imagine what it could have been like to have polio. That void has been bridged, in a powerful and moving way, by personal accounts of living with polio, and I am indebted to those who were prepared to share their experiences with us: John and Margaret Johnston, Bristol; Margaret Scrimgeour, London; and Zoltán Lipey Rózsa and Iván Szánti, Budapest. Zoltán and Iván were interviewed by Anna Tóth and Magda Pribojszki, third-year medical students at Szeged University, and my thanks go to them and to Ferenc Bari for having arranged this.

In November 2011, I visited the United States and had the good fortune to meet an outstanding array of experts in the fields of polio, vaccines and public health. My grateful thanks are due to the Wellcome Trust for the travel grant that made the trip possible, and to the following, for their time and wisdom: the late Hilary Koprowski (and his team, led by the ever-energetic Sue Jones), DA Henderson (with additional thanks to DA and Nana, for their hospitality), Neal Nathanson, Neal Halsey, Jason Schwartz and Al Sommer.

In this digital age, much of the scientific literature is available within milliseconds through the Internet. Happily, though, there is still a need for real archives and those who guard them. My gratitude, for their warm welcome as well as their expertise and help, goes to the late Marshall Barr, Tim Smith and Carol Barton, Berkshire Medical Heritage Centre, Reading; Annie Brogan, Mutter Museum and Library, College of Physicians, Philadelphia; Barry North, British Polio Fellowship; Katerina Petruskova, World Health Organisation, Geneva; Jane Sweetland, Burden Neurological Institute, Bristol; Danielle Seilhean and Réni Sabathier, Bibliothèque Charcot, ICM, Paris; Jean-François Vincent, Bibliothèque Interuniversitaire de Santé, Paris; Jonas Christian Ryborg, Ny Carlsberg Glyptotek, Copenhagen; Nicole Babcock, Mayo Clinic Historical Unit,

Rochester, Minnesota; Anna Dysert, Osler Library of the History of Medicine, McGill University, Montreal; Ed Jackson, University of Georgia, Athens; Georgia; David Wencer, Hospital for Sick Children Archives, Toronto; Colin Brown and Selina Hurley, Science Museum, London; Keith Manchester and Alan Humphries, Thackray Medical Museum, Leeds; Rebecca Winson, *British Medical Journal*; Patsy Williams, Gladstone Library, Hawarden; Sandra Landberg, Sister Kenny Rehabilitation Institute, Minneapolis; Margaret Hogan, Rockefeller Archive Center, New York; Kay Peterson, National Museum of American History, Smithsonian Institute, Washington DC; Crestino Forcina, Wellcome Library, London and Jan Leach and Glenys Hannam, Dr Jenner's House, Berkeley.

Many friends have shared the suffering while the book was in gestation and have greeted its safe delivery with relief. My thanks to them for nudging me along: June, my mother-in-law; Tim and Julie Mann; Colin and Angela Gardner; Santi Rodriguez, Sue Kriefman, David Jackson and Chris Cooper; Tim Jones and Alison Paton; and the other 80% of the Rockhampton Wind Quintet: Sophie Cottrell, Charlie Tomson, Margaret Putin and Chris Pople.

It is always sensible to say thank you to your publishers, even if you're not sure what they have done or will do. In this instance, there is no doubt whatsoever. The team at Palgrave Macmillan did a brilliant job with *Angel of Death*, and exceeded my expectations with this book. They have provided encouragement, inspiration and the sort of gentle reminders that have you sweating at your desk before dawn – and all with great humour and a sense of fun that has kept up the momentum through the inevitable dark moments. I'm indebted to Jenny McCall, publisher extraordinaire, wise counsel and diplomat; Holly Tyler, Claire Morrison and Verity Holliday, for keeping me on the straight and narrow; and Oscar Spigolon, for superhuman patience and a fantastic cover design; and the excellent production team led by Matt Aldridge and Vidhya Jayaprakash.

Finally, my thanks are due to my parents, Joan and Alwyn, for nature, nurture, curiosity and a love of writing. Sadly, neither of them saw this book. My mother read an early draft and (having taken her red pen to the manuscript) said that she'd enjoyed it. However, she thought it probably would not sell as well as that other book about 'shades of grey'.

A Gentle Introduction

It was S– who first gave me the idea of writing this book, on 16 February 2011. We were in Geneva and I had just given a talk about the history of smallpox. 'You should look into polio', he said. 'It's a gripping story, full of twists and turns. And if you're quick, you might get a book out just before it disappears forever. It should be eradicated by the end of next year'.

As you will know, that prediction was close, but not close enough; S– was certainly right about those twists and turns. His broader view has also turned out to be accurate. This is a great story, with a powerful storyline, a rich cast of heroes and villains and a nail-biting final chapter that really deserves to resolve into a happy ending – but may not.

Polio is now in an odd state of limbo. Vaccination finally wiped it out from the Western Hemisphere a generation ago, and it is a rapidly receding memory across most of the rest of the planet. Even in its last boltholes, in northern Nigeria, Afghanistan and Pakistan, polio paralyses only a handful of people each year. When I qualified in medicine in 1977, polio was already a fine-print item of largely historical interest. Most Western-trained doctors of my generation never saw an acute case, or laid eyes (let alone hands) on an iron lung. And for the public at large, polio is just a name from the ritual of childhood vaccination that means as little as tetanus or diphtheria. So what's all the fuss about?

Wind the clock back 60 years to America in the early 1950s – before Jonas Salk's vaccine proved that polio could be beaten – and things look entirely different. Here, polio is a terrifying plague that cuts down tens of thousands of children every summer. Nobody's home is safe, and nobody can predict where polio will strike that year; the only certainty is that it will. What keeps Americans awake at night? Top of the list: the fear of nuclear warfare. Close behind: polio.

But here lies a paradox. Other countries, hit just as hard as the United States, seem to take polio in their stride. After all, in terms of headcount, polio lies far down the list of infections that kill or maim. So what is going on in America?

The answer has little to do with the polio virus. Instead, the fear of polio is deliberately played up, by an organisation prepared to steamroller its way through American society to achieve its goal of conquering polio. This is not an organisation to argue with, because the man at the top is in the White House. He is another of the paradoxes of polio: one of the most powerful men on the planet, yet he refuses to let himself be photographed in the wheelchair to which he was consigned by polio.

America was the centre of gravity for the science and medicine of polio during most of the 20th century, but a much broader canvas is needed to paint the picture as a whole. Following the history of polio takes us all over the world: from the Canadian Arctic in mid-winter to the Australian outback, via Peru, the Dutch Bible Belt and the cold-war USSR. Along the way, we will hear stories that enthrall and move us, from a 12-year-old girl in Copenhagen with just minutes to live, to the 2-year-old Pakistani boy who was paralysed because the Taliban had threatened to kill anyone in his village who had their children vaccinated against polio.

Science has – in theory at least – solved the problem of polio. The process of discovery that led to effective polio vaccines is often portrayed as logical and linear: brilliant scientist has an idea, does experiments, finds the answer, cures polio and wins Nobel Prize. However, medical and scientific journals from the first half of the 20th century tell a different story – tortuous, unexpected and fascinating. The gems of scientific endeavour are scattered across an odd terrain of good, bad and dreadful research. Here, we find blind alleys, red herrings and experiments that should never have been done, let alone published. Among those with feet of clay (which in some cases extends to well above the knee) are some of the most powerful scientists of the day – including men who could, and did, block research on polio vaccines for nearly 30 years.

It is easy to be critical in hindsight. In fairness, we have to acknowledge that these were desperate times that demanded desperate remedies. Why else would an American professor take a red-hot poker to the back of a child paralysed by polio?

We are now on the brink of ridding the planet of polio for all time, but it is distinctly possible that we shall fail. The final obstacles are acts of man, rather than anything to do with the poliovirus. The virus is blameless, as all it does is to follow the orders stamped into its genetic code. People are infinitely more sophisticated.

At the time of writing, the vaccination campaigns to clear polio out of Pakistan, Afghanistan and northern Nigeria had all been halted, following the murder of vaccination workers by Islamist extremists. Most of the victims were women; several were schoolgirls. Even against a background of daily bombings and shootings, these atrocities stand out for their brutality and vindictiveness. As collateral damage, millions of children in those regions have been robbed of the chance to be protected against an infection that is both cruel and eminently preventable.

This brings me back to S–. His presence in Geneva in February 2011 was no coincidence, as he works there with the World Health Organisation's polio eradication initiative. He has to remain anonymous because he regularly visits the front lines of the campaign, and cannot afford to take chances. When I thought up the title for this book, I had no idea that it would have such a strong resonance in 2013.

Finally, I have a personal interest to declare. My family has a brief walk-on role in Chapter 10, including (characteristically) a speaking part for my mother. What happened then was trivial beside the rest of the story, but it was one of the things that nudged me into medicine.

Returning to the subject over half a century later, I can only agree with S–. This is indeed a gripping story, full of twists and turns. I hope I have done it justice.

1

A Plague from Nowhere

The victim, known only by his initials, 'GS', was ten years old, sporty and in perfect health. For the family holiday during the summer of 1909, his parents had rented a lakeside cabin in the Laurentian Mountains, four hours' drive north-west of their home in Montreal. The boy was active and happy, spending his days swimming and boating with friends.

On the morning of 5 August, he complained of a headache and feeling hot. He stayed indoors for the rest of the day, but seemed better when he went to bed that evening.

The next morning, his mother was woken by panic-stricken screaming from his bedroom. Both his legs were completely paralysed. By the time the doctor arrived, his left arm was also losing its strength. There was little that the doctor could do, other than confirm the parents' greatest fear and chart the progression of the illness. The next day, the boy couldn't lift either arm off the bed. His breathing became more and more laboured as the paralysis crept up to the diaphragm and the muscles of the rib-cage. During the afternoon of 9 August, he fought for his life, now with only the muscles of the shoulder girdle and neck struggling to drag air into his lungs. That evening, exhaustion set in, and his lips turned blue with the lack of oxygen. He died quietly a couple of hours later, four and a half days after developing his first symptoms.

His parents drove him home during the night. A post-mortem was carried out in their house the following morning, ten hours after death. The boy's brain was bloated and looked too large to fit back inside the skull, but the main abnormality was in the spinal cord. It was swollen, bulging through the filigree of membranes on its surface as though it had been trussed up with cotton thread, and it felt hard, almost like a fat lead pencil. When it was sliced across with a scalpel, two symmetrical pinkish areas could be seen on either side of the midline, towards the front.

Later that day, thin sections of the boy's spinal cord were examined under the microscope. This confirmed the diagnosis that his mother had made as soon as she had heard him screaming that he couldn't move his legs. The pinkish areas visible to the naked eye were patches of inflammation, stamped across the butterfly pattern of the grey matter that occupies the centre of the cord. The forewings of the butterfly – the 'anterior horns' of the grey matter – had been obliterated, together with the large 'motoneurone' nerve cells which normally reside there. The motoneurones give rise to the motor nerves which power the muscles.

The diagnosis was acute anterior poliomyelitis, better known at the time as 'infantile paralysis' and familiar to us today as 'polio'.[1]

This family tragedy was repeated thousands of times that summer, because polio was tightening its hold on North America. Just 25 years earlier, polio had been a fine-print medical curiosity which featured mainly in obscure articles from Sweden, Germany and France. Americans first took notice in the summer of 1894, when a polio outbreak in Vermont paralysed 130 children and killed 18 of them.[2] After the turn of the century, polio gained momentum and settled into a pattern of annual outbreaks during the 'polio season' of the summer months. These culminated in 1916 with a major epidemic that hit New York City, then swept down the East Coast and swung west towards the heart of the continent. In all, 27,000 people were paralysed, and 6,000 died, most of them previously healthy children.[3]

Polio rapidly became every American parent's worst nightmare: a horrific disease which killed and maimed children, which could break into any house without warning and which could not be prevented or treated. The result was widespread fear which boiled up into panic at the approach of each year's polio season and continued to paralyse the American public for the next half-century.

In numerical terms, polio was never one of the great scourges of mankind. Even the common childhood infections such as measles, whooping cough and diphtheria regularly eclipsed the death toll of the great New York epidemic of 1916. And a couple of years later, influenza tore through every continent on the planet and killed 25 million people, roughly three times more than all the casualties of the World War which ended in that year.[4]

Yet polio punched far above its weight in its power to terrify, because it could pick off children from inside the cleanest, most secure household. If smallpox was a mass murderer, then polio was a sniper, and all the more menacing because nobody could see where the fatal shots had come from. Panic drove rational people to do bizarre things to protect themselves and their loved ones. Swimming pools, churches and cinemas were closed during the polio season; children were prevented from using public transport or even leaving their homes; and public funerals of polio victims were banned. The medical authorities also reacted with desperation. To try to prevent polio from spreading, doctors and terrified parents sprayed toxic chemicals up the nostrils of thousands of American and Canadian children, while American cities were doused with DDT.

Fear of the disease was all the greater because its cause and means of spread were a mystery. Some scientists were confident that they had found a virus that caused it, while others claimed that bacteria, toxins and even milk and fresh fruit were responsible. On the street, people blamed swimming, Italian immigrants and cats – prejudices that were shared and spread by doctors. The only solution was to get away from all possible sources of infection, and the wealthy began to desert the cities at the start of each polio season. But as the family from Montreal found out, even the idyllic isolation of the Laurentian Lakes was no guarantee of protection.

Naming of parts

> The name of a disease is always a matter of some importance. It should be short for the sake of convenience in writing, and euphonious for ease in pronunciation.
>
> Henry Veale, Scottish military physician, 1866

'Polio' is short for 'poliomyelitis', which slips less easily off the tongue. The name is a fusion of two Greek words, *polios* meaning 'grey' and *myelos* meaning 'spinal cord' (*myelos* is also used for 'bone marrow'). The ending '–itis', familiar from appendicitis and tonsillitis, indicates inflammation. 'Poliomyelitis' therefore describes the abnormalities seen down the microscope – 'inflammation of the grey matter of the spinal cord'. The term was coined in 1847 by the German pathologist Albert Kussmaul and was rapidly adopted. Twenty-five years later, the great French neurologist Jean-Martin Charcot tried to introduce his own term, 'tephromyelitis', from *tephros*, another Greek word for 'grey'. If it had caught on, this book would have been subtitled *The story of tephro*.[5]

Polio first entered the medical literature in 1789 with the cumbersome title of 'debility of the lower extremities'.[6] This term slipped neatly into German half a century later as *Lähmungzustände der unteren Extremitäten*, but was soon replaced by *Kinderlähmung*, which translated back into English as 'infantile paralysis'.[7,8] At the time, this designation seemed entirely appropriate, as older people appeared to be miraculously spared; in 1858, an otherwise classic case in a 50-year-old Swiss man was written up as the first ever recorded in an adult.[9] The name remained fashionable into the 1930s – as in America's National Foundation for Infantile Paralysis, founded in 1938. But by then, polio was obviously attacking adults as well as children, and a non-ageist alternative was needed. 'Poliomyelitis', the literal description of the lesions seen down the microscope, gradually took over as the name of the clinical disease as well. This was inevitably contracted to 'polio', especially when survivors of the infection began calling themselves 'the polios'.

Polio has had other, more figurative names. 'Morning paralysis' was a reminder of how abruptly a healthy child could be struck down by the curse which apparently dropped out of the night sky.[10] In 1907, the Swedish epidemiologist-detective Ivar Wickman used the name 'Heine-Medin disease' to honour his two personal heroes, Jacob Heine and Oscar Medin.[11] More bluntly, polio was known as 'The Crippler'. This term grates on us today, but the word was used widely and non-pejoratively up to the 1950s; researchers fought to have papers published in the *Journal of the Crippled Child*, while rich benefactors were proud to perpetuate their names through institutions such as the Betty Bacharach Home for Crippled Children. One of those who referred to The Crippler was someone who knew all about it: Franklin D. Roosevelt, President of the United States and polio victim. But it could have been worse. The North American Indians' name for smallpox was 'rotting face'.[12]

Polio causes paralysis because the inflammation in the spinal cord kills off the 'motoneurones' which are packed into the anterior horns of the grey matter. The motoneurones are large nerve cells which power the 'motor' nerves that supply the muscles and make them contract. They can be damaged by several viruses and many natural or synthetic poisons. By far the commonest culprit is an 'enterovirus', which often infects the human bowel but only rarely causes any problems. The enteroviruses are among the smallest and simplest of all viruses; they are also some of the prettiest, as they have a complex symmetrical shape like tiny mineral crystals.

The guilty enterovirus is now called the 'poliovirus'. The suffix 'hominis' is sometimes added as a reminder that this virus only infects us (and a few subhuman primates such as the chimp and gorilla). During the 1950s, the sturdy souls who classify viruses came up with a cryptic, fake Latin name for the virus: *Legio debilitans*, meaning 'the legion that weakens'.[13] This looked posh but added nothing. Mercifully, it was quickly abandoned.

Not noticed by any medical writer

For such a striking disease, polio was slow to impress medical men. Twenty-five years elapsed before the first description was followed up, and the next half-century saw barely 100 cases reported. To give the polio detectives their due, they were dealing with a suspect which covered its tracks remarkably well, and which adapted its pattern of attack in response to changes in the behaviour of its prey.

To begin at the beginning: the history of polio is virtually blank until the eighteenth century. The Bible mentions weakness of the limbs, but there are no clear accounts of children being suddenly struck down and unable to walk.[14] This seems a wasted opportunity, as polio would have made an ideal weapon for the Almighty to terrorise the opposition and keep His own flock under control. Similarly, the collected writings of the physicians of Ancient Arabia, Greece, Rome, India and China do not contain convincing descriptions of polio-like illnesses – in contrast to their accounts of afflictions such as diabetes, plague and measles, which are still recognisable today.

Indirect evidence suggests that a paralytic disease resembling polio might have cropped up early in human history, even if rarely. Skeletons of adults, some dating back to the Neolithic and the Bronze Age (respectively 8,000 and 4,000 years ago), have been found that show marked shortening and underdevelopment of one leg (Figure 1.1). This is the hallmark of complete paralysis of the limb in early childhood, although diseases other than polio could be responsible.[15] The most famous ancient image believed to depict polio is in a painted Egyptian frieze, dating from the Eighteenth Dynasty (1570–1342 BC), and now in the Ny Carlsberg Glyptotek in Copenhagen.[16] The frieze shows Ruma, guardian of the Temple of the goddess Astarte in Memphis, supporting himself on a stick; his right leg is withered and shortened, with the foot dropped in the 'equinus' posture characteristic of polio (Figure 1.2). However, retrospective diagnosis can be tricky, especially after 3,500 years, and the appearance could be due to something else. The clubfoot of the mummified Pharaoh Siptah from the Nineteenth Dynasty (1342–1197 BC) was originally diagnosed as polio, but is now thought to be a birth defect.[17]

The first convincing reports of polio did not appear until the eighteenth century, by which time infections such as plague, smallpox and tuberculosis were all deeply embedded in medical practice around the world. In December 1734, Jean-Godefroy Salzmann submitted a 50-page dissertation to the University of Strasbourg for the degree of Doctor of Medicine.[18] Entitled *A defect of many muscles of the foot*, it described a diagnostic conundrum: a previously healthy boy who suddenly lost power in both legs and was left with a paralysed and wasted right leg. Lapsing into depression and alcohol abuse, the patient came to post-mortem at the age of 40. Salzmann noted that the muscles of the right leg were shrivelled and replaced by pale fatty tissue; he blamed an imbalance in 'tension' between the legs, but noted that the patient's fondness for brandy, liberally applied externally and internally, would not have helped. Salzmann's report highlighted the key features of polio, but he did not invent a name for the condition. His case was cited in a textbook on orthopaedic surgery in 1743 and then forgotten.[19]

The man generally credited with putting polio on the map was Michael Underwood (1738–1810), a multitalented London physician.[20] Underwood became famous for his comprehensive *Treatise on diseases of*

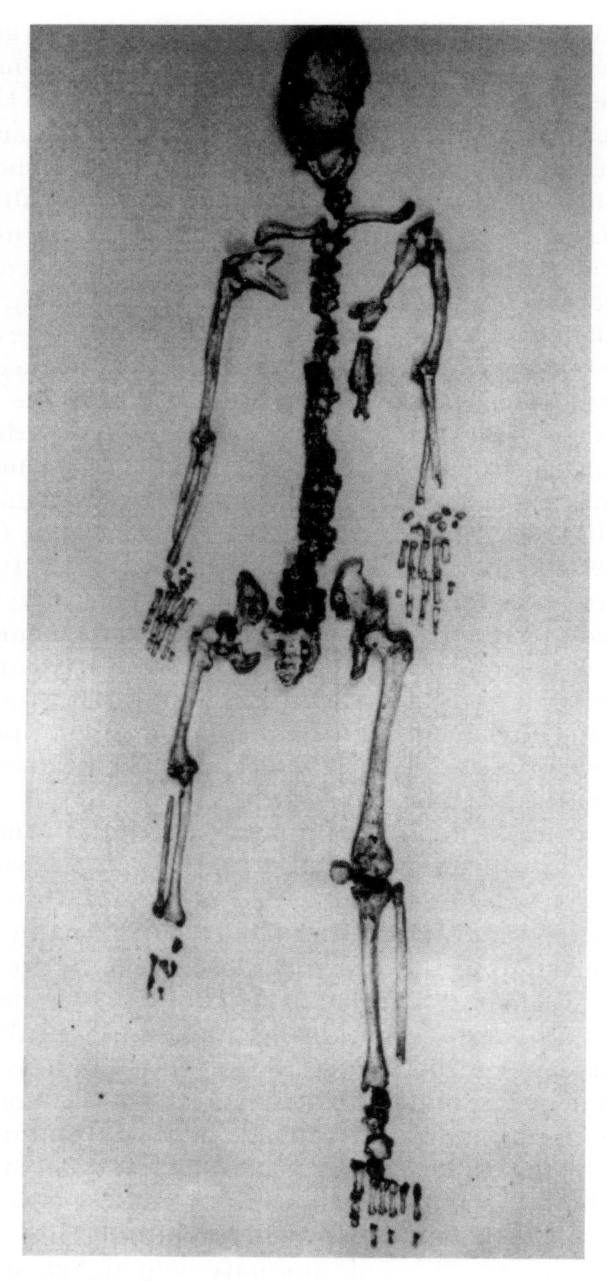

Figure 1.1 Adult human skeleton, showing marked shortening and growth failure of the right leg, consistent with polio or another paralysing disease of childhood. Medieval (8–10 centuries AD), from Raunds, Northamptonshire, England. Reproduced by kind permission of Dr Keith Manchester, Bradford

Figure 1.2 Ruma, guardian priest of the Temple of Astarte in Memphis, whose right leg shows deformities typical of polio. Egyptian funerary stele from the Eighteenth Dynasty (1570–1342 BC). Ruma is depicted with his wife, Amao, and their son, offering a prayer to Astarte to rest Ruma's soul. Reproduced by kind permission of the Ny Carlsberg Glyptotek, Copenhagen

children, but also strayed into the forbidden territories of surgery and midwifery; he was fined 20 guineas by the Company of Surgeons for daring to write a book about operations to treat leg ulcers. The second edition of Underwood's *Treatise* (1789) contained a new entry, 'Debility of the lower extremities', describing paralysis of the legs in children aged between one and four years.[6] This was a rare disease, especially in London, and as far as Underwood knew, it 'is not noticed by any medical writer within the compass of my reading, or is not so described as to ascertain the disease'.

Underwood noted that paralysis was often preceded by fever and thought that it arose from 'debility'. There were no signs of worms or 'other foulness of the bowels', but he had not had the chance to 'examine the body of any child who had died of this complaint', and so was not prepared to speculate about its possible cause. He believed that a couple of cases had regained some strength, thanks to his treatment. This included the routine therapeutic miseries of the day, notably 'caustics' to blister the skin over the sacrum and hips. Underwood's description crossed the Atlantic in 1793 and passed unchanged through over 20 editions of the *Treatise* – even though some of his cases were probably due to tuberculosis of the spine.[21,22]

Further intelligence about this 'new' disease was slow to come in over the next half-century. It seemed to be sporadic and rare, even to those specialising in diseases of childhood. During his 30-year career as a surgeon, physician and obstetrician, Giovanni Battista Monteggia (1782–1815) managed to collect several cases. In his eight-volume *System of surgery* (1813), Monteggia described a form of paralysis of one or both legs that appeared suddenly in infants, heralded by fever and diarrhoea. Like Underwood, Monteggia thought that this disorder had not previously been reported.[23]

The first coherent description of polio, which cemented its reputation as a real disease, came in 1840 from Jacob Heine (1800–79; Figure 1.3). Heine practised as an orthopaedic surgeon in Stuttgart and had built up a cottage industry specialising in braces and walking machines for the lame. Heine's 80-page dissertation on 'Debility of the lower extremities' was based on only 14 cases, but all meticulously described. Victims were generally aged six months to three years and in good health until paralysis struck, after a couple of days of mild fever. Usually, both legs were affected. There were no symptoms to suggest a problem in the brain, and Heine concluded that the spinal cord had been attacked.[7]

Like others of his time, Heine blamed teething. He was particularly impressed by Dr Fliess of Neustadt, who suggested that diseases of the teeth caused swelling of the spinal cord. Dr Fliess had dissected a five-year-old boy who suffered severe left-sided toothache and then woke with his left arm completely paralysed, 'like a corpse'. The opportunity for 'a very careful post-mortem' arose when the boy fell off the back of a cart taking him to a relative's house and hit his head on a stone. 'Commotion of the brain' was the cause of death, but there was also dramatic dilatation of the blood vessels supplying the upper spinal cord – and around the rotting teeth that were deeply impacted in his left jaw.

Figure 1.3 Jacob von Heine (1800–79), German orthopaedic surgeon and manufacturer of walking machines, who wrote the first comprehensive clinical account of polio in 1840. Reproduced by kind permission of the History of Medicine Archive, National Institutes of Health, USA

Heine updated his treatise 20 years later, under a punchier title that pinned the blame firmly on the spinal cord: *Spinale Kinderlähmung*.[8] This second edition (200 pages, with 15 lavish lithographs) was weightier, and so was its author, who by now had acquired the aristocratic prefix 'Von', together with the Order of the Württemberg Crown and the Russian

Imperial Order of St. Vladimir. Von Heine's monograph transformed 'spinal infantile paralysis' into a disease that doctors could now look out for. It was therefore entirely reasonable for Wickman to flatter his memory in 1907 by calling polio 'Heine-Medin disease'.[11]

By this time, small clusters of polio cases had been described, although the notion of an infection which spread from person to person was still years away. The earliest outbreak, back in 1835, had involved just four cases from Nottinghamshire in northern England. The author was John Badham (1807–40), who left Glasgow University in 1828 with a medical degree and the tuberculosis which eventually killed him. Seeking therapeutic warmth, Badham worked in the West Indies and Nice, but had to return to the cold greyness of England and the mining town of Worksop. There, during one week in August 1835, he saw four cases of paralysis that struck down previously healthy children overnight. Arms, legs and even the tiny muscles which move the eyeball were variously affected. All the victims were under three years of age. Some were drowsy, suggesting to Badham that the cause was in the brain rather than the spinal cord. There was a partially happy ending, as one girl had regained some strength in her leg after two months. Badham's paper was published in 1835 in the *London Medical Gazette*, which carried his own obituary just five years later.[24]

In 1843 and across the Atlantic in Louisiana, George Colmer described over a dozen instances of paralysis, all hitting previously healthy children in and around the town of Feliciana. The clinical features appeared broadly similar to Badham's cases. Like Von Heine, Colmer blamed teething.[25]

Over the next 40 years, a smattering of small outbreaks of sudden-onset paralysis affecting children were reported from Sweden, Norway and France. These had little impact in England, let alone America, as the papers were in foreign languages and appeared in journals with cryptic titles and little penetration overseas. However, they included some gems. In 1881, Nils Bergenholtz, a provincial doctor from northern Sweden, wrote up 13 cases of paralysis in Umea and outlying villages.[26] His conclusion that these were causally linked was ahead of its time, even if the translation reads rather clumsily: 'Poliomyelitis acuta anterior and its appearance as an epidemic is because the same cause of disease has simultaneously affected several persons'. This account was particularly slow to break out of its native Sweden. Bergenholtz presented his research in 1881, but then lost the text and had to reconstruct it from memory (a copy eventually turned up in 1949).

Four years later, in 1885, a paralytic illness broke out among in children in the isolated silver-mining village of Sainte-Foy l'Argentière in Rhône-Alpes, eastern France. There were only 13 cases, but enough to devastate the tiny community. The episode was not investigated until 15 months later, when Dr S. Cordier of Lyon was so struck by what had happened – and especially the similarity of the cases to seven-month-old 'petit Benoît Villars', the first to be paralysed – that he used the word 'epidemic' in the title of his paper.[27]

Meanwhile, over in the United States, the 1886 edition of William Pepper's massive *American system of medicine* included a 50-page chapter on infantile paralysis by the energetic and accomplished Mary Putnam Jacobi.[28] She had trained in neurology in Paris and now brought European authority to the topic. Somewhat ahead of her time, Jacobi speculated that polio might be an infection, caused by an 'as yet unknown member of the great class of pathogenic bacteria'.

All this set the scene for polio to make headlines at the Tenth International Medical Congress, held in Berlin in August 1890. The conference attracted a record 5,000 participants, squeezed into the theatrical setting of the old Market Hall. The auditorium usually housed the renowned Circus Renz, which had recently outgrown its original Circus Olympic building, where Johann Strauss had premiered his *Blue Danube* waltz. Excitement ran high from the opening day, when Robert Koch, the charismatic evangelist for 'germ theory' unveiled tuberculin, his novel 'remedy' for tuberculosis. This treatment had worked wonders in tuberculous guinea pigs, and Koch had injected it into Hedwig, his newly acquired second wife, without obvious ill effects.[29] Those who gave Koch a rapturous reception were not to know that tuberculin would turn out to be useless for treating tuberculosis in humans.

Another sensation was created by a 43-year-old Swedish paediatrician, applauded for his painstaking detective work and the rare ability to paint a coherent picture of an elusive disease. Karl Oscar Medin (1847–1927) held the Chair of Paediatrics in Stockholm during the three decades which straddled the turn of the twentieth century (Figure 1.4). Medin knew polio as a sporadic illness which paralysed one or two children in southern Sweden each year. Then, in the autumn of 1887, dozens of cases appeared around Stockholm – an astonishing step up in incidence which prompted Medin to investigate the outbreak in detail as it evolved.[30]

Visiting the patients' homes, Medin realised that the outbreak was much more than a chance clustering of sporadic cases. Like Bergenholtz and Cordier, Medin used the word 'epidemic' in the title of his presentation to the Congress – the first time that the concept was driven home before a large, international and influential audience. What really made them sit up was Medin's view that polio affected the whole body, and that paralysis only occurred if it happened to attack the spinal cord. Medin's evidence was the occurrence of vague 'minor' symptoms, a few days before the onset of paralysis. The symptoms included fever, malaise, diarrhoea

OSCAR MEDIN

Figure 1.4 Karl Oscar Medin (1847–1927), Professor of Paediatrics in Stockholm, whose study of the 1887 Swedish epidemic showed that polio was a systemic illness that did not necessarily paralyse. Reproduced by kind permission of the Swedish National Archives

and headache, none of them anything to do with the spinal cord. Such symptoms had previously been noted by others in passing, but Medin's careful questioning now showed that all cases of paralysis had suffered them. And for the first time, Medin showed that these 'minor' symptoms also affected some patients who did not go on to develop paralysis. Medin's vision of polio was much broader: a systemic illness that might or might not turn on the spinal cord and, crucially, one that appeared to be changing its behaviour into an 'epidemic' disease alongside cholera, plague and smallpox. Suddenly, polio was a disease to watch.

In 1895, another polio outbreak hit the countryside around Stockholm. Only 21 children were paralysed, but Medin was able to confirm his earlier conclusions. The epidemic also helped to train up Medin's ambitious young assistant, Ivar Wickman (1872–1914; Figure 1.5). The 25-year-old paediatrician was determined to make his career in polio and was writing a 300-page thesis about the disease. Before long, he also had his eye on his boss's job.

Medin had convincingly recast polio as an epidemic disease and encouraged everyone to think laterally. Strangely, though, he failed to follow his own observations to the logical conclusion that polio was contagious. Medin shied away from any notion that polio could spread from person to person and years later – in a move which would have infuriated Robert Koch and his germ-theorist disciples – he seemed to imply that polio was due to 'miasma', the toxic emanations of decomposing matter in the soil.[31] But by then, Medin was beyond reproach and safely coasting on his reputation as a grand old man of microbiology.

The eager young Wickman had to wait nearly a decade for the perfect moment to pick up where Medin had left off and blaze his own trail. In early July 1904, reports filtered in of polio cases rapidly accumulating around Trästena, a small village in the middle of nowhere (more accurately, about 250 kilometres south-west of Stockholm). New cases appeared in nearby settlements, and then the epidemic really took off. Over the next three months, outbreaks flared up in virtually every county in Sweden, with cross-border incursions into Norway. The numbers swelled into an epidemic which was several times larger than anything that had gone before. August saw a massive 360 cases, and when the outbreak finally ground to a halt at the end of the year, Wickman had recorded 1,031 cases.[32]

Wickman made even better use of his material than Medin had done. He got his hands dirty and his boots muddy, following up 300 cases personally and extracting blow-by-blow histories for the others from the doctors who had attended them. In the process, he put together a jigsaw puzzle that showed polio as it had never been seen before, filling in crucial gaps with some pieces handed to him by Medin and some that he had to make himself. He focused on small populations such as the 500 residents of Trästena. Wickman was helped by the sparseness of human habitation across this stretch of Sweden, which made both the foci of polio cases and the lines of communication between them – roads and railways – stand out clearly. He took for granted Medin's notion

Figure 1.5 Ivar Wickman (1872–1914), Swedish paediatrician who concluded from a polio outbreak around Trästena that polio was an infection that spread from person to person

that polio cases included those with minor symptoms who had escaped paralysis. In most of his outbreaks, these non-paralytic cases accounted for up to half of those affected.

Wickman's records were so precise that he was able to reconstruct the Trästena outbreak, case by case and day by day, rather like time-lapse photography. His analysis showed a clear pattern. Polio had started in Trästena and had radiated out to invade other communities, following the obvious routes of human traffic. Trästena's school, the focus of life for the village's children and where the schoolmaster's six children slept each night, turned out to be a hub for the transmission of polio. In other words, polio was contagious and spread from person to person.

However, the pattern could only be created if Wickman made some big assumptions. Polio had to be spread not only by the obvious paralysed cases but also by those with Medin-style minor symptoms – and by others who were unknown because they had no symptoms at all. He reached this conclusion because there were many instances where polio must have tracked from community A to community C via community B, but nobody in community B showed any clinical evidence of polio, not even the minor symptoms. This was the first hint that polio could infect and be passed on by 'asymptomatic carriers'. Wickman's inspired hunch turned out to be the key to understanding how polio was spread, although it took several more years before the hypothetical asymptomatic carriers were shown to exist.

Wickman's detective work still stands out today as one of the great deductive exercises in epidemiology. His findings proved (to the satisfaction of most, but not all) that polio was an infection spread by personal contact. He laid down the concept that all those infected, whether symptomatic or not, were in turn contagious. From his freeze-frame montages of outbreaks, he could calculate the incubation period of polio: 3–4 days after exposure to the onset of minor symptoms, and a further 6–8 days for paralysis to take hold in those whose luck was destined to run out.

Wickman wrote up his findings in another monumental paper, initially published in German in 1907. He dedicated to it to his mentor, with a further gesture of homage in its title, 'Contributions to the knowledge of Heine-Medin disease'. The paper was well received, especially after its translation into English in 1911, and rapidly revolutionised thinking about polio on both sides of the Atlantic. Back in Sweden, Wickman successfully argued that polio cases should be notified to the public health authorities, as with major infections such as smallpox and tuberculosis. This was in 1908, and set the trend for other countries.[33]

Unfortunately, all this did not boost Wickman's career as much as he had hoped. He was not seen as Medin's natural successor to the Chair of Paediatrics in Stockholm. Stung by the criticism that his research was too focused on one disease (polio), Wickman toured European centres of paediatric excellence to try to bolster his curriculum vitae. He ended up in Strasbourg with the all-powerful Adalbert Czerny. From there, Wickman applied for Medin's Chair when the old man retired in 1914, but he sent out mixed messages. Claiming a debilitating stutter, he refused to give a public lecture, which all the applicants were required to do. Instead, Czerny sent a note to the appointment committee confirming that Wickman was a fine teacher. The committee were unmoved and appointed somebody else. Wickman, always up and down in mood, took their decision badly. On 20 April 1914, he went to his room with his revolver and shot himself through the heart.[34]

It took some years before Wickman's work was fully recognised as the decisive study which unmasked the true nature of polio. Today, Wickman's bronze face stares out from the Polio Wall of Fame in Warm Springs, Georgia. He is flanked by his heroes, Jacob von Heine and Oskar Medin – no mean feat, as there are only four non-Americans in the line-up of seventeen.

Wickman was the right man in the right place at the right time. 'His' epidemic marked a watershed in the behaviour of polio, which would transform it into one of the headline diseases of the twentieth century. Originally, it had grumbled away sporadically in the background of childhood infections. Then, at the turn of the twentieth century, it shifted up a gear. The Swedish epidemic of 1905 was the first to break through the 1,000-case barrier. The numbers were partly inflated, as Wickman included all the minor cases as well as those paralysed, but this revealed for the first time the true extent of an outbreak.

And this was not a blip, but the start of a trend. Polio began to break out in epidemics that soon became an annual fixture in Scandinavia. This pattern was picked up in other regions around the world: America in the early 1900s, followed over the next 40 years by South Africa, Australia, continental Europe and Great Britain.

In his paper, Wickman referred to an epidemic that hit the green and pleasant agricultural landscape of Rutland County, Vermont, during the summer of 1894. This was the first substantial epidemic in North America, and it paralysed 300 people. This outbreak was much smaller than the one around Stockholm a decade later, but it too proved to be hugely significant.

The man in the hot seat in Vermont was Charles Caverly, the state's public health officer. Like Wickman, Caverly was a field worker rather than an armchair detective, but he did not dig as deeply as Wickman had done. Caverly found cases who had minor symptoms without paralysis, but did not appreciate their significance. As for the cause, he simply parroted the impressions of the local physicians, which included falls on the head, chills and drinking birch beer from a travelling circus. However, Caverly made two striking observations: many victims were adults, and the death rate of 14 per cent was substantially higher than previously recorded.[2]

These changes in predilection and mortality turned out to be the shape of things to come. Polio was evolving into an epidemic illness that extended its reach through adolescence and into adulthood and was becoming more dangerous.

The Vermont epidemic of 1894 was momentous in another way. It marked the start of the reincarnation of polio – until then, a minor clinical oddity from Europe – as an American disease, and its takeover by the American scientific and medical establishment. The 'Americanisation' of polio would have far-reaching implications. Polio was turned into an enemy which had unwisely chosen to target the American people, and therefore set itself up for revenge through the might of American science.

Shades of grey

The 41 years of Ivar Wickman's life saw two major scientific milestones laid down in polio research. The first, while he was still a child, was the evidence that paralysis was caused by damage to a tiny area of the grey matter in the spinal cord. The second, just a few years before his death, was the revelation that the spinal cord lesions were caused by a 'filterable virus', a new class of infectious microorganism.

This was a period when neurology was highly developed as a clinical art, but often teetered on shaky scientific foundations. The gross anatomy of the brain was already well described, and thanks to accidents of both nature and man, the connections and functions of its main structures were becoming clear. Lesions that destroyed specific areas or cut through particular tracts of nerve fibres could be highly informative, especially when whatever had gone wrong with the patient could be tied in with a thorough post-mortem to localise the site of the problem. These tell-tale lesions included strokes, brain tumours and abscesses, gunshot wounds and – in one celebrated case – an iron rod which had ill-advisedly been used to tamp gunpowder while blasting out a railway cutting.[35]

Meanwhile, neuropathology – the science of disease processes in the nervous system – was struggling through a difficult childhood. Doctors were familiar with the brain, both in and out of the skull, as illustrated by Dr G. Thompson in his paper of 1875, 'On the physiology of general paralysis of the insane, and epilepsy':

> Who has not, in slicing down such a brain, felt the hard resilient touch given to the knife, almost amounting to the sensation given in cutting down a hard-boiled potato?[36]

However, the causes of most diseases of the central nervous system were entirely mysterious and the subject of fierce controversy. When it was suggested that 'general paralysis of the insane' was end-stage syphilis, many neurologists refused to accept the notion, even when the corkscrew-shaped 'spirochaete' bacteria of syphilis were clearly shown in victims' brains.

So what was happening in polio? At this point, we need to divert – briefly and, with any luck, painlessly – into the anatomy of the spinal cord.

The spinal cord is a downward continuation of the brainstem, to which are attached the large paired hemispheres of the cerebrum and, towards the back, the smaller cerebellum (Figure 1.6). The cord emerges from the foramen magnum ('big hole' in Latin) in the base of the skull and runs down through the spinal canal, like a thick string in an unhooked necklace made up of vertebral 'beads'. The cord is mostly about the size

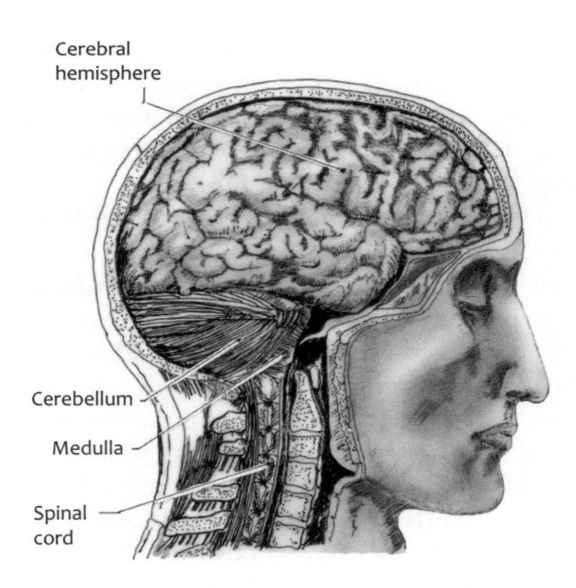

Figure 1.6 The brain and vertebral column, with the spinal canal removed to show the inside of the spinal canal that runs through the middle of the vertebrae. Illustration by Ray Loadman

of the index finger, but is expanded towards its top and bottom to contain all the extra wiring that supplies the arms and legs.

Grey matter occupies the centre of the cord and runs right through its length, like the lettering in a stick of rock (Figure 1.7). As with the grey matter of the brain, it owes its colour to tightly packed nerve-cell (neurone) bodies. The surrounding white matter consists of tracts of nerve fibres, and is rich in the fatty myelin which insulates the axons, the

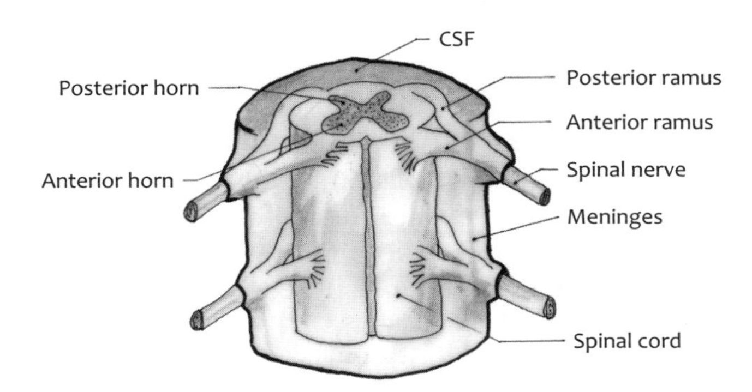

Figure 1.7 The spinal cord. The protective meninges are partly cut away to show the anterior and posterior rami on each side merging to form the spinal nerve. Illustration by Ray Loadman

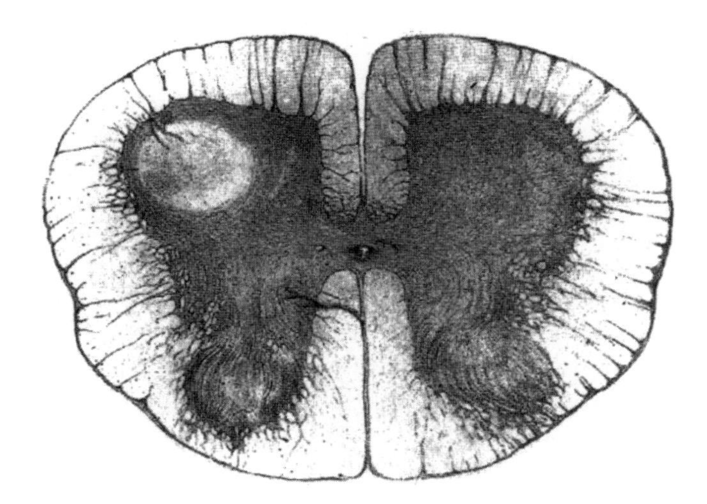

Figure 1.8 Cross-section of spinal cord, showing the anterior and posterior horns of the grey matter. The left anterior horn (top left in the image) contains a lesion of poliomyelitis. Reproduced from Roger H, Recherches anatomo-pathologiques sur la paralysie de l'enfance, Paris: Adrien Delahaye (1871), © Bibliothèque Interuniversitaire Santé, Paris

long extensions of the neurones. The axons conduct electrical impulses away from the neurones, to switch other neurones on or off or to make muscle fibres contract. In cross-section, the grey matter of the spinal cord is shaped like a butterfly with its wings pinned out. The anterior horns, at the front, represent the forewings of the butterfly, while the posterior horns are its hindwings (Figure 1.8).

In the gaps between the vertebrae, large nerve bundles (rami) run into the front and back of the cord, on each side. The rami have tightly demarcated job descriptions. Each posterior ramus gathers together all the sensory nerves on that side from that level, and channels all the information about touch, temperature, pain and the positions of joints into the corresponding posterior horn.

By contrast, the front half of the cord – the anterior horn and the anterior ramus – governs motor function, giving power and movement to the muscles. The ultimate drivers are the motoneurones, large nerve cells with a prominent nucleus, packed into the anterior horns. Their axons, bunched together like the strands in a fibreoptic cable, run out through the anterior ramus.

On each side, the posterior and anterior rami run together, and the combined bundle is then re-divided into the nerves that supply the parts of the body at that level. These 'peripheral' nerves therefore include both outgoing (efferent) motor fibres running out to the relevant muscles, and incoming (afferent) fibres bringing in sensory information about what is happening inside the body and its immediate environment.

It took years to prove that paralysis in polio results from destruction of the anterior horns, partly because the lesions were often small and patchy. Also, there were other places to look, as the characteristic floppy paralysis and muscle wasting could also be caused by diseases of the peripheral nerves. However, peripheral nerve damage – for example in diabetes and leprosy – usually develops much more slowly. Furthermore, because these nerves carry sensory fibres, the ability to feel touch, pressure, temperature and pain is also lost. A diabetic patient with severe nerve damage may not realise that he has stepped on a nail and transfixed his foot until he tries to take his shoe off. By contrast, polio patients may suffer painful muscle spasms in affected limbs, but sensation is intact.

Careful clinical examination of polio victims led Jacob von Heine to label 'infantile paralysis' as 'spinal' in the 1860 edition of his treatise.[8] However, others begged to differ. In 1843, Frédéric Rilliet, chief physician in Geneva, had been persuaded by a few perfunctory post-mortems on young polio victims that there was nothing wrong in the cord.[37] Rilliet dismissed the disease as '*essentielle*' or 'idiopathic', which translates into the shorthand of today's doctors as 'GOK' ('God only knows').

The dispute was eventually settled by Jean-Martin Charcot and Alfred Vulpian, the two greatest neurologists of the late nineteenth century in France, if not the world. Charcot's interests ranged from hysteria to multiple sclerosis (which he first diagnosed in his housemaid), while Vulpian had discovered adrenaline in the adrenal gland. The two men co-directed the Salpetrière Hospital in Paris, so named because the original building had housed Louis XIII's gunpowder store. In 1870, Charcot and Vulpian published separate papers in the same issue of the *Archives of Physiology and Pathology*.[38,39] These had virtually identical titles, and both demonstrated lesions in the anterior horns of the grey matter of the spinal cord (Figure 1.8). Embarrassingly, Charcot spotted the lesions in the same microscope slides which one of his underlings had previously passed as normal.[40]

The finding was rapidly confirmed by others. In an acute case, inflammation could be caught in the act: the anterior horns were stuffed with white blood cells, gobbling up the remains of the motoneurones, which had mostly disintegrated. If the victim survived, the acute inflammation settled down, leaving a puckered scar towards the front of the cord where the anterior horn used to be. Typically, the posterior horns and the rest of the cord were spared – indicating how precisely the poliovirus targets the motoneurones.

Pinning down the lesions to the anterior horns did nothing practical for polio victims. However, this insight was crucial in understanding how paralysis arose. And now that the characteristic pathological signature of the disease was known, this could be used to detect and ultimately identify the agent responsible.

Which takes us forward 30 years, from Paris in the 1870s to the Vienna of Gustav Mahler and Gustav Klimt, and the next great milestone in polio research.

Cause and effect

As the twentieth century got under way, it was accepted that infantile paralysis was due to anterior poliomyelitis – but what actually damaged the spinal cord was anyone's guess. Suggestions such as teething, cold exposure and a bang on the head were based on anecdote and propagated through the medical literature by serial plagiarists who could not be bothered to check the original source.

Mary Putnam Jacobi's hunch that bacteria might be responsible was closer to the mark than most.[28] That was back in 1886, 50 years after the word 'bacterium' was coined, but two decades before Ivar Wickman's study indicated that polio was an infection. Many researchers set off on the trail of the infectious agent, and by 1905, there was a bewildering assortment of 'poliomyelitic bacteria' of various shapes and sizes, each backed by someone convinced that he alone had found the true cause. As described later in this book, confusion reigned.

All uncertainty should instantly have been swept aside on the evening of 18 December 1908, when three microscope slides were projected by magic lantern before the Royal and Imperial Society of Physicians in Vienna. All three showed sections of spinal cord disfigured by the anterior horn lesions characteristic of acute polio. The first slide was entirely routine, from a boy who had died of classic paralytic polio a few weeks earlier. But the two other slides created a sensation, because they were from monkeys.

The speaker was Karl Landsteiner (1868–1943), the 40-year-old Professor of Pathology at the Wilhelmina Hospital in the city (Figure 1.9). Landsteiner had already blazed a brilliant trail across the firmament of Viennese medicine, with breathtakingly clever experiments and a publication output that averaged one paper every three weeks. He also had a

Figure 1.9 Karl Landsteiner (1868–1943), Professor of Pathology in Vienna, who showed in 1908 that polio could be transmitted from man to monkey, and that the infectious agent was a filterable virus. He was awarded a Nobel Prize in 1930, for having discovered the blood groups. Reproduced by courtesy of the Rockefeller Archive Center, New York

voracious appetite for new challenges. In 1900, he had worked out why blood transfusions often killed the recipient; his discovery of the major blood groups had, within a few years, revolutionised the procedure and saved hundreds of thousands of lives around the world.

Then he turned his attention to infections. He started with the enigma of syphilis, an obviously contagious disease whose causative germ had not been cultured in the laboratory or even seen. In 1905, Landsteiner successfully transmitted syphilis to a monkey, and a year later managed to visualise the organism responsible for the infection using a new method that he had developed. This was 'dark-field illumination', which lit up even elusive bacteria against a black background, like a moth caught in a torch beam against the night sky. And there it was: a tiny silver corkscrew, the first sighting of the spiral *Treponema* bacterium.[41]

Landsteiner became intrigued by the cause of polio, and this took him into the mysterious new world of 'filterable viruses'. Viruses had been around for centuries, but only in an abstract sense. 'Virus' means 'poison' in Latin, and the word was widely used in the late eighteenth century as a vague term for the cause of any medical unpleasantness. It acquired a specific meaning around the turn of the twentieth century, when the term 'filterable virus' was adopted to cover a new kind of infectious agent that broke the fundamental rules of engagement for bacteria. Somehow, these agents attacked and killed living cells, yet could not be grown in laboratory cultures or seen down the microscope.

The first virus to be isolated was the agent of tobacco mosaic disease, an ugly brown blotching of tobacco leaves that quickly kills the plant. This was an urgent research priority because outbreaks regularly wrecked the American tobacco harvest. Sap from infected plants could infect healthy plants, even though no bacteria could be isolated. Curiously, the infection could be spread with an extract of sap which had been filtered through porous porcelain that held back all bacteria. The first animal disease proved to be caused by a filterable virus was foot-and-mouth disease (1898), and the first human infection was yellow fever (1900).[42] Viruses are much smaller than bacteria, which is why they could pass through the porcelain filter. They are far too tiny to be seen under the conventional light microscope, and remained invisible until the 1930s, when the invention of the electron microscope pushed the limit of visualisation down into the nanometre range (a nanometre is one millionth of a millimetre).

Landsteiner's search for the infectious agent of polio began in an acutely inflamed spinal cord, which he believed must harbour the organism. The opportunity came when a nine-year-old boy came to post-mortem (one of the 5,600 which Landsteiner personally performed) four days after being admitted to the Wilhelmina with acute paralytic polio. Part of the boy's spinal cord was preserved and sectioned to confirm the diagnosis by microscopy, while the rest was ground up in sterile water. The cord suspension grew no bacteria, and did nothing obvious when injected into guinea pigs, rabbits and mice. But Landsteiner and his assistant, Erwin Popper, had also injected the sterile cord extract into the abdominal cavity of two monkeys left over from other experiments. Monkeys were expensive and in short supply, and it was a further stroke of luck that both belonged to species that are easily susceptible to polio.

Landsteiner struck gold. One monkey died suddenly after eight days and the second after seventeen days, having become paralysed in both legs. Under the microscope, the spinal cord of both monkeys showed the diagnostic lesions of poliomyelitis.

These were the images that stunned Landsteiner's audience at the Royal and Imperial Society in the week before Christmas 1908. They proved that infantile paralysis was an infection that, like syphilis, could be transmitted to monkeys. Even more exciting was Landsteiner's conclusion that the infective agent was not a conventional bacterium because none could be isolated from the boy's cord. Instead, the culprit must be one of the newly described filterable viruses.[43]

Confirmation followed over the next few months, although Landsteiner had to team up with Constantin Levaditi at the Institut Pasteur in Paris, where monkeys were more easily available. Landsteiner and Levaditi showed that filtrates of poliomyelitic cord suspension could transmit paralysis to healthy monkeys and chimpanzees, which then showed identical anterior horn lesions.[44,45]

Some of the key messages from Landsteiner's work were soon lost in the noise as polio research became both fashionable and fundable, and began to attract other, less expert players onto the stage. By then, the ever-restless Landsteiner had moved on to his next big question – how antibodies react only with the antigens they were generated to recognise – which would obsess him until the end of his life. Perhaps order might have prevailed if Landsteiner had kept up his interest in polio. Instead, polio research drifted off beam and, as explained later, spawned some of the oddest notions in the story of disease.

To many people, Landsteiner epitomised the perfect scientist. In 1930, the American novelist Sinclair Lewis said, 'I shall be happy on that day I can write a book the hero of which should personify the ethical and material value to humanity of Dr Landsteiner'. He had just heard the news that he and Landsteiner were to be awarded Nobel Prizes. Landsteiner received his Prize, for Medicine or Physiology, to recognise his discovery of blood groups; this was seen as his greatest contribution to science, and the Nobel citation did not even mention that he had proved the viral cause of polio.

Landsteiner's career also rehearsed some of the tragedies that would soon tear middle European medicine apart. Despite the weight of his curriculum vitae, his application for promotion was sabotaged, because his recent conversion from Jew to Catholic was not quite convincing enough. He read the omens of what was to come more astutely than some of his friends, who stayed on in Vienna and ended up in Auschwitz. In 1919, Landsteiner took his wife, Helene, first to The Hague, and then in 1923 to New York, where he was snapped up triumphantly by the Rockefeller Institute.[41]

The Rockefeller was powerful and well funded, and gave Landsteiner everything he needed to maintain his trajectory, but it was also an idiosyncratic and dysfunctional workplace. Simon Flexner, the director, ruled with a firm and often stifling grip; also, he had claimed polio as a research priority for his Institute and his own pet subject. Flexner was highly intelligent and charismatic but also dogmatic and with an unfortunate blind spot for evidence that contradicted his own opinions. The fear of arguing with Flexner sent many of his underlings chasing up blind alleys in pursuit of some of his worst ideas. The result, told later, makes a good story, but held back progress in combating polio for over 20 years.

Photographs of Landsteiner show a stern and apparently humourless man, in line with his dictatorial reputation in his laboratory. However, he was also modest and self-doubting. He did not bother to include his Nobel Prize address in the collection of personal documents that he put aside for his family towards the end of his life. He was also locked into a bizarre relationship with his mother, who had been widowed when he was seven years old. Landsteiner concealed his marriage from her and publicly announced his wedding only after her death. Even then, her presence still hung around in the conjugal home; Landsteiner nailed her death mask up on the wall above his bed.[41]

But above all, Landsteiner is most appropriately remembered as a brilliant scientist of Olympian achievements. He died on 26 June 1943 as he had lived, still working on in retirement right up to the instant when a massive heart attack felled him at his laboratory bench.

This is a good moment to put the science of polio to one side. We know what the poliovirus does to the spinal cord, and that the infection had mutated from a rare and sporadic disease of childhood into one that was breaking out in explosive epidemics around the world.

It is now time to see what polio did to people, individually and en masse.

2

The Crippler

Oh Lord:
Restore me to that state of mind and body.
In which I was, when you created me.

The Polio Child's Prayer, quoted by Philip Lewin, *Infantile paralysis*, 1941.

Mia Farrow and Francis Ford Coppola were both nine years old when they caught it. So were Joni Mitchell and Johnny Weissmuller, who took up singing and swimming, respectively, to help their recovery. Itzhak Perlman and John Slessor were only four years old; both lost the use of their legs, but this did not stop them from realising their ambitions, on the international concert circuit and as Air Chief Marshal of the Royal Air Force. Others whose biography features a brush with polio include Arthur C. Clarke, Alan Alda, Sir Walter Scott (probably) and, with a twist, President Franklin D. Roosevelt – one of the many who used the politically incorrect name for polio which gives this chapter its title.

Hundreds of millions of others were killed or paralysed by polio. Their numbers are not known with any precision. Across most of the developing world, polio was until recently just one of the many rivers that all children must cross – and beside the raging torrents of cholera, measles and malaria, it was a mere trickle. Even in Western countries, the first step towards accurate counting of cases was not taken until the 1910s, when polio was accorded the status of a 'notifiable' infectious disease.[1]

Polio was one of the diseases that defined the twentieth century. Its transformation from a grumbling endemic curiosity to explosive outbreaks that left thousands paralysed occurred at different times in different regions: the 1910s in the United States, the years after the Second World War in Great Britain and Europe, and the 1960s in Russia. The worst times for the Western world were during the decade leading up to the introduction of the first effective polio vaccine in the mid-1950s. Great Britain suffered major epidemics in 1947, 1949 and 1950, with 6,000–8,000 paralysed and several hundred deaths in each of those years.[2] In 1949, the US Surgeon General, Leonard Scheele, warned that this would be the nation's worst year yet for polio and that the future looked even grimmer. He was right on both counts: 1949 saw 42,000 Americans paralysed and 2,700 deaths from polio, rising in 1952 to 58,000 paralysed and 3,000 fatalities.[3]

We can only guess whether or not the saw-tooth pattern of ever-greater epidemics would have continued if the brake of vaccination had not been applied. After a couple of false starts, vaccination quickly gained traction and, as well as protecting individuals, began to sweep away the infection itself. Within a few years, polio had become as much of a rarity in the Western world as it had been a century earlier. In 1961, barely 1,300 cases were reported in the United States.[4]

By 1988, polio was endemic in only 125 countries (about one-third of the total number), and caused about 350,000 cases worldwide each year. Flushed by its success in eradicating smallpox in 1980, the World Health Organisation (WHO) announced a campaign in April 1988 to rid the entire planet of polio by 2000.[5] The Global Polio Eradication Initiative has been remarkably successful, and the number of cases reported across the world has fallen by over 99 per cent – but the 2000 deadline for eradication was missed, as were a couple of its successors.

By the start of 2013, only three regions – Northern Nigeria, Afghanistan and Pakistan – remained in the grip of endemic polio, and just 150 cases were reported worldwide in 2012.[6] But like Sisyphus, forever pushing his boulder uphill, we are not there yet. This particular boulder has fallen back several times, with polio reinvading previously cleared countries where the defences thrown up by vaccination have crumbled.

The stories of two young victims of polio bring the triumphs and failures of the eradication campaign into sharp focus. One is assured a place in history, while the other will soon be forgotten. Luis Fermín Tenorio Cortez, a two-year-old boy from Junín, Peru, lost the use of his left leg in August 1991. The vaccination net had been tightening around South America, but he slipped through. He was the last victim of natural polio in the whole of the Western hemisphere.[7] A different kind of failure ensnared two-year-old Fahad Usman, who now faces a precarious future with his parents in the refugee camp at Jalozai, south-east of Peshawar, in Pakistan. In early 2012, polio paralysed both his legs. Fighting along the border with Afghanistan had prevented vaccination teams from reaching the area. Even if they had, Fahad and the other children in his village would probably have remained unprotected. The Taliban had threatened violence to anyone giving or receiving the vaccine, on the pretext that it contained a poison developed in the West to sterilise Muslims.[8]

At the bedside

The horrific dramas of paralysis and death by suffocation are the hallmark of polio. In fact, these are rare complications of an infection that usually runs a surprisingly benign course – a reminder that the poliovirus normally lives in the gut and causes no problems if it passes straight through. Out of every hundred patients who catch polio, at least 95 will not notice anything untoward (Figure 2.1).[9]

Of the 5 per cent of cases who become ill, most also escape paralysis. Symptoms generally begin about 10 days after meeting the virus, although the incubation period can be anywhere between two days and five weeks (Figure 2.2). These 'minor' symptoms include fever, headache, sore throat, feeling sick and unwell, and constipation or diarrhoea. The temperature often reaches 40°C (104°F), and occasionally hits 43°C (110°F), a level that can be life threatening in its own right. Fever can be unbearable, as illustrated by a ten-year old boy from La Crosse, Wisconsin, who was found lying under a garden sprinkler in a desperate attempt to cool down.[10] These symptoms correspond to the 'viraemic' phase, when the virus circulates in the bloodstream after breaking through the immune defences in the wall of the gut. In most cases, the fever and minor symptoms usually melt away after a few days, marking the end of the episode.

Most other patients go on to develop symptoms which indicate that the virus has invaded the central nervous system but has got no further than attacking the meninges, the multi-layered membranes which line

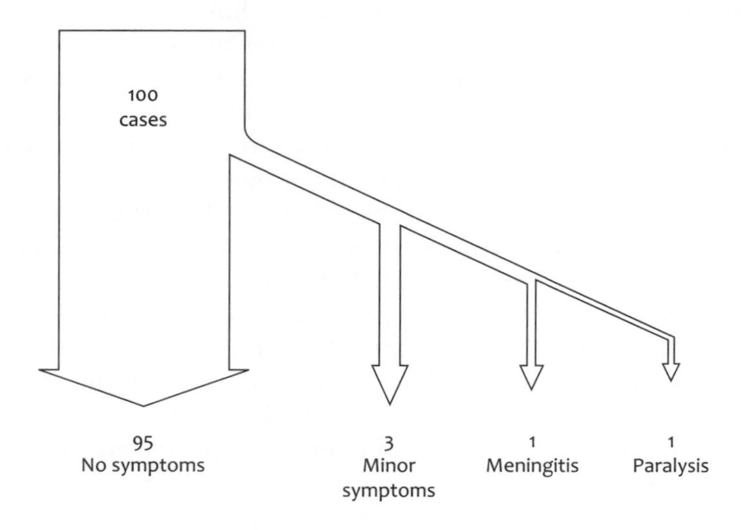

Figure 2.1 Possible outcomes of an acute attack of polio. The vast majority of cases have no symptoms; only 1 per cent or fewer develop paralysis

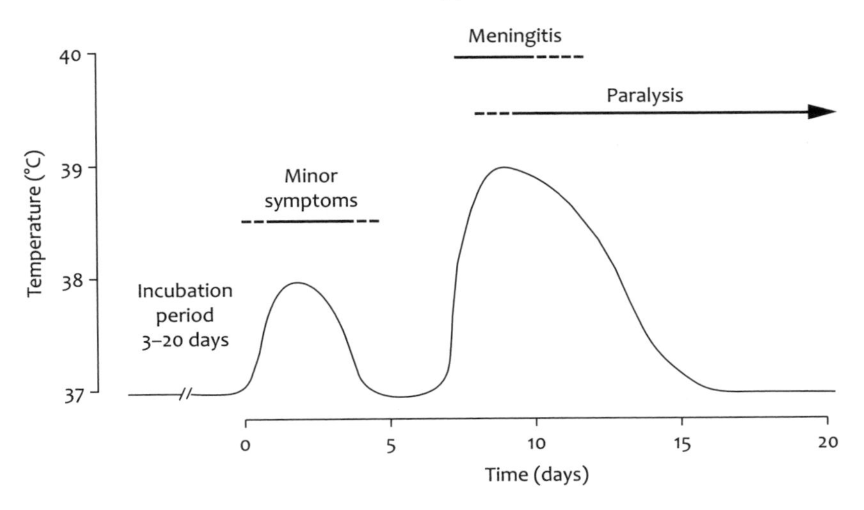

Figure 2.2 Time-course of symptoms during an acute attack of polio. The classic 'dromedary' pattern shows two peaks of fever, the first coinciding with 'minor' symptoms and the second with meningitis and paralysis. Adults may show only the second 'hump'

the skull cavity and the spinal canal and protect the brain and spinal cord. These patients are not paralysed, but can be very ill with symptoms of meningitis (inflammation of the meninges), notably vomiting, severe headache and agonising pains in the neck, back and limbs. Meningitis triggers a reflex spasm of the muscles which normally keep the spine straight. The neck is stiff; characteristically, patients cannot put their chin on their chest, even when helped by the doctor (a diagnostic test that causes intense pain). In extreme cases, the spine is rigid 'like a bent bow', with the head and feet thrust back, and a child can be lifted with one hand under the neck and the other under the heels.[11] Meningitic symptoms in polio may be mild or severe, but usually resolve completely within a few days. By contrast, bacterial meningitis due to the meningococcus may be life threatening or leave permanent brain damage, if not treated promptly with high-dose antibiotics.

On average, fewer than 1 per cent of all those who catch polio go on to develop the classic acute flaccid (floppy) paralysis. In outbreaks of the milder Type 2 polioviruses, the incidence of paralysis is as low as one in 1,000. In children, the minor symptoms settle into a deceptive lull for a couple of days before the fever picks up again with a vengeance, and paralysis sets in (Figure 2.2). This 'biphasic' profile of fever and symptoms was known as the 'dromedary' pattern (even though the dromedary has only one hump). Adults tend to omit the first 'hump' and present directly with fever, pains in the limbs and paralysis.[12]

Paralysis can strike overnight or develop gradually over days. It can be subtle: a mild squint, due to loss of one of the tiny muscles that moves the eyeball from side to side, or a crooked smile, because one side of the face no longer responds. One or both arms and/or legs can be affected, causing monoplegia (one limb paralysed), paraplegia (both legs) or tetraplegia (all four limbs). Children usually suffer monoplegia, especially of a leg, or paraplegia. Tetraplegia is commonest in adults. In some cases, weakness marches steadily up from the feet, paralysing the legs, then the ribcage, arms and diaphragm.

Persistent fever is an ominous sign, likened to lava continuing to pour from a volcano[9], and indicates that the spinal cord is still under attack. Conversely, a fall in the patient's temperature is an encouraging sign that paralysis has reached its peak.

Not all paralysis is permanent, presumably because the motoneurones have been temporarily incapacitated by inflammation nearby rather than being destroyed by the poliovirus. Even tetraplegic patients who had to be kept breathing by the iron lung have been known to make a full recovery. However, the extent of paralysis after a month generally predicts the final outcome, and further recovery is unlikely after nine months.

The breath of life

Not being able to breathe is the most potent of our primal fears, and rightly the stuff of horror fiction and nightmares. Suffocation is the main cause of death in the 10–15 per cent of patients with paralytic polio who die during the acute attack. This is because the poliovirus has destroyed the motoneurones which maintain breathing.

Respiration is one of the 'autonomic' activities that we take for granted, and we really only notice it when it goes wrong. It is regulated in a top-down hierarchy, ultimately controlled by the 'respiratory centre'. This is a cluster of specialised motoneurones in the medulla, the expanded upward extension of the spinal cord (see Figure 1.6). The fat, rounded medulla was formerly known as the 'bulb' – hence the term 'bulbar paralysis', indicating that the medulla has been attacked and that respiration is likely to be affected.

The motoneurones of the respiratory centre have a very demanding brief. They operate continuously for every second throughout life, day and night, whether awake or asleep. They must also ensure that breathing responds urgently to emergency signals such as a fall in the blood oxygen level or a rise in carbon dioxide.

The respiratory centre controls the motoneurones which directly drive the muscles that do the mechanical work of breathing. Motoneurones in the upper spinal cord supply the diaphragm and the intercostal muscles between the ribs. The diaphragm acts like the plunger in a bicycle pump, pushing the viscera down and pulling air into the lungs, while the intercostals expand the chest by levering up the rib cage. Less important are the 'accessory' muscles of respiration in the neck (trapezius at the back and sternomastoid at the front), which pull up the rib cage from above; these are driven by motoneurones in the medulla. The vital importance of the connections running down from the respiratory centre is demonstrated by the cause of death in hanging. Thanks to a design fault in the human spine, a peg-shaped part of the second cervical vertebra snaps backwards when the body drops. The bony peg severs the cord above the motoneurones which supply the intercostals, and the diaphragm and breathing stop instantly.

Polio can paralyse respiration at various levels. Destruction of the motoneurones in the upper spinal cord will take out the intercostals and/or the diaphragm. This leaves only the accessory muscles in the neck, which shift relatively small volumes of air and tire quickly; alone, they cannot keep the patient alive for long. Bulbar paralysis, due to a direct hit on the respiratory centre in the medulla, is much more dangerous. The attack effectively smashes the control box and rips out the wiring to all the muscles which keep the patient breathing.

Respiratory paralysis often develops gradually over a day or more, although bulbar paralysis can kill in an hour or two. The inexorable descent into respiratory failure is horrific for all concerned. The early stage is described by Lewin:

> One sees a sleepy baby become all at once awake, highly strung ... the whole mind and body appear to be concentrated on respiration. The child gives the impression of one who has a fight on his hands and knows perfectly well how to manage it ... The child is nervous, fearful, and dreads being left alone.[13]

As the respiratory muscles weaken and the movement of the chest decreases, the victim breathes faster to try to compensate. This tires the muscles more quickly and heightens the feeling of imminent suffocation. The victim looks scared to death, with nostrils dilated in a futile attempt to drag in air, and eyes 'staring in an unmistakable and unblinking terror'.[14] Eventually, oxygen levels in the blood fall so low that the lips and tongue turn dusky blue, and the victim becomes ominously calm before slipping into a coma. Shortly after, all respiratory efforts cease.

Collateral damage

Occasionally, the poliovirus floods into the entire brain, causing widespread lesions known as 'polioencephalitis'. Compared with a precisely targeted attack that picks off the motoneurones of the anterior horns, polioencephalitis is carpet bombing. Clinical features indicate diffuse brain damage and include confusion, epileptic fits and coma. Most victims are infants, and the few who survive are generally left with permanent brain damage. Henry Frauenthal and Jacolyn Manning, in their famous *Manual of infantile paralysis* (1914), list the main outcomes as blindness, spasticity and feeble-mindedness, and offer their opinion that 'a majority of all morons, idiots and imbeciles are victims of polioencephalitis affecting the frontal lobes'.[15,16]

In the time-honoured tradition of doctors who attach their names to the misfortunes of their patients, polioencephalitis used to be known as Strümpell-Marie disease. Paul Marie was Jean-Martin Charcot's brightest pupil in Paris, while Adolph von Strümpell was a leading light in Vienna – a city which later witnessed an even more horrific destiny for some victims of this cruel condition.

Aftermath

For many survivors of polio, the acute attack was just the start of a grim life with paralysis, deformities and disability.[17]

The deformities of polio result from a conspiracy between paralysed and unaffected muscles. Anterior horn damage is virtually always patchy, even within one segment of the spinal cord, hitting some muscles in a limb while leaving others intact. For example, paralysis commonly affects the peroneal muscles on the front of the shin, which cock the foot up at the ankle. This action is normally counterbalanced by the fat-bellied gastrocnemius on the back of the calf, controlled by motoneurones a few centimetres lower down the spinal cord. If the peroneals are paralysed, the foot drops into the 'equinus' (horse-like) posture nicely shown by Astarte's priest, Ruma (Figure 1.2). This is accentuated by the still-healthy gastrocnemius, pulling through the Achilles tendon that runs into the top of the heel bone (calcaneum).

The paralysed peroneals begin to shrivel away, because muscles need constant prompting by motor nerve impulses to synthesise the proteins which give them bulk and power. Within weeks, wasting (atrophy) can reduce a plump, healthy muscle to a thin strap of fatty gristle. Without treatment, the deformity becomes permanent, with the ankle joint effectively locked in the equinus position. The patient can only walk with a heavy limp, leaning towards the good side and swinging the paralysed leg out so that the dropped foot clears the ground.

In children, normal development throws in another cruel twist. In order to grow, bones need to be stimulated by the everyday pull and push of the muscles attached to them. The long bones in the paralysed limb fall progressively behind their normal growth curve and end up thinner and shorter, sometimes by several inches (see Figure 1.1). Having a paralysed leg that is several inches shorter than the other makes walking even more difficult.

The effects of paralysis are not limited to the arms and legs. The vertebral column is normally held vertical, with gentle curves towards the back (kyphosis) in the thorax and forwards (lordosis) in the lumbar spine. The posture is maintained by hefty muscles that run up on either side of the vertebrae; their cross-sectional anatomy is well demonstrated in a lamb chop. If one set of these muscles is paralysed, the unopposed pull of their counterparts will bend the spine towards the good side. The result depends on the distribution and extent of paralysis, and includes an exaggerated kyphosis or scoliosis (curvature to the left or right). Combinations of these deformities that twist the spine so badly that it interferes with breathing.

Textbooks from the early twentieth century contain galleries of photographs of polio cases, often with children posed naked to show their deformity to its best advantage. This was an age when the doctor ruled supreme, and before the stirrings of real respect for patients and their

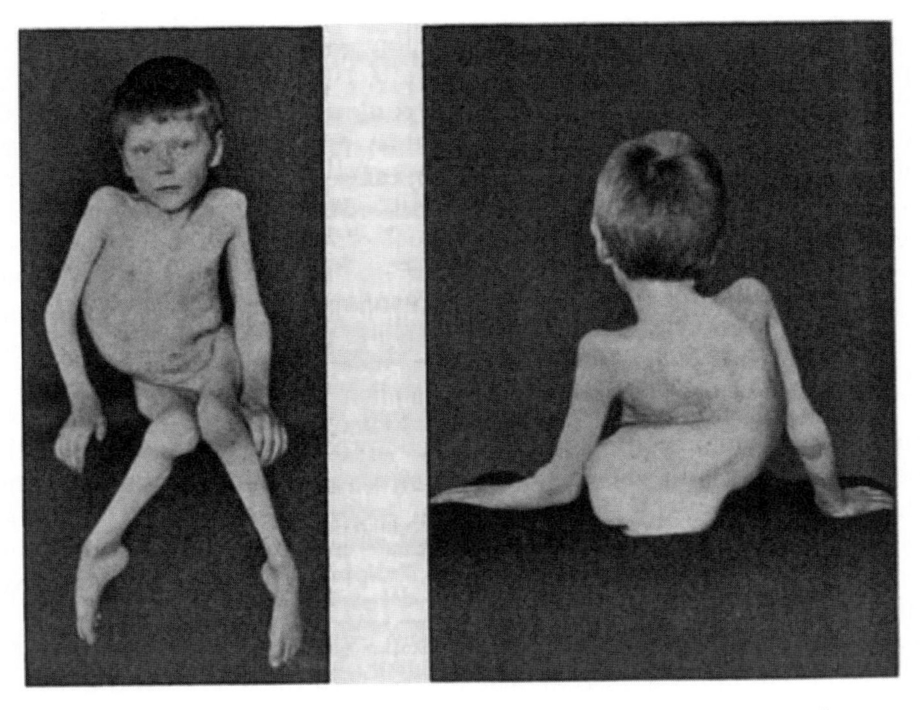

Figure 2.3 Permanent deformities of polio in a boy. From: Wickman I. Beiträge zur Kenntnis der Heine-Medinschen Krankheit: Poliomyelitis acuta und verwandter Erkrankungen. S. Karger; 1907

feelings. Today, the images of grotesquely contorted bodies, laid out like a medical freak show, make uncomfortable viewing (Figure 2.3).

Occasionally, there are glimpses of the person behind the 'clinical material': ribbons in a girl's hair, or a cheeky grin for the camera. These insights are also disturbing. They remind us that – because of other people's reactions to disability as much as the damage inflicted by the poliovirus – many Western polio victims from a century ago faced a life that matched Thomas Hobbes' description of 'nasty, brutish and short'.[18] The same applies today to millions of polio survivors in the developing world.

Recurring nightmare

Until recently, it was assumed that polio survivors were stuck with the disabilities which remained after a year or so. They had been dealt a bad hand, but the good news was that they were still alive and could not get any worse.

This assumption has turned out to be wrong. A sizeable proportion of polio survivors eventually suffer worsening or reawakening of muscle weakness. This can begin decades after the original attack, and in some cases can be more disabling. This strange condition is now known as 'postpolio syndrome' (PPS) or the 'late effects of polio' (LEP).[19]

Like polio itself, PPS seems to have come from nowhere, and it took a long time – a whole century – to establish a foothold in the medical literature. A handful of suggestive cases were reported by Charcot's colleagues during the 1870s,[20] but were lost in the flood of neurological syndromes which poured out from Paris at that time. For the next 80 years, there was a void in the medical literature. Finally, in 1969, 80 cases of 'late motor neuron degeneration following polio' were brought together,[21,22] but it was not until the 1980s that this vague entity mutated into a respectable diagnosis.

The 80 cases reported in 1969 were the tip of a massive problem whose full extent remains uncertain. In the United States, PPS is currently believed to be the single greatest cause of disabling motoneurone disease.[19] PPS probably affects 20–40 per cent of patients who had polio 30 or more years earlier. The risk increases with age and time after the original attack, and is higher in women and those who suffered severe paralysis at the outset – even if this subsequently resolved.[19]

PPS appears to result from a healing process which works initially, but is destined to fail when ageing catches up with the spinal cord decades later.[23] After the acute attack settles, surviving motoneurones try to take control of fibres in the affected muscles that are otherwise doomed because their original motoneurone has been killed. New terminals sprout from the endings of viable motor nerves in the muscles and make contact with the denervated muscle fibres, bringing them back to life. The muscle may show a 'tabby-cat' pattern of healthy red fibres interspersed with the fatty yellow streaks of others that are beyond rescue. Strength may improve and be maintained for years. However, the overstretched motoneurones eventually lose the battle with their extra workload. The motor nerve endings shrivel away, followed by the muscle fibres which they served. Degeneration speeds up after the age of 60, when the number of healthy motoneurones in the spinal cord begins to fall off, even in normal subjects. Loss of motoneurones in PPS may also be accelerated by overuse of the muscles.

The cardinal features of PPS are the *new* onset of muscular weakness, pain and fatigue, together with physical and mental tiredness affecting the whole body. Muscles other than those originally paralysed are often involved, sometimes with deep, burning pain and tenderness over tendons and other 'trigger points'. A flu-like exhaustion can be permanent and debilitating. Symptoms usually develop insidiously and are often made worse by exertion.[19]

The diagnosis rests solely on the patient's account of his or her symptoms and can be contentious. There are no diagnostic physical signs, nor any laboratory tests which can distinguish a polio survivor with PPS symptoms from one who is unaffected. Some symptoms of PPS – such as exhaustion, tender 'trigger points' and poor enjoyment of life – are common in the general population, especially in older people and those with anxiety and disturbed sleep.[19]

The impact of PPS is highly variable, as is the way it is perceived by the medical profession. One the one hand, PPS is seen as important enough to figure in a chapter title from a major textbook on infections of the central nervous system: 'Poliomyelitis, polio vaccine and the postpoliomyelitis syndrome'.[24] On the other hand, another authority dismisses PPS as 'an indolent condition that rarely leads to disability or death'. This view might not be shared by the many who coped well for 20 or 30 years after surviving polio, but who now can no longer climb the stairs, run their homes or hold down a job.

Why did it take so long for the condition to be recognised? Cynics would argue that this was another example of doctors failing to listen to their patients, especially those written off by an illness that had defeated medical science. Yet the symptoms of PPS are non-specific and shared with many other conditions, and the long latency of PPS meant that many earlier polio victims did not live long enough to develop the complication.

Pattern recognition

Great diagnostic acumen was not needed to work out what was wrong with a child laid low with fever, a stiff neck and paralysis at the height of a polio epidemic. However, it was not always that easy, and before reliable laboratory tests became widely available, it was surprisingly common for polio to be mislabelled as something else, and vice versa. In 1944, Donald Johnstone, the superintendent of the Isolation Hospital in Plymouth, complained that many polio cases were misdiagnosed as brain tumours, multiple sclerosis or 'drunkard's palsy' – but he had the humility to confess that he felt 'quite incompetent to discuss the differential diagnosis in nervous diseases'.[25] Conversely, of 104 cases of suspected polio admitted to the Middlesex Hospital in London's West End during the great epidemic of 1947, 44 turned out to have other diseases, including bacterial meningitis and glandular fever.[26] Further confusion came from people with hysterical tendencies who believed that they had caught polio; many cases had symptoms so convincing that they ended up in hospital with a lumbar puncture needle in their back.[11]

The greatest confusion reigned during the pre-paralytic phase, when the minor symptoms such as fever, headache and vomiting could masquerade as many other conditions, including colds, heatstroke or food poisoning (an understandable mistake if vomiting was forceful enough to hit the ceiling). According to Frauenthal and Manning, many cases of suspected polio turned out to have 'intestinal autointoxication with retarded elimination' (i.e. constipation). Luckily, diagnostic confusion could be settled if the symptoms improved after an enema that induced a 'thorough flushing of the bowels'.[27]

Polio is only one of many infections that can produce meningitic symptoms. Other viruses include several different enteroviruses and the Epstein-Barr virus of glandular fever. All cause 'aseptic' meningitis which usually resolves within a few days. The crucial distinction to be made is from bacterial meningitis, which often kills and leaves survivors deaf, blind or otherwise brain-damaged – hence the need for early lumbar puncture to identify the cause and begin antibiotic treatment if bacteria are responsible.

Even the classic picture of 'acute flaccid paralysis' is not exclusive to polio. Various enterovirus cousins of the poliovirus can attack the anterior horn cells, causing an illness that is assumed to be polio until the laboratory identifies the guilty party. The main culprits are Coxsackie A7 and the drably named enterovirus 71, responsible for outbreaks of paralytic disease around the world.[28] Recently arrived in the Western hemisphere is West Nile fever, a virus endemic in birds and transmitted to man by mosquito bites.[29] This disease appeared in the United States in

1999, possibly having hitched a lift in a bird imported from Israel, and is now endemic across America and in parts of Canada. Ominously, West Nile fever has now cropped up in Greece and elsewhere in Southern Europe. Polio-like paralysis can also result when a monkey virus simply called 'B' gets into man.[30] The enigma of its name, and the tragic circumstances under which it helped to advance the scientific career of vaccine pioneer Albert Sabin, are explained later.

Lead, arsenic and other chemical poisons can occasionally produce paralysis, as they are toxic to the anterior horn motoneurones. In these cases, paralysis usually develops slowly and is accompanied by other signs of poisoning, but diagnostic blunders have been made. One notorious compound is used as gun oil; the curious story of how it found its way into bootleg liquor during the Depression and paralysed tens of thousands of Americans is told in the next chapter.

Testing times

Until the 1950s, polio was diagnosed by the doctor's clinical impression, supplemented by some rudimentary investigations. Lumbar puncture (spinal tap) was part of the ritual of hospital admission with suspected polio, uncomfortable if it went smoothly and agony if the needle went a few millimetres off course. The purpose is to sample the spinal fluid, known more accurately as 'cerebrospinal fluid' (CSF), as it bathes the outside and inside of the brain as well as the spinal cord.

A lumbar puncture is done by pushing a fine needle into the middle of the back, between the spines of the third and fourth lumbar vertebrae (see Figure 2.4). This level is chosen because the spinal cord ends several centimetres higher up, and the chances of spearing a spinal nerve are small. The patient lies on the left side with the knees pulled up to the chin, to open up the gap between the vertebrae. The needle is angled backwards about 45°, and should pass through the ligament bridging the vertebrae and then the meninges, to enter the CSF. All these structures are exquisitely pain sensitive, so generous local anaesthesia and warm encouragement to relax are the rule. Frauenthal and Manning provide helpful tips on how to strap up the patient to prevent the back from arching during the procedure, which can snap the needle deep under the skin. Other hazards of lumbar puncture which they mention include infection, sudden death and friends of the patient attributing paralysis to the procedure.[31]

The first drops of CSF to emerge from the needle may give useful clues about the diagnosis. The pressure of the CSF is usually slightly raised in polio and viral meningitis, but greatly elevated in bacterial meningitis. Nowadays, CSF pressure is measured using a slender manometer tube

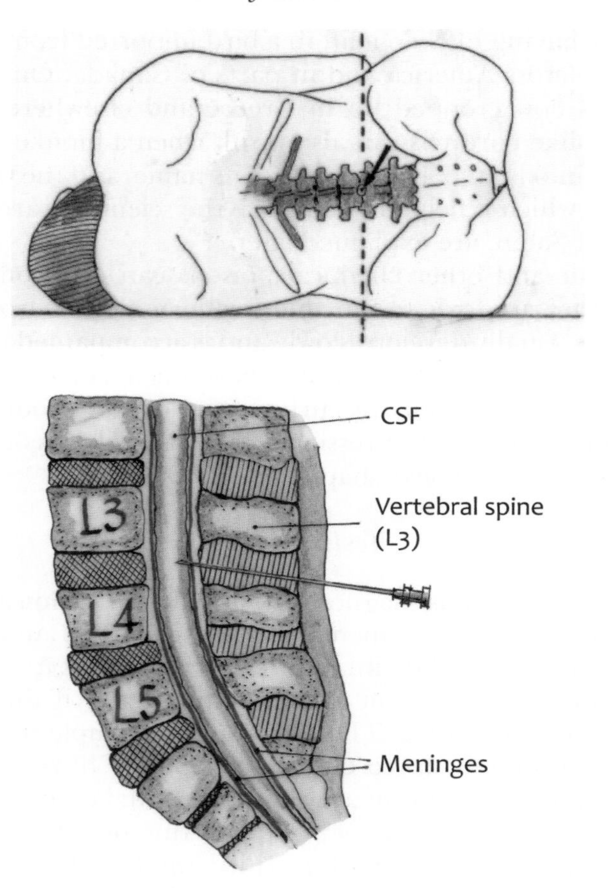

Figure 2.4 Lumbar puncture. With the patient lying on the left side, the needle is pushed in between the spines of the third and fourth lumbar vertebrae, penetrating the meninges and entering the spinal fluid. This level is chosen because the spinal cord ends several centimetres higher up

plugged into the end of the needle. This gives a reading quickly and without wasting the precious sample. Only a few millilitres of CSF are collected, to reduce the headache that often follows the procedure.

In Frauenthal's day (1914), CSF pressure was estimated from how much CSF flowed out before the tap ran dry.[31] In polio, the CSF sometimes spurted several inches out of the needle, and about 50 millilitres would follow. With bacterial meningitis, the high pressure typically pushed out over 100 millilitres. This practice would have made most patients feel worse, especially as lumbar punctures were often repeated every couple of days to see how things were going. There was also a vogue for repeated lumbar punctures to relieve the pressure on the brain supposedly caused by polio. This idea was carried to bizarre extremes by the 'brain washout therapy' devised during the 1930s by Dr George Retan. As explained later, CSF was drained off continuously through a lumbar puncture needle, while weak saline solution was run into a vein.[32]

Normal CSF is 'gin clear', rich in glucose and low in protein. It contains very few white blood cells that have wandered in across the meninges, the barrier which keeps microorganisms out of the central nervous system. The calm clarity of the CSF is quickly disturbed if viruses or bacteria invade the meninges. Viruses such as polio drag in a few lymphocytes from the bloodstream; the protein level rises slightly, but glucose is unaffected. Bacteria such as the meningococcus (which causes the most feared form of meningitis) attract masses of white blood cells, mostly the 'polymorphs' which kill and ingest bacteria. The protein level rises dramatically, while the glucose plummets, consumed by the energy-hungry polymorphs in their battle with the invaders. The self-limiting meningitis caused by polio and many other viruses is called 'aseptic' because, unlike bacterial meningitis, no organisms grow when CSF samples are put into bacterial culture media.

On looking down the microscope, the difference between viral and bacterial meningitis is stark. In aseptic viral meningitis such as with polio, the CSF shows scanty lymphocytes; in bacterial meningitis, there are vast sheets of polymorphs, many containing the offending bacteria. The microscope is superfluous in extreme cases of bacterial meningitis, when the CSF looks like thin mayonnaise and smells like pus from a boil.

One diagnostic elephant trap is Guillain-Barré syndrome (GBS), a peculiar condition that damages the spinal nerve roots where they run into and out of the spinal cord. In 1916, Georges Guillain and Jean-Alexandre Barré attached their names to a strange paralysis that began in the feet and advanced steadily up the body, cutting off breathing if it reached the diaphragm. This was the same as the 'ascending paralysis' reported 60 years earlier by Jean-Bernard Landry, one of the protégés of the great Charcot (who was at Landry's bedside when he died of cholera in 1865).[33,34] The causes of GBS include *Campylobacter*, a bacterium which causes food poisoning, and glandular fever. GBS must be diagnosed without delay, as paralysis may be slowed or halted by intravenous immunoglobulins – but only if given early.[35]

GBS can usually be distinguished from polio by clinical examination and lumbar puncture. In GBS, sensation is impaired as well as motor function, and loss of feeling marches up the body in tandem with paralysis; in polio, the muscles can be painful but sensation is intact. GBS is symmetrical in its attack, whereas polio is usually patchy. The spinal fluid usually shows diagnostic abnormalities in GBS, with very high protein levels and few white blood cells. However, it can be difficult to decide at the bedside whether a patient has GBS or polio. Recently, it was suggested that one famous man assumed to have been paralysed by polio actually had GBS[36] – an intriguing twist, as the man in question led the battle against polio and was instrumental in changing the course of its history.

As late as the 1950s, proving that the poliovirus was responsible for a patient's illness took several days and required the irreversible use of many monkeys. In 1947, Dr Douglas McAlpine of the Middlesex Hospital, London, expressed regret that 'facilities in this country for this type of work are at present extremely limited'.[26] By the time the results came through, most patients were either on the mend, in the life-saving embrace of the iron lung or in transit to or from the post-mortem room.

Nowadays, the diagnosis of polio is plain sailing.[37] The virus can be quickly identified in throat swabs, stool samples or CSF, from its ability to kill particular cell types in culture. Many other viruses do this too, but polio can be pinned down as the culprit by running the test with and without specific antibodies against each of the three types of poliovirus; the cultured cells survive only when the correct antibody is added. A recent encounter with the poliovirus can be confirmed indirectly by comparing antibody levels against the three types in blood samples taken from the patient during and after the illness. As antibodies take several days to be generated, a 'convalescent' sample collected 3–4 weeks after the illness will show higher levels of the specific antibody than in a baseline sample taken at the outset. For absolute proof, the DNA sequence of the guilty poliovirus can be read out by molecular biological techniques. This is essential to work out whether the virus is a 'wild type' or, in a regrettable twist of fate, a paralysing mutant derived from the supposedly protective vaccine.[38]

Luck of the draw

The vast majority of those who met the poliovirus were unaware that they had brushed up against Fate. For most of the rest, polio was an unpleasant but short-lived illness that was soon buried among the other afflictions of childhood. It was only the unfortunate few who saw 'The Crippler' in its full, terrifying glory.

The clinical outcome of polio has always been a lottery. In time, the molecules that determine the risk of paralysis and death will be identified; until then, it will remain a question of luck.

Both the poliovirus and the victim determine what happens. Some strains of the virus are more virulent than others, such as the notorious Type 1 strain known as 'Mahoney', which was particularly likely to cause paralysis and death. The individual's ability to fight off the virus is crucial. There may be some pre-existing immunity from an earlier encounter with the virus or from vaccination. Otherwise, protective antibodies against the virus have to be generated from scratch once the virus is picked up by the immune system. The process is a frantic scramble against the timetable of invasion. Ideally, the poliovirus is confined where it belongs, in the gut. However, defences may not build up fast enough to prevent the virus from breaking into the bloodstream and finally the central nervous system.

The impact of polio depends crucially on other characteristics of the victim, ranging from general health and nutrition to the unmeasurable qualities of resilience and optimism. Different people develop different ways of coping (or not) with being brutally wrenched out of a healthy childhood and dumped into a life sentence with leg braces, a wheelchair or the iron lung. At one end of the spectrum is Paul Bates, who caught polio at the age of 20 while on patrol in the Malaysian jungle in 1954.[39] He was paralysed from the neck down and kept alive by artificial respirators, but refused to be beaten. He nicknamed himself 'the Horizontal Man', and initially communicated by pointing to letters on a card one by one with a stick held in his mouth. Later, he ran a radio station, acted as an ambassador for polio sufferers and even planned his route for taking part in the London to Paris air race. In stark contrast was the unnamed American high school student, also tetraplegic, who cried inconsolably because his paralysis had robbed him of the means to kill himself.[40]

Family and friends can also limit or enhance the damage done by the poliovirus. Positive support can make all the difference, such as being taken out to follow the fortunes of a favourite sports team – quite an undertaking if an iron lung (in the back of a specially converted truck) has to come too.

Others were much less fortunate. In the past, polio carried a stigma in many societies – although not as damning as with the greater scourges of smallpox, syphilis or cholera (as late as 1832, cholera sufferers were beaten to death on the streets of Paris).[41] Even in the civilised United States, many families simply hid away their polio victims. Paralysed or brain-damaged youngsters were packed off to be among their own kind, in one of the thousands of long-term residential homes for 'crippled children' that were dotted around American cities.

Europeans did the same. On the outskirts of Vienna stands an imposing turreted building, Am Spiegelgrund. During the 1920s and 1930s, this was a long-stay 'psychiatric hospital' to which were consigned mentally handicapped children with hereditary conditions such as Down's syndrome, or severe brain damage acquired from cerebral palsy or polio. During the late 1930s, Am Spiegelgrund was one of the centres in the

Nazi diaspora that pushed through the 'T4 Aktion' directive, so called because orders were sent out from the offices at No. 4 Tiergartenstrasse, near the Zoo in Berlin. All that was needed to enter a child into the T4 programme were two doctors' signatures, certifying that this was *lebensunwertes Leben* ('life unworthy of life'). Some were starved to death in their hospital beds, carefully monitored by doctors and nurses; others were murdered by injecting petrol or phenol directly into the heart.[42] All this took place barely five miles from the hospital where Adolf Strümpell first described polioencephalitis.

Close up and personal

Books have been written about the personal experience of polio, from different viewpoints: a direct hit by the virus, the family caught up in the collateral damage, and those swept along in the furore of an epidemic or trying to cope with the aftermath. Many of the stories are gripping and profoundly moving. They often tell us more about human nature than the actuality of mankind's battle against the poliovirus. They also provide fascinating and often sobering insights into the state of medical and scientific knowledge, and attitudes to health and disability at the time.

This is an anthology that continues to grow, but very slowly now. It is possible that the last entry will have been added by the time you read this.

A.J.
Cornwall, England. 22 September 1911

A previously healthy 25-year-old man, holidaying on the Cornish coast. He felt perfectly well on going to bed the previous evening but awoke the next morning with a severe frontal headache and retching. Nonetheless, he joined a pleasure boat for a day trip around the harbour. Although the sea was calm, he felt increasingly unwell and began vomiting.

On returning to the dock at 5 p.m. that evening, he felt too ill to walk home and took a taxi directly to the hospital. On admission, he was drowsy, and by 6 p.m. had fallen into a coma. He stopped breathing just before 8 p.m.

Post-mortem showed congestion of the entire brain, the appearances of polioencephalitis.[43]

Akearok
Chesterfield Inlet, Nunavut, Canadian Arctic. February 1949

Outside, the temperature was –30° C. Inside, 10-year-old Akearok and his parents huddled together on the bench of compacted snow that served as the family bed, heavily wrapped in caribou furs. Their igloo was 35 miles away from the Hudson Bay Company's trading post at Chesterfield Inlet, high up the eastern coast of Hudson Bay. They had been cut off for months and were unaware that polio had picked off a string of Inuit settlements hundreds of miles to the south.

On 11 February 1949, Akearok's father braved the cold and made the day-long trek into Chesterfield to collect his family allowance. Everyone at the trading post seemed well, and he returned home the following day. A few days later, he fell ill with a high fever, headache and a stiff neck, and his son quickly developed the same symptoms. Within a week, the father was recovering, but Akearok was paralysed and rapidly losing ground. He died on 22 February, still enveloped in his furs and lying on the bed of snow.[44,45]

By then, polio had broken out in Chesterfield and was being carried to the surrounding settlements by the constant traffic through the community's hub, the trading post. The epidemic lasted just three weeks, from 14 February to 7 March, but was vicious and devastating. Of the 275 Inuit in the area, half developed polio symptoms and 39 were left paralysed. Akearok was one of 14 who died.

It took months of dogged detective work by Dr Joseph P Moody, Field Medical Officer for the Eastern Arctic, to work out how polio had found its prey in 40,000 square miles of desolation that were home to just 4,800 Inuit. Moody retraced the epidemic on foot and horseback, accompanied by constables of the Royal Canadian Mounted Police and often facing 'considerable hazard and personal danger'.[46]

The first case had occurred seven months earlier and 500 miles south of Chesterfield, in Churchill, Canada's main Arctic port on Hudson Bay (see Figure 2.5). On 8 July, a white Canadian working at the air force base developed paralytic polio and was flown to Winnipeg for treatment. The infection may have been brought in by the crew of a ship from Liverpool. Polio then began to track north through Inuit settlements along the coast of Hudson Bay, beginning with an Inuit man and a Chipewyan Indian girl just north of Churchill. Three cases of paralysis followed in September in Nunella, 75 miles north, followed in October by eight more cases in Eskimo Point, 150 miles further on.

Polio travelled at roughly walking speed across this difficult terrain, and Moody soon had his man: an Inuit called Tutu, who had left Churchill on foot in late July for Eskimo Point. Tutu had visited all the settlements where polio later broke out, and was always received with the traditional Inuit welcome of a shared bed and food. As he remained well throughout his journey, Tutu must have been an asymptomatic carrier.

From Eskimo Point, polio moved inland, flaring up in December in Padlei, 125 miles to the east, and then in February in Kazan, 75 miles south

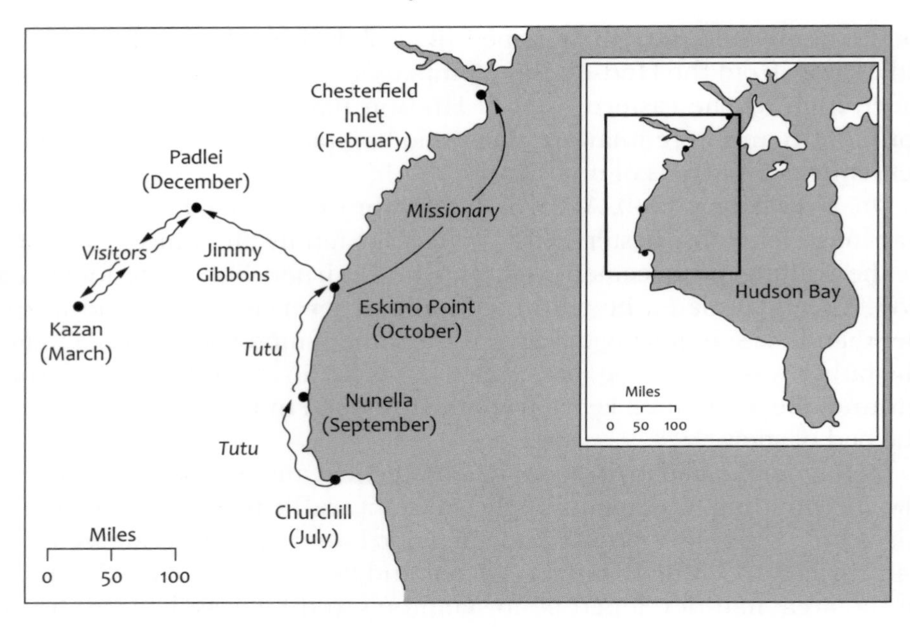

Figure 2.5 Map showing the Hudson Bay polio outbreak of 1948–49. Redrawn from JD Adamson et al. (47) and AFW Peart (48) by Ray Loadman

of Padlei. For the first time, there were deaths: two in each settlement, with ten people paralysed in Padlei and four in Kazan. Moody again tracked down the bearers of polio. Jimmy Gibbons, an Inuit special constable with the RCMP, had been paralysed in both arms during the Eskimo Point outbreak. He still had residual weakness when he set off on horseback for Padlei in late November. Traders from Kazan who visited Padlei in January took home polio as well as essential supplies.

All this was the prelude to the disastrous outbreak in Chesterfield Inlet. Polio jumped 180 miles from Eskimo Point with a white missionary who flew in on 28 January. He had visited a polio victim in hospital in Eskimo Point and, like Tutu, was an asymptomatic carrier. The missionary divided his time between the two magnets for visitors to Chesterfield – the Mission and the trading post – and so was strategically placed to spread the infection. The first cases appeared 16 days after his arrival; Akearok was one of the earliest to die.

The Hudson Bay outbreak was catastrophic because polio was unleashed on a 'virgin' population. Thanks to their isolation, these Inuit had never met polio before and therefore had no individual or herd immunity to limit its spread.

This 'virgin soil' epidemic provided Dr Moody and his colleagues with an epidemiological wonderland to explore. They noted that the

Inuit way of life encouraged faecal-oral transmission of polio: excreta deposited immediately outside igloos, 'intimate' sleeping arrangements that included visitors like Tutu, and the mothers' habit of chewing tough meat before passing it on to their children. Moody complained how difficult it had been to pin down the sequence of events because 'no-one could be more indifferent to the passage of time or more vague in temporal orientation than the Eskimo'. Other challenges included doing a post-mortem on a cadaver that had been kept outside, then accidentally thawed and refrozen. Moody was excited by the old notion (by then discredited) that certain facial features – 'mongoloid' eyes, big front teeth and freckles – predicted susceptibility to polo, but did not manage to put the theory to the test:

> Although no specific anthropomorphic observations were made on the group under discussion, it is well known that Eskimos show all of these characteristics.

The Chesterfield Inlet polio outbreak made minor celebrities of Dr Moody and his team. It also went into the record books for having the highest attack, paralysis and death rates ever reported for polio. Of the 275 Inuit at risk, 50 per cent developed clinical polio, 14 per cent became paralysed and 5 per cent died. If New York had suffered the same incidence during the great epidemic of 1916 which crippled the city and terrified the American nation, there would have been 800,000 New Yorkers paralysed and 280,000 deaths, rather than 9,000 and 2,500, respectively.

The Field Medical Officer for the Eastern Arctic added a grim footnote to the tragedy of Chesterfield Inlet. On 21 August 1949, a Canso seaplane of the Royal Canadian Air Force arrived to take seven Inuit polio victims for rehabilitation at King George Hospital in Winnipeg. The plane crashed during a thunderstorm at Bigstone Lake, 200 miles north of Winnipeg. All on board were killed.

The bodies of the white Canadian crew and passengers were flown home to their families. On the orders of Dr Moody, all the Inuit were wrapped in canvas and bundled together in a single coffin. This was buried in an unmarked grave at Norway House Inuit Reserve, near the crash site and 700 miles away from home.[47]

Vivi Ebert

Copenhagen, Denmark. August 1952

The Blegdams Hospital was already stretched to breaking point when 12-year-old Vivi was rushed into the isolation unit on 26 August. Copenhagen had been hit by the biggest and most vicious polio outbreak

Figure 2.6 Parents on the outdoors observation platform, looking at their child in the polio isolation ward at Blegdams Hospital, Copenhagen, August 1952

in Denmark's history. Over 1,000 patients had been admitted since early July; half were paralysed and 150 needed respiratory support in the iron lung. This did not guarantee a happy ending: 80 per cent of the patients put into the machine died in it, mostly from chest infections.[48]

Vivi was one of 50 patients admitted that day and it was soon obvious to her parents – standing outside in the sunshine on the wooden platform below the observation window of the little girl's isolation cubicle (Figure 2.6) – that she was losing the battle. Over the next day, paralysis closed in on her respiratory muscles, and by the morning of 27 August, she was barely breathing. Her parents were told that she was going to die because, regrettably, she was beyond help.

Later that morning, someone looked at the terrified, suffocating girl and thought that she – and he – had nothing to lose. This was Bjorn Ibsen, a Danish anaesthetist who had recently trained at Massachusetts General Hospital in Boston and was helping out at Blegdams. Ibsen was only there by chance. His wife had fallen into conversation with the hospital's deputy director, who happened to be on board the ship bringing the Ibsens back to Copenhagen from America.[49]

Ibsen now decided to try a method used to ventilate patients who were deliberately paralysed by curare-like drugs to relax their muscles during major surgery. He put the girl to sleep, and a surgical colleague cut a small window in the front of her trachea. Ibsen pushed in a rubber tube with a cuff around its end that could be inflated to make an airtight seal inside the trachea. A small crowd had gathered to watch him perform this operation; they all quickly left the room when Ibsen attached a bag filled with oxygen to the tube and discovered that he could not inflate her lungs. The muscles of the rib cage had gone into terminal spasm. In desperation, Ibsen gave her an intravenous injection of barbiturates, which relaxed her ribcage enough for him to begin artificial respiration by squeezing the bag of oxygen (Figure 2.7).

Ibsen's bold idea saved Vivi's life, and those of over 100 other polio victims during that outbreak. Once started, the repeated squeezing of the bag – 20 times a minute for adults, faster for children – could not be stopped until the patient was ready to breathe unassisted again. The solution: a huge team of volunteers, including 1,500 medical and dental students, working in shifts of several hours each. They kept the patient breathing for as long as it took; in some cases, this was over three months.

By the time the epidemic ran out of momentum in the spring of 1953, just under 200 polio victims had been treated in this way for a total of 165,000 hours. The mortality rate during artificial respiration had fallen to just over 20 per cent.[48]

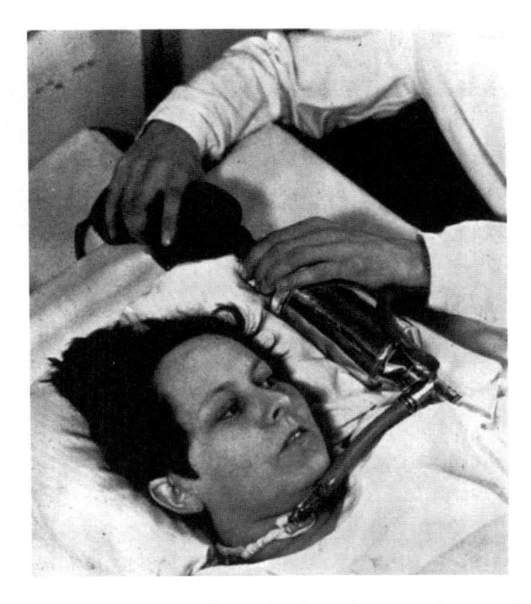

Figure 2.7 Young woman being ventilated by hand through a tracheostomy, Blegdams Hospital, Copenhagen. Working in relays, over 1,500 medical students and other volunteers kept nearly 200 polio victims with respiratory paralysis alive during the outbreak in the summer of 1952. More patients survived with this method than with the iron lung. Reproduced by kind permission of the World Health Organisation

And Vivi was back home, her tracheotomy healing nicely, and on the road to recovery.

Margaret Scrimgeour
Essex, England. August 1955.[50]

It was a very hot summer, and I was five and a half. The afternoon before, I had pains in both legs, and my mother walked me from school to the doctor's – he couldn't find anything wrong – and then home again.

The next morning, I couldn't move my arms or legs. I managed to crawl off the bed, but fell on the floor, trying to call to my mother. The doctor came immediately, and this time, he looked worried. He lifted one of my arms and told me to keep it up; I couldn't, and it fell back onto the bed. The other arm was the same, and both legs. And by then I couldn't lift my head off the pillow.

The doctor told my mother it was infantile paralysis. He didn't say anything to me, though, and neither did the doctors or nurses in the Rush Green Fever Hospital in Romford. They were all dressed in white and wearing masks, with just their eyes showing. I was terrified.

After they'd examined me, they said they were going to do something to me. They didn't explain what it was for or what it would be like. I thought they said 'punch', but of course it was a lumbar puncture. They put me on my side and doubled me up, with three or four of them holding me down in case I wriggled. Then they stuck the needle in my back. I was scared stiff, and it was agony. It's nearly 60 years ago now, but I can still hear my screams echoing off the walls.

My mother remembered those screams too, although it upset her so much that she couldn't tell me for many years. She was sitting outside the closed door, by herself. They hadn't told her either what they were going to do to me.

I was on the unit for several months. It was a frightening place to begin with, although I got used to it in the end. There were several children with polio, in beds or cots or the iron lung, with just screens to separate us if the doctors or nurses needed to do anything.

The iron lungs scared me. My first sight of them, through the bars of my cot when they wheeled me into the ward, was horrific: great metal tanks with a child's head sticking out at one end. They didn't tell me what they were for, and most of the children in them either couldn't speak or didn't want to. I thought this was what they did to everyone and used to wake up in a panic in case it was my turn to go in.

One boy I remember was in an iron lung next to my bed. He had curly fair hair and could only move his eyes and the muscles in his face. He loved reading. The nurses would put a book in a special frame on the end of the machine above his face, but were often so busy that they then forgot about him. Later, when my strength began to come back, I used to crawl out of bed and across the floor, then pull myself up beside him. I could follow his eyes as he read, and turned the page for him when he reached the bottom. He could still smile, and used to do that when I wound up my doll and let it dance on his bedside table. The doll was bright blue, and it played 'Waltzing Matilda'. I don't know what happened to him.

The place played tricks with my mind, especially after dark. During the day, with the nurses bustling around, you didn't notice the constant sighing noise that the iron lungs made. At night, when it was still, the machines sounded like monsters breathing, and they gave me nightmares. And then one day, I looked out of the window and saw my father driving past. Sometime later, I realised that we didn't have a car. It was a hallucination. So were the nun and my grandmother from Ireland, both of whom came in and sat down on my bed and had a conversation with me.

The doctors never said anything, just looked at the charts on the end of my bed and walked away. As I remember them, the nurses were mostly too busy to spare me any comfort or warmth. Once they gave me an enema, on a sheet of red rubber and with the screen pulled around my bed. As usual, they didn't say why they were doing it, just arrived and got me ready to push the tube in. By then, my strength was coming back, and I fought them off. So more of them came – five or six of them in the end – to hold me down until they'd done it.

But they weren't all like that. After months in bed, I finally took my first wobbly step, and the young nurse who helped me cried with happiness. That memory still serves me today.

I remember being horribly lonely. My parents were allowed to visit just once a month when I was in isolation. Like everyone else, they wore white gowns and masks, and I could just see their eyes. They had to sit away from my bed and couldn't give me a kiss or a cuddle. They brought news from outside, including our neighbours who had put up high fences around their gardens when they knew what I had. Talking to my parents, I also realised how much I missed my friends.

When I got out, I needed regular physiotherapy and operations on my legs. The worst treatment was electrical stimulation, which made your muscles jump. When nurses were new, they always seemed to set the dial far too high, and it felt like being electrocuted. Years later, I found out that the electrical stimulation treatment was completely useless.

The first time we saw a doctor after my discharge, he told my mother, 'She'll end up in a wheelchair like the rest of them'. He made my mother cry, but I suppose I'm grateful to him because that's when I made up my mind to keep walking for as long as I could.

And I still am, 57 years later, although it's taken ten operations and a lot of determination.

John Johnston
Bristol, England. November 2012.[51]

I caught polio in 1935, when I was two and a half years old. We lived in a small village in Hampshire, and they took me to the hospital in Alton. When I came out, my right leg couldn't take my weight, and I had to wear a leg-iron that ended in my right shoe and was strapped around my thigh so that the knee wouldn't buckle. But I got around and managed pretty well at primary school, thanks to the headmistress who was quite protective of me; the other children were fine, too.

In 1938, we moved with my father's job to Gillingham in Kent, not far from London. By then, my right leg was a couple of inches shorter than the left. I went up to London to the Royal National Orthopaedic Hospital in the West End. With the rigid leg-iron, my right leg stuck out straight and blocked the aisle on the bus. I needed an operation on my ankle and a shoe with a built-up heel to walk satisfactorily.

This was wartime, and the after-effects of polio made the Blitz all the more challenging. Our house was on the way in for air-raids on London. Many children there were evacuated to safer parts of England, but we didn't have that option; because of my leg, it was felt that another family would not be able to look after me. So I stayed at school in Gillingham. The rule was that if you could run home and get into the Anderson air-raid shelter within three minutes of the air-raid siren starting, then that's what you did. The rest of us simply hid under desks in the classroom while the bombers went over. Luckily, we were never hit.

Back then, my paralysed right leg didn't really hold me back that much. I played table tennis and was a not-very-good goal-keeper in the school hockey team. I never learned to swim, though; in the pool, my right leg went blue and cold and dragged me down.

After school, I worked first in local government. By law, the County Council had to fill a quota, with 4 per cent of its employees being 'green card' holders, registered as disabled. Later, I went to college and trained as a history and geography teacher, and that's what I did until I retired. There, I met Margaret; we married in 1963 and have two sons.

It was only much later, in my early sixties, that polio really caught up with me. My muscles started losing strength – and not just in the right leg, which had been paralysed since the original attack. Margaret and I used to love exploring castles and churches; before long, I couldn't do that any more.

The muscle weakness got worse quickly after I reached 70. It took the doctors a long time to work out what was wrong. Most of them, even the specialists, had never seen a patient with acute polio, and to them the disease simply didn't exist any more. I remember a doctor at Frenchay Hospital, the big specialist neurology centre in Bristol, confessing that he didn't know much about polio and suggesting that we should look it up on the Internet. The next time we went back, he said he'd printed off some information for us – it was a pile of paper about an inch thick – but he still didn't know what was going on.

We went to another specialist in Bath, and at least he investigated me. A scan suggested that the bottom end of my spine was 'all mangled' (his words). This gave us hope that the weakness was due to something else that might be remediable by surgery. Unfortunately, the operation didn't help.

They finally told me that I had post-polio syndrome. Both my arms and both legs are partially paralysed now, and I have trouble swallowing and speaking. The last time I walked on my own was seven years ago, and two years ago I could no longer manage the stairlift. Since then, I've been confined to an armchair in our living room. My bed is in there too. Margaret can't get me out of the chair now, but they've put a hoist on the ceiling which means that we can mostly cope on our own.

I suppose I was lucky that polio let me off so long. Looking back, it's little things that strike me now. The first set of leg-irons with a hinged knee joint was a godsend. This meant simply that I could bend the knee while sitting down, and lock it straight for walking. It sounds trivial, but it made such a difference to me. I had to wait until I was 20 to get one. It still weighed a ton; they told me that it was perfectly feasible to make a lightweight version, but they couldn't afford to.

We visited Prague during the 1980s, and that was memorable too. We went to a concert in the opera hall and had to climb up to our seats, right up in the Gods. We saw the usherette watching us. In the interval, she came to find us and took us down in the lift. We had rather better seats for the second half: in the Presidential Box.

Zoltán Lipey Rózsa and Iván Szánti

Long-term residents in the Baba Utca Rehabilitation Home, Budapest, Hungary. January 2013.[52]

Zoltán:

I fell ill with polio in 1957, when I was 18 months old. I was severely affected, and a tracheotomy was immediately performed on me so that I could be ventilated. I have been dependent on the ventilator ever since, that is, for 56 years; I am one of the longest survivors on ventilation in Hungary. I have to be on the ventilator for 16 hours a day.

I have been living in Baba Utca since 1959. When I was young, I attended classes held here in the home and did homework and exams like everyone else. In the end, I got my secondary-school diploma. My mother adapted more easily to my illness and tried her best to support me in everything. It proved much harder for my father to deal with my situation and all the burdens it entailed.

I can only move my legs; all the other muscles in my body are paralysed, unfortunately. Using my first two toes, I can drive a disabled car, use a phone, write and even play the zither (I was taught to play by nuns who were hidden in the home during the socialist era). I have a knack with machines. Doing repairs is one of my hobbies, and I am skilled at repairing ventilators. I know the ventilation settings of almost all of my fellow patients by heart – hence my nickname, The Expert. It's an official nickname, printed on my nameplate.

These days I like surfing the net, though unfortunately I can make out the display less and less. I very much enjoy talking to people, and when the weather is good, I like to get about in the streets in my special car and do the shopping. I still get some malicious remarks; they can hurt me, but I try to ignore them.

I like socialising. Some events are held here in the Home, such as the carnivals before Ash Wednesday, although less frequently now than in my youth. I like music, and dance in my own way; having healthy people dancing around me makes me happy, not miserable. I've had relationships with women, but now I'm single.

Some of my friends in Baba Utca are no longer with us. When I was younger, I couldn't cope with the thought of dying, but now have come to accept death as a part of life. All the same, it still hurts me to witness grief and suffering.

Iván:

I am 54 years old, and was six months old when I caught polio during the big Hungarian epidemic of 1959. I was the last polio patient in my village of Budaörs, just outside Budapest. Initially, I spent a few hours in an iron lung, but my condition deteriorated and I had to have a tracheotomy. I have been on a ventilator ever since.

All my muscles except those in my right arm and my head and neck are paralysed. My parents took my illness very badly; they dragged me from doctor to doctor and experimented with alternative treatments on me, hoping to find someone that could heal me. My mother now lives in Canada; unfortunately her illness prevents her from coming to visit me.

I spent the first ten years of my life in Saint László hospital in Budapest. I very much liked being there, and I felt torn apart when I was moved here: I rebelled at first, but eventually came to like it. This is my forty-fourth year here.

School was hard for me in the beginning. I was obstinate and didn't study, but eventually I got my secondary-school diploma and took a degree in theology at Péter Pázmány University. I made a lot of friends in the university and still keep in contact with them. I am religious; faith plays a very important role in my life. At the same time, I enjoy the company of beautiful women.

It's not easy, but I love travelling and flying. I have been to Austria, Germany, the Netherlands, Luxembourg, and even to Tunisia and Israel. People abroad, especially in Western Europe, accept my condition much more readily and behave much more naturally in my presence than those here in Hungary.

Nowadays I spend my time watching television and surfing the Internet, or receive visitors. I usually stay up late, until two in the morning. I have come to accept my condition, and if I am allowed to say so, I am content with my life.

3

The Virus That Never Was

Even five years ago, if anyone had suggested that polio was an infectious disease, it would have been looked upon as a joke.

Dr L. Emmett Holt, commenting in 1910 on Simon Flexner's paper confirming that polio was transmissible from humans to monkeys.

We have already seen the steps leading to a fact that we take for granted, namely that polio is an infection. The cluster of paralysed children around the silver mine at Sainte-Foy l'Argentière in 1885, which persuaded Cordier to write of 'an epidemic'. Then the network of polio cases, visible and invisible, which fanned out from the village school in Trästena in 1904 and convinced Ivar Wickman that this must be an infection, transmitted from person to person. And finally, in the run-up to Christmas four years later, Karl Landsteiner's magic lantern show in Vienna, showing that a filtered extract of spinal cord could transmit polio from a dead boy to a monkey.

Until that moment, there had been plenty of room for uncertainty about what might cause polio. Throughout the nineteenth century, polio was a medical curiosity that was unknown to most doctors and gave away few clues about its nature. This was also a period of turmoil and wild ideas in the history of infectious diseases in general. The action centred on the spirited and occasionally violent standoff between two warring factions, the 'germ theorists' and the 'miasmatists'. Their energies were focused on the greatest human pestilences, such as cholera and smallpox, which routinely rampaged through entire cities and killed thousands at a stroke. Polio, this fine-print rarity that occasionally picked off a few children, was never one of the prize pieces in this conflict. Nevertheless, early notions about its possible cause were inevitably moulded by both sides of the germ-miasma debate.

Theories about what we would recognise today as 'infections' go back a long way. The idea that invisible particles could cause diseases like measles and smallpox was originally planted by Girolamo Fracastoro during the 1550s, while bacteria were first seen down the microscope in 1674 by Antonie van Leeuwenhoek, who was captivated by the minute 'animalcules' swimming around in the slime scraped off his teeth. They were duly classified as tiny life forms, but another two centuries rolled by before bacteria were shown to cause human diseases.

In the meantime, some believed that diseases could be spread from person to person by 'contagion' or by 'infection'. The vagueness of these notions is nicely illustrated by the *American Medical Lexicon* of 1811–

> *Contagion*: 'a poisonous humour secreted by a living vascular surface that can cause disease'.
> *Infection*: 'effluvia or particles, shed by distempered bodies which mix with the juices of others [and] occasion the same disorders as in the bodies they come from'.[1]

– and by Noah Webster's helpful definition of 'contagion' as a 'septic acid' that could hit a man at ten paces.[2]

By contrast, the miasmatists believed that many diseases were due to 'miasma', or poisons released into the air by putrefaction. Pus in war wounds was produced by vapours from the decaying corpses of those slain in battle. The torrential diarrhoea of cholera was nothing to do with 'germs'; the stink of human excrement in open sewers made it blindingly obvious that miasma was responsible. Moreover, malaria (from 'bad air', in medieval Italian) was caused by the emanations of swamps, and yellow fever by the fumes from sacks of coffee left to rot on the quayside.

Miasma theory flourished throughout the nineteenth century, even though it had no credibility other than a history which went back to the Middle Ages.[3] Miasmatists were notoriously unimpressed by evidence, such as John Snow's famous demonstration in 1854 that a cholera outbreak in London's Soho could be blamed on drinking water contaminated with the 'putrid effluvia' of cholera victims. Snow found the pump responsible and stopped the outbreak by removing its handle so that it could not be used.

'Miasma' is not even mentioned in the 1892 edition of William Osler's massive *Principles and practice of medicine*.[4] This was because 'microbe hunters' such as the German Robert Koch had by then proved that certain bacteria caused specific infections. Koch put germs decisively on the map by characterising the bacteria of tuberculosis and cholera in 1882 and 1883. Exploiting his methods, others rapidly nailed the causes of diphtheria (1883), typhoid (1884) and plague (1894). But this did not prevent some doctors from continuing to deny that epidemic diseases such as smallpox, cholera and polio could be caused by microscopic, infectious agents transmitted from person to person.

Those unconvinced that polio was an infection included respected authorities on both sides of the Atlantic. In Sweden, the great Medin sat on the fence following his monumental study of the great epidemic of 1887.[5] He clearly regarded polio as 'contagious', but did not believe that it spread from person to person. He may have thought that it was miasmatic, or perhaps belonged in the strange no-man's-land for diseases with a combined germ-miasma origin, which briefly separated the two camps.[6]

Seven years later and over in America, Charles Caverly plotted the course of the first major American epidemic, in Vermont. He spotted non-paralysed cases and deduced that these too 'belonged to the epidemic', but somehow concluded that 'it is very certain that [polio] is noncontagious'. Caverly had no original thoughts about the possible cause; instead, he regurgitated the clinical impressions of the attending doctors, such as overheating and minor bangs on the head.[7]

Doubts about the infectious nature of polio persisted in the United States, thanks to influential people such as Caverly and the orthopaedic surgeon and polio expert Dr Robert W. Lovett, later famous as the man who diagnosed and treated the paralytic illness of Franklin D. Roosevelt.

On 9 June 1908, the Massachusetts Medical Society assembled in the Harvard Medical School in Boston for its annual meeting. Top of the bill was Lovett's state-of-the-art address on polio, with 'especial reference to etiology'.[8] Lovett acknowledged the 'commonly received opinion' that polio was an infection, but insisted that an infectious cause 'cannot be regarded as established by bacterial evidence so far collected'. He gave a long list of possible causes of the characteristic spinal cord lesions, including 'sepsis, trauma, specific infections such as measles, chilling of the body and over-exertion'. Toxins such as lead and arsenic could also be responsible. It was difficult to argue with Lovett, who was powerful and persuasive and had a plethora of references to back up his claims.

Within a year, however, Lovett had changed his mind, together with most polio experts. Just six months after the Boston meeting, Landsteiner's landmark article on the filterable virus of polio was published, followed rapidly by a clutch of confirmatory papers from Landsteiner and Levaditi in Paris and – closer to home and American consciousness – by Simon Flexner's group at the Rockefeller in New York. Polio was an infection. QED.

Except for those who were convinced that the 'so-called polio virus' was a bandwagon going nowhere. To them, polio was caused by foods, poisonous weeds, electrical appliances or chemical toxins. Surprisingly, the poliovirus agnostics were not just the usual suspects, such as the anti-vaccinationists and natural therapists who had a vested interest in undermining orthodox medicine. Some were doctors and scientists, respected enough for their views to be published in reputable scientific journals.

And this was not simply a passing phase while the poliovirus was establishing its credibility. Papers denying the existence of the poliovirus were still appearing in 1954, when the virus had been seen under the electron microscope and was the key ingredient in the new polio vaccine which Jonas Salk was injecting into 650,000 American children.

Conspicuous consumption

> Foods must be in the condition in which they are found in nature, or at least in a condition as close as possible to that found in nature.
>
> Hippocrates, quoted by Dr Benjamin Sandler, 1951[9]

Eating is at the centre of human life and health, and the notion that particular foods caused polio was guaranteed to seize attention and induce fear. The menu of incriminated foods stretched from fresh fruit and milk to sugar and seafood, and brought in some spectacularly poor science along the way.

It had long been noted that the polio season coincided with harvest time, at least in North America and Europe. Many fruits and vegetables were blamed: bananas, blueberries and English mulberries in Massachusetts in 1908; assorted produce sold by Greek and other foreign peddlers in America during the 1920s; strawberries in an English boarding school in 1939; green salad leaves forced down reluctant French children during the German Occupation[10]; and up to the 1950s, windfall apples in Sweden.[6] All this led Dr Ralph Scobey, president of his own Polio Research Institute Inc. (1140 Salinas Avenue, Syracuse, New York) and a leading poliovirus agnostic, to claim that polio 'only occurs in those countries which raise the same type of agricultural products'.[11] This intelligence might have come as a surprise to victims of polio epidemics in Northern Africa or the Canadian Arctic.

How could eating fruit cause polio? To believers in the poliovirus, there was no mystery: the virus was simply carried on unwashed fruit like any other organism responsible for food poisoning. However, an eight-year research programme to find the poliovirus in washings of fresh fruit (and in well water and dog stools) from Northern Ohio finally drew a complete blank in 1943.[12] This left the door open for Scobey and others who believed that the real culprit was agricultural pesticides, described below.

Sugar was another 'dietary toxin' that caused polio, at least according to anti-vaccinationist Eleanor McBean.[13] Writing in *The Poisoned Needle* in 1957, McBean pointed out that sugar had a chemical formula ($C_{12}H_{22}O_{11}$) and must therefore be 'classified as a drug' rather than an 'article of food as was formerly believed'. Sugar was a 'destroyer of health', and polio was one of its most serious consequences. Polio was so common in the United States because the average American went through 100 pounds of sugar each year; it was 'practically unknown' in China, where the annual sugar intake averaged only 3 pounds. Her explanation of how sugar caused polio was pithy –

> Sugar is converted into alcohol almost immediately after it is taken into the body and does the same damage as alcohol. It dehydrates the cells and leeches [*sic*] the calcium from the nerves, muscles, bones, teeth and all tissues that are supplied with calcium ... A serious calcium deficiency is a forerunner of polio.

– and factually wrong in every respect.

Cows' milk was another popular culprit, beginning with Ivar Wickman's observation in the Swedish epidemic of 1913 that all ten polio cases in the parish of Ukla drank milk from the same supplier, whose son had been paralysed.[14] Similarly, a polio epidemic 'of unusual severity' which struck down 62 people in the English town of Broadstairs in 1926 fell within the delivery area of a single dairy.[10] Poor hygiene, with poliovirus hitching a lift in the milk?

Not necessarily, claimed those who were convinced that milk contained a polio-inducing toxin. There were precedents. Formaldehyde, better known as embalming fluid, had been used to sterilise milk in Australia in 1897, and had apparently caused some cases of paralysis.[15] The practice had been abandoned, but the notion persisted. Natural poisons were known to find their way into milk. The mysterious 'milk sickness', which regularly killed infants in the Smoky Mountains of North Carolina until the 1920s, turned out to be due to the toxin tremetol, found in a woodland plant called white snakeroot.[16] Wayward cows that ate the plant developed muscle tremors and produced highly toxic tremetol-laced milk that, according to William Osler, could kill a dog in six days.[17] People who drank contaminated milk also developed muscle trembling, vomiting and coma.

During the 1940s, Dr Marsh Pitzman, a physician in St. Louis, Missouri, set out to prove that polio was due to poisoning by the thorn apple, known in North America as 'jimson weed'.[18] Pitzman knew about the alleged virus of polio, recently imaged by the 'super-dooper' electron microscope, but thought it was too small to be believable. Instead, the cause was atropine, the ingredient that put the 'deadly' in 'deadly nightshade' and that caused the toxic effects of jimson weed. There were promising hints, such as the ten-year old boy with polio who was the only milk drinker in an otherwise unaffected community; while visiting him, Pitzman spotted the family cow browsing on hay contaminated with jimson weed.

Pitzman's jimson weed hypothesis was published in June 1928, in the *Journal of the Missouri Medical Association* (edited by 'my good friend, Eddie Goodwin'), but all his experiments to prove the case failed spectacularly. Atropine could not be detected in milk from cows fed jimson weed, and polio-like paralysis could not be produced even with massive doses of atropine injected into dogs and cats. Pitzman refused to let negative results spoil a good idea, however, and over the next 16 years collected over 20 rejection letters, beginning with the *Journal of Experimental Medicine* and ending in 1945 with *Readers' Digest*.

In 1951, disgusted by the prejudice among the 'strong virus proponents', Pitzman bowed out with a self-published a monograph, 'The cause and prevention of infantile paralysis'.[18] The copy in the archives of the Philadelphia College of Physicians is inscribed by the author, in red crayon: 'An old-timer talks out in good old-fashioned style. And maybe solves the mystery of justly dreaded poliomyelitis'. He adds a plaintive PS: 'Maybe to your Book Reviews Editor (If any?)'. His hint was not followed up.

In his various diatribes against the poliovirus, Ralph Scobey mentioned the ability of seafood and shellfish to paralyse.[10] Here, at least, there was some substance. Shellfish can occasionally paralyse and kill, because they concentrate potent nerve toxins produced by the algae on which they feed.[19] These toxins can also accumulate in larger species, including the Pacific puffer fish (*fugu*, a traditional Japanese delicacy), sea slugs and the Oregon rough-skinned newt. Massive 'blooms' of toxic algae can poison rivers and even the open sea, turning the water red, green or blue. The 'Red Tides' of the Bible, which killed fish and Egyptians alike, may have been early records of this phenomenon.

The algal toxins, saxitoxin and tetrodotoxin, are the most lethal human poisons known; the fatal dose is about 0.2 milligrammes (the weight of a fine grain of salt), making it 1,000 times more deadly than the nerve gas sarin. The toxins kill by paralysing the respiratory muscles. Death can also result from eating *fugu* fish or even (and this is serious) an Oregon rough-skinned newt.[20] Tingling of the lips often precedes paralysis, and is Nature's way of telling you to put your fork down – although with the quickest death on record just 17 minutes after the first mouthful of *fugu* fish, it may already be too late. But if you survive, you will have no residual paralysis and will be back to normal within days.[19]

In his review of polio in Boston on 9 June 1908, Robert W. Lovett drew attention to a mysterious outbreak of acute paralysis on the coast of Labrador the previous autumn.[8] Five cases were affected; three died within hours, while the survivors had recovered completely the next day. The history was unobtainable in each case, and no post-mortem examinations were performed. However, all cases had been seen gorging themselves on rotting herrings, heaving with maggots, which were lying on the beach.

In fact, the clinical features were nothing like human polio, which is not surprising, as all five victims were dogs. The diagnosis was probably acute 'scombroid' poisoning, due to massive levels of histamine released by decomposing fish proteins – a condition not recognised until decades later.[21]

There are unlikely to be any lessons for the human disease, even though rotting herring is a traditional dish in Sweden. It is eaten more as a test of manhood than for the delicacy of its flavour, and does not travel well (many airlines refuse to carry it). Side effects include nausea and vomiting, which can affect onlookers. However, perhaps because maggots are absent, paralysis does not occur.

Choose your poison

Willie and three other brats,
Ate up all the Rough-on-Rats,
Papa said, when Mama cried,
'Don't worry, dear, they'll die outside'.

Pelican, University of California, reprinted in
Los Angeles Herald, 11 March 1906

Before the discovery of the poliovirus, it was quite reasonable to suggest that polio was due to toxic metals. In particular, lead and arsenic were commonly used around the home, and were known to cause muscular weakness and spinal cord damage. Lovett devoted much of his Boston lecture to the notion, citing examples from Ancient Rome to the *crème de la crème* Parisian neurologists of the day.

It had been known for centuries that lead miners could lose the use of their limbs; more recently, so had a painter-decorator who was overfond of lead-based pigments.[22] In 1888, 500 people in Hyères, Provence, suffered paralysis of their legs after a major health and safety lapse caused wine to be clarified with white arsenic oxide instead of harmless gypsum.[23] Two years later, paralysis struck several hundred beer drinkers in Manchester who had drunk a 'fourpenny brew' which turned out to be unexpectedly rich in arsenic.[24] The microscope also suggested that poisonous metals could destroy the motoneurones in the spinal cord – at least in cases that came to autopsy, such as the unfortunate painter mentioned above, or dogs given massive doses of lead or arsenic.[22]

Up to the 1950s, arsenic and lead commonly cropped up in the home.[23] Paris Green, a brilliant emerald pigment popular in wallpapers, was a

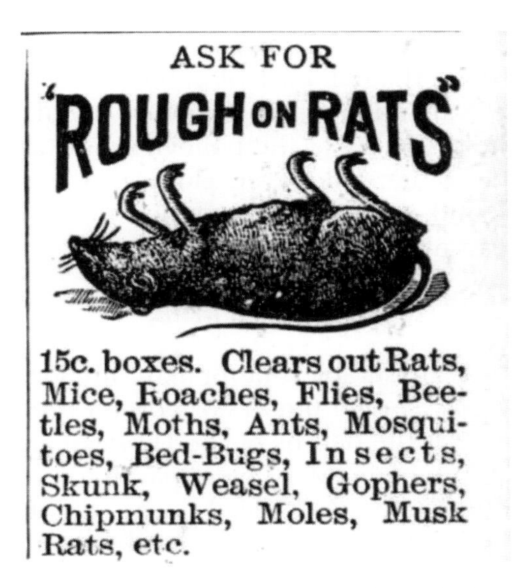

Figure 3.1 'Rough on Rats', an arsenic-based rat poison which was widely used in the United States during the early twentieth century. It inspired a song, with words by with words by W.J. Boston and music by Jules Juniper. Reproduced by kind permission of the Hagley Museum and Library, Wilmington, Delaware, United States

copper-arsenic salt. Also arsenic based were rodent poisons such as 'Rough on Rats', so widely used that it figured in a popular song (to the tune of 'Little Brown Jug') as well as the grimly witty stanza quoted above (Figure 3.1).

Lead oxide was the time-honoured opacifier in white paint, which tended to flake off woodwork and toys. Metallic lead was a major ingredient in the shields which nursing mothers used to protect their nipples, and in toy 'tin' soldiers. By the early twentieth century, lead poisoning was well known in toddlers who enjoyed oral exploration of their surroundings.[25]

It all sounds plausible, but toxic metals do not cause a disease recognisable as polio. Apart from acute poisoning (accidental or deliberate), metal toxicity builds up over weeks or months, and each poison also has its own diagnostic hallmarks. Chronic lead poisoning can cause severe muscle weakness, but this is a late symptom which follows months of constipation and abdominal colic; other features include blue-grey 'lead lines' on the gums and, in children, hydrocephalus (producing a dull sound like a cracked pot when the skull was tapped) and delayed mental development.[25,26] In arsenic poisoning, paralysis is accompanied by abdominal pains, agonising pins and needles in the hands and feet and scaly skin growths that can turn cancerous.[27]

The case also falls apart under the microscope. Toxic metals can kill the anterior horn cells, but only at extremely high doses that also destroy many areas of the brain as well as organs such as the liver and kidney. Selective 'poliomyelitis', the hallmark of polio, does not occur.

Lovett and others abandoned the idea that metal poisoning caused polio soon after the discovery of the poliovirus was announced in time for Christmas 1908. Others, however, clung to the notion that toxins were responsible – and shifted their attention to more plausible candidates, in the shape of man-made organic chemicals.

Dying for a drink

The spring of 1930 was a miserable time for many inhabitants of Cincinnati, Ohio, caught between the twin pincers of poverty and Prohibition. The city, struggling through the depths of the Depression, provided a grim backdrop for a puzzling outbreak of acute paralysis.

The first cases cropped up well ahead of the traditional polio season, but otherwise seemed unremarkable. Profound weakness developed over a day or two, usually in both legs and often the arms, and sometimes with respiratory paralysis. Cases which ended up on the post-mortem table showed typical poliomyelitis in the spinal cord. However, differences from the classic disease emerged as that year's polio epidemic got under way. Most victims also had severe pain and pins and needles in their limbs, while fever and general symptoms were absent. Strikingly, this new variant spared children, and instead targeted a particular sector of society: young to middle-aged men, including armed services veterans and recent immigrants, often alcoholic and living rough.[28]

By the time the Cincinnati outbreak had claimed its five-hundredth victim, it was clear that this was part of a wider epidemic sweeping through the Southern and Midwestern states and preying particularly on the citizens of Skid Row. The final tally was estimated at 150,000 cases.[29]

This was long before routine laboratory tests were available to diagnose polio. It took some inspired detective work by an usual alliance of public health doctors and US Customs and Excise to work out what was going on.[30]

Careful questioning showed that all the victims had drunk a 'medicinal tonic' called Jamaica Ginger ('Ginger Jake', to its many friends), in the ten days before paralysis set in. Jamaican Ginger had been around for decades, but had gained popularity during Prohibition because it had a high alcohol content yet remained on sale (and cheap) thanks to its 'medicinal' status (Figure 3.2). Evidently a new neurotoxic ingredient had found its way into Ginger Jake. The culprit caused surprise when it was finally nailed: an industrial organophosphate compound, triorthocresyl phosphate (TOCP).

TOCP is a heavy, inert liquid with high boiling point, ideal for gun oil and hydraulic fluid. It is also tasteless and odourless, and happened

Figure 3.2 'Ginger Jake', an alcoholic tonic that evaded Prohibition restrictions because of its 'medicinal' status. Adulteration with TOCP caused an outbreak of polio-like illness that killed 2,000 people and paralysed 35,000, mostly in the southern American states, in 1930

to interfere with the tests which the analytic chemists of US Customs used to pass alcoholic drinks as 'medicinal'. The use of TOCP was the brainchild of two amateur chemists and bootleggers from Boston, Harry Gross and Max Reisman. TOCP was a perfect adulterant: it did not give itself away, and left Ginger Jake as delicious and alcoholic as always, and made a fortune for Gross and Reisman.[30]

It is unfortunate that TOCP is a powerful neurotoxin which kills off many neurones in the nervous system, notably the motoneurones of the anterior horns. The 1930 outbreak of TOCP-induced paralysis killed 2,000 Americans and left 35,000 permanently paralysed.[29]

It also set the stage for a more pervasive chemical toxin that had daily access to hundreds of millions of people around the world.

Silent Spring

The saga of dichloro-diphenyl-trichloroethane, better known as the insecticide DDT, is part fairy tale, part horror story. It was discovered in 1873 by a teenaged apprentice chemist, and then forgotten for nearly 70 years. In 1939, the Swiss chemist Paul Muller synthesised it again and ran some experiments to see if it did anything. It did.

Muller's discovery that DDT was a staggeringly potent contact poison for insects won him the Nobel Prize for Medicine or Physiology in 1948, and transformed this neglected molecule into one of the industrial marvels of the twentieth century. Weight for weight, DDT was several hundred times more powerful at killing insects than the best existing pesticides – and miraculously, seemed to be harmless to humans and other mammals. This is because of its selective action on the 'sodium channels' that control the electrical activity of nerves. DDT is a spiky, crab-like molecule that scuttles inside the sodium channel in insect nerves, wedging it open. The nerves fire continuously, paralysing the muscles in contraction and killing the insect within seconds. Mammals are spared because their sodium channels have a tight entrance that keeps DDT out.[31]

DDT was initially mobilised in 1940 to defeat the insects that were killing hundreds of thousands of Allied troops. It was sprayed on swamps in the Far East to eradicate the mosquitoes that spread malaria, and on soldiers to rid them of the lice that carried typhus. After the War, its use mushroomed. Photographs of the time show cotton fields being sprayed from planes, and fruit trees, cattle, poultry, houses, children in swimming pools and sunbathers on beaches, all being doused in a heavy grey mist of DDT. Housewives squirted it around with impunity, because it said on the tin, 'Harmless for warm-blooded animals'. Homes could even be protected aesthetically with DDT-impregnated wallpapers and paints (Figure 3.3).

But things were set to turn sour. A link between DDT and polio was first suggested in 1953 by Dr Morton J. Biskind, a physician at the prestigious Beth Israel Hospital in Boston. Biskind believed that DDT was responsible for 'the most intensive campaign of mass poisoning in human history'.[32] He was struck by the worsening polio epidemics since the advent of DDT and the spread of polio into countries such as Mexico, the Philippines and Israel, where 'vast quantities' of DDT were now used. Biskind also dug up 'neglected' studies from 1944, showing that massive doses of DDT injected into experimental animals caused anterior horn damage 'often enough to be significant'. He threw down the gauntlet:

> When the population is exposed to a chemical agent known to produce in animals lesions in the spinal cord, resembling those in human polio, and thereafter the disease increases sharply in incidence and maintains its epidemic character year after year, is it unreasonable to suspect an etiologic relationship?[32]

Figure 3.3 Advertisement for DDT-impregnated wallpaper, as carried in many popular magazines in the United States during the 1950s

His claim was promptly denied by US government officials who, in his words, 'relied solely on the prestige of government authority and sheer numbers to bolster their position'.

Meanwhile, however, signs were emerging that large-scale use of DDT might kill other species: ominously fish-free lakes, Florida beaches piled high with dead crabs, plummeting populations of pelicans and eagles. Worryingly, it was discovered that DDT concentrations rose exponentially through the food chain to dizzying levels in fish and shellfish – up to 70,000 times those in water – while DDT broke down so slowly that it would remain in soil for decades.[33] Rachel Carson's *Silent Spring*, published in 1962, made an eloquent case for the environmental catastrophe threatened by the use of DDT. The book quickly became a bestseller and signalled the start of the modern environmental campaigns.[34]

DDT also got into humans and stayed there; it was still easily detectable in the body fat of normal Americans in 2001, nearly 30 years after its use was banned.[31] But was it dangerous, and could it cause polio?

DDT could kill, as a few people had died after drinking large amounts by accident or with suicidal intent.[35] However, most of these deaths were probably caused by the kerosene used to dissolve DDT. The lethal dose of DDT in adults was estimated to be 20–30 grammes – making it only half as deadly as paracetamol.[31]

Nonetheless, toxic effects were sought energetically. DDT was fed to American prisoners for 21 months to a total dose that was well over half the lethal threshold, while volunteers ate food drenched in DDT.[31] Skin absorption was investigated in stoical British soldiers who wore DDT-impregnated underwear for up to a month without changing, and (for undisclosed 'Service reasons') in two semi-naked volunteers who pressed their backs against a DDT-painted wall soaked with machine oil and sweat.[36] Following a bet, an officer with the British Army in Germany in 1946 ate six pancakes made of DDT instead of flour and 'enjoyed his meal greatly without any untoward effects'.[37] If he had taken a similar amount of paracetamol, he probably would have died of liver failure.

There was no hint of polio-like symptoms in any of these subjects, and none of the cases that came to post-mortem showed any microscopic signs of poliomyelitis.

All the evidence was resoundingly negative, but Biskind was undeterred. He produced graphs showing that DDT use and the incidence of polio rose together in the United States through the 1930s to the 1950s – but glossed over the massive epidemics of 1916 and the 1920s which had marched across America decades before the first puff of DDT hit the continent. Had he plotted other data, he would have discovered that the rise of polio in mid-twentieth century America was also paralleled by sales of Coca Cola, ownership of televisions and cars, and other vital signs of the American lifestyle.

The unavoidable conclusion: a good story, but with nothing in it.

Time to move on

Drs. Biskind and Scobey each had their 15 minutes of fame (actually several hours) in 1950, petitioning the House of Representatives to axe DDT and other industrial toxins because of their risks to human health.[38,39] They failed then, but DDT eventually worked its way to the top of the 'dirty dozen' list of industrial pollutants. It was banned in the United States in 1972, ten years after the publication of *Silent Spring*, on account of its environmental hazards, and has since been withdrawn around the world.[33]

Meanwhile, another organic compound – and a genuine neurotoxin – has had a lasting impact. This is TOCP, famous for helping Ginger Jake to sneak through an analytical loophole during Prohibition. The survivors of the American outbreak of TOCP poisoning in 1930 carried a huge burden of disability for the rest of their lives. Many remained trapped as 'stick' or 'chair' cases. Those who could still walk unassisted were easily recognised (and stigmatised) by their characteristic 'Jake walk': each leg had to be lifted high so that the dangling, useless foot could clear the ground.[29]

These unfortunates received little or no compensation, and precious little sympathy in those judgmental times. Their plight and their disabilities – useless legs, impotence and social exclusion – were commemorated in verse and song, including the plaintive Blues number by Asa Martin, 'Jake Walk Papa'.[40]

TOCP poisoning has cropped up repeatedly since then, often thanks to a surprising tendency of army cooks to confuse machine-gun lubricant with cooking oil. Over 10,000 locals were paralysed in Morocco in 1959 following a blunder by the American forces; the outbreak nearly provoked a major diplomatic incident.[41] When TOCP found its way into army food during the Vietnam War, victims were identified by the spot diagnosis of 'Jake walk'.[42] And in 1992, the long-standing mystery of the 'Saval cripples' was finally solved. These workers on a farm near Verona had been paralysed during 1942, together with their animals and even a blackbird in a cage. The cause: TOCP-contaminated engine oil from army vehicles, dumped on the farm.[43]

All these cases suffered severe paralysis, and the impact on them was every bit as devastating as if they had been attacked by the poliovirus. But it was not polio.

In the library of the College of Physicians in Philadelphia, there is a vintage wooden cabinet containing index cards which catalogue the older books and pamphlets in the collection. Under 'Poliomyelitis' is a card for Dr J.G. Woolley's self-published 'Report on the possible relationship between electrical appliances and the incidence of anterior poliomyelitis'. The card carries a hand-written disclaimer, making it clear that the College wished to dissociate itself from Dr Woolley's views. And so it did: the College's copy has vanished without trace.

However, Dr Marsh Pitzman's monograph on jimson weed and polio, also self-published, is still in the archives. So are the bound volumes of the *Archives of Pediatrics* and other reputable medical journals containing the publications of Drs Biskind and Scobey. And a couple of minutes spent searching the Internet will confirm that the myth of the toxic cause of polio is still out there today, as pervasive and persistent as DDT itself.

4

Germs of Ideas

A bacterial cause has been definitely ruled out and there is now undisputed evidence that the infectious agent belongs to the group of so-called filterable viruses.

Francis W. Peabody, George Draper and A.R. Dochez, *A clinical study of acute poliomyelitis*, 1912

The story so far: polio is not caused by teething, pesticides, ptomaine poisoning, electrical appliances or misaligned vertebrae in the neck. It is an infection, as was confirmed in 1908 by the grisly experiments in which Karl Landsteiner transmitted polio to live monkeys with an extract of spinal cord from a dead boy.[1]

Landsteiner also provided vital information about the nature – or at least the size – of the infectious agent. The spinal cord extract injected into the monkeys had been filtered through a 'Berkefelt candle'. The candle was a hollow cylinder of baked diatomite, a whitish silica clay made up of billions of skeletons of microscopic marine organisms. The channels through the filter were so minute and tortuous that they trapped bacteria. Whatever passed through was therefore even smaller – one of the 'so-called filterable viruses' mentioned in the quotation above, from one of the great contemporary monographs on polio.[2]

The discovery that polio was caused by a virus rather than a bacterium was not just scientific fine print. This was the essential first step towards accurate diagnostic tests and properly targeted treatments – remembering that antibiotics, our most effective weapons against bacteria, are powerless against viruses.

Landsteiner's paper came out at a time when the pace of discovery was brisk, and it seemed likely that the 'filterable virus' of polio would be pinned down within a few years. In the event, the task took nearly forty years. This was a momentous journey which generated a clutch of Nobel Prizes and pulled in some of the greatest scientific names of the twentieth century. But there were also sideshows and distractions along the way – blind alleys, red herrings and charlatans, and many instances where evidence was not allowed to stand in the way of a good story.

Perhaps surprisingly, many scientists did not accept that polio was due to a virus. In their well-known *Manual of infantile paralysis*, Henry Frauenthal and Jacolyn Manning gave less space to the filterable virus as a possible cause of polio than to 'Dixon's protozoon' (which fell from grace a few years later).[3] Their *Manual* was published in 1916, four years after Peabody's classic monograph, by which time Landsteiner's findings had been amply confirmed by several other groups. Years later, an eminent microbiologist was invited to talk about the cause of polio at the biggest medical conference in America. He devoted his entire lecture to the 'poliomyelitic' bacterium named after himself.[4] This was not in the dark ages of the 1920s but in 1952, long after the electron microscope had revealed the poliovirus for all to see.

To make sense of this story, we need to go back a hundred years, to an age when polio was just one of many diseases waiting to be solved by germ hunters who hoped to follow in the giant footsteps of Robert Koch and Louis Pasteur.

Causes and effects

The first decade of the twentieth century was a boom time for bacteria, genuine and otherwise. Journals devoted to infectious diseases flourished and were full of reports of weird and wonderful bacteria, previously unknown to science. One new species was *Streptothrix interproximalis*, rescued from mouths tersely described as 'habitually unclean'. It certainly looked dramatic – rods, scrolls or branches, with or without spots – but whether it did any harm was anyone's guess.[5]

The discoverers of new bacteria were also searching for a role for them, especially in diseases which had long refused to reveal their secrets. The pressure of competition, helped along by dubious experiments and wishful thinking, often stretched the imagination more than the boundaries of knowledge. Particular bacteria were identified as the causes of smallpox, rabies, measles and rubella – all of which are viral infections – and malaria, caused by a protozoon.

Other bacteria had even more to live up to, being blamed for diseases that are not even infections. Several authors claimed to have found bacteria that caused scurvy, the scourge which made teeth fall out of the bleeding gums of sailors on long-haul ocean voyages.[6] An as yet unidentified germ, reportedly spread by insect bites, was believed to cause pellagra. This was a disfiguring condition with a skin eruption so ugly that some American hospitals refused to admit sufferers, and nurses went on strike to avoid looking after them.[7] Another presumed infection was beriberi, which struck down millions in the Far East with agonising pains in the limbs and massive fluid accumulation in the legs.

In *Principles and practice of medicine* (1892), William Osler maintained, 'It is probably due to a microorganism'.[8] The more exotic candidates included Pekelharing's staphylococcus, Taylor's spirillum and Durham's looped streptococcus. The elusive bacterium of beriberi was perhaps the germ hunters' holiest grail – and one of those who joined the chase was the greatest of them all, Robert Koch himself.

Many researchers became convinced that polio was due to 'poliomyelitic bacteria', which they found in compromising situations in tissues or body fluids of polio patients. These included rod-shaped bacilli, spherical cocci and more bizarre forms such as the branched, root-shaped rhizopods. The commonest were streptococci, similar to those that cause tonsillitis and blood poisoning. Most of the purported 'poliomyelitic bacteria' soon slid back into obscurity when other researchers could not find them, or showed them to be contaminants. For example, Ivar Wickman, epidemiological detective extraordinaire, set off after the fabled 'tetracoccus' identified by Magnus Geirsvold, only to find that it was indeed a fable. No convincing case had been made by 1908, prompting the American orthopaedic surgeon B. Sachs to write, 'The infectious microorganism, whatever its character may be, has not yet been revealed'.[9] This set the stage neatly for Landsteiner to usher in his filterable virus a year later.

Landsteiner's paper did not persuade everyone to think small. The filtration candles were relatively new and were throwing up some odd results, including the heretical claim that tuberculosis was caused by a filterable virus rather than the bacillus that had won Koch his Nobel Prize in 1905.[10] Sceptics, some of them well respected and powerful, therefore clung to the belief that bacteria caused polio – and some continued to do so long after their pet microbe had been discredited.

The believers in poliomyelitic bacteria included Ludvig Hektoen, Professor of Pathology at the University of Chicago and founding editor of the *Journal of Infectious Diseases*. A prolific researcher, Hektoen had notched up his hundredth paper ('An anatomical study of a short-limbed dwarf') by his fortieth birthday.[11] In 1918, Hektoen reported in the *Journal* that he had found 'cocci' in stained sections of brain and spinal cord from polio victims. These bacteria looked 'quite like' those which others had claimed to have cultured from patients' tissues and fluids. Hektoen concluded that 'such cocci occur constantly in the central nervous system in epidemic poliomyelitis, and their presence here is not explainable as due to accident or contamination'.[12]

Hektoen's findings were a direct challenge to Landsteiner's paper, published a full decade earlier, but Hektoen had authority. His obsession with accuracy was legendary, and his *Journal* was top in the field. And he had examined a vast number of samples, including archive material from a Norwegian epidemic in 1906, and spinal cords that had been collected with disconcerting speed, within an hour of death.

Within a few years, Hektoen's cocci were destined to follow many other poliomyelitic bacteria into oblivion. For now, though, the door was still open for those who remained unconvinced about the filterable virus. Meanwhile, those who did believe in Landsteiner were still scrabbling for ideas about what could have crept through the Berkefelt candle.

All this uncertainty provided the way in for two intriguing red herrings: Rosenow's poliomyelitic streptococcus and the globoid bodies of Noguchi and Flexner. Both should have been quickly consigned to the deep, but the power of the scientific hierarchy ensured that they continued to swim against the tide of evidence for many years.

Downsizing

The issue of the *Journal of Infectious Diseases* which contained Hektoen's article on poliomyelitic cocci was dominated by a clutch of papers about an even more promising germ. The author was Dr Edward C. Rosenow, Head of Experimental Microbiology at the Mayo Foundation in Rochester, Minnesota, one of the top research centres in the United States (Figure 4.1). Rosenow had made his reputation studying infections of the brain, but had recently set off on a crusade to prove that streptococci caused important human diseases – beginning with polio.

The massive American polio epidemic of 1916 and a smaller local outbreak the following year had provided Rosenow with fresh experimental

Figure 4.1 Edward C. Rosenow (1875–1966), tireless promoter of the 'poliomyelitic streptococcus'

material and the chance to search for streptococci. He found lots of them, in the body fluids, brain and spinal cord of polio victims.[13,14] Tonsils, removed by a friendly surgeon from patients 'who were not convalescing satisfactorily', were also a rich source. These streptococci were tricky to grow in the laboratory, but flourished in a special culture medium which Rosenow had dreamed up. The recipe began with live twenty-day-old chicks still in their shells and proceeded – badly for them – with the intervention of a mortar and pestle.

To be confident that his streptococci were responsible for polio, Rosenow had to show that they fulfilled 'Koch's Postulates'. Robert Koch had laid these down in 1884 as the essential criteria to prove that Germ X genuinely caused Disease A, and was not just an innocent fellow traveller. Germ X had to be isolated from all cases of Disease A, and never from healthy subjects. A pure culture of Germ X, grown in the laboratory, must produce Disease A when given to healthy animals. Finally, Germ X had to be isolated again from the animals with experimentally induced Disease A, and these isolates must in turn induce Disease A in fresh animals. This exercise had been straightforward in the case of cholera: easily seen, cultured and transmitted, and quick to induce the diagnostic diarrhoea. Rosenow's task was much harder, because the streptococcus was fussy, and the detection of infection inside the central nervous system was so laborious.

But he persevered, and Koch would have been satisfied by the papers which Rosenow published in 1918 – the product of four years' work and some 2,000 experimental animals. Rosenow reported that he had grown streptococci from the brain and cord of polio victims; that these caused poliomyelitis and paralysis when injected directly into the brains of monkeys or rabbits; and that the same bacteria could then be recovered and produce paralysis in healthy animals. As proof, there were photographs of unhappy rabbits trailing their useless limbs. Rosenow concluded that his 'poliomyelitic streptococcus' normally lived in the tonsils, but sometimes invaded the bloodstream, when it homed in on the spinal cord and caused paralysis.[13,14]

Being a bacterium, Rosenow's streptococcus was far too bulky to be filterable. However, it was 'pleomorphic' and could change its shape. Specifically, it underwent an extraordinary transformation when grown in the 'chick-mash' medium, forming minute 'micrococci' that were one-tenth of its normal size. The micrococci could slip easily through the Berkefelt filter – and also through the 'blood-brain barrier' which normally keeps bacteria out of the central nervous system.

Rosenow reported that the micrococci were just visible under a powerful light microscope which magnified about 1,000 times (Figure 4.2). He later described them in greater detail, thanks to the ingenuity of

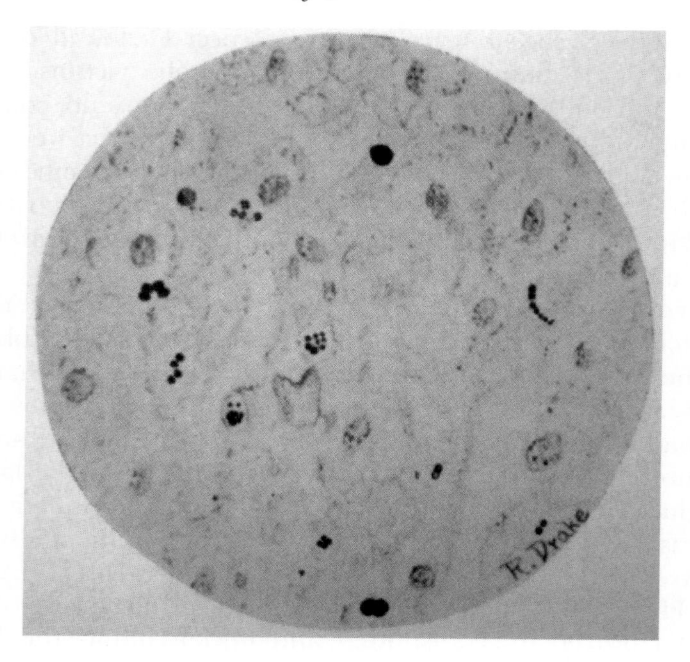

Figure 4.2 Rosenow's 'filter-passing micrococci', drawn for a paper in 1918.[13] They are marked by the arrows; the larger paired structures at the bottom of the field (circled) are normal streptococci. The 'micrococci' looked like tiny bacteria, but were artefacts. Reproduced by kind permission of the *Journal of Infectious Diseases*

Royal R. Rife, a Californian inventor known chiefly for his colourful personality and boundless energy for self-promotion. Using optical wizardry that involved polarised light and rotating quartz prisms, Rife had built a 'virus microscope' which somehow stretched the power of magnification to an incredible 8,000 times. This instrument had recently shown the causative organism of typhus (a notoriously tiny bacterium) to be perfectly round and bright blue – a claim greeted with derision by most of the medical fraternity. In July 1932, Rosenow visited Rife in San Diego, taking filtrates of his streptococcus. To his delight, the 'virus microscope' confirmed the missing link: round blobs, grey-brown in colour, small enough to be 'filter-passing'.[15]

There was more. Back in 1917, Rosenow made the stunning claim that his pleomorphic streptococcus held the key to a miracle cure for polio[16]. Animals which survived infection with the streptococcus developed antibodies against it, which then blocked subsequent attempts to infect them – mimicking the natural immunity induced by an attack of polio. Rosenow harvested large amounts of 'immune serum', rich in protective antibodies, from a horse which he had immunised with the streptococcus. This was a stoical horse, as it was injected with extracts from 40 litres of culture broth over six months, and then bled repeatedly. The immune serum extracted from its blood was magical stuff: it could snatch back from certain death monkeys which had been given usually fatal doses of polio brain extract.

Rosenow then extended his trials to humans, treating 58 patients during a polio outbreak that hit Davenport, Iowa, during the summer of 1917.[17] He screened 'tentative' cases with lumbar puncture, even though this could not diagnose early polio with any certainty. Likely patients, such as those whose spinal fluid spurted dramatically out of the needle, were immediately given the immune serum; this was injected into an arm vein, or in babies, the jugular vein in the neck. Some subjects were already moribund when treated and, as expected, soon died. Otherwise, Rosenow reported, 'Paralysis did not develop in a single instance when treatment was begun before its onset, and all recovered'.[17] Rosenow's excitement at these astounding results was shared by the American drug company Eli Lilly, which set up cultures of the streptococcus and began manufacturing its own antiserum.

Rosenow's poliomyelitic streptococcus certainly looked the part, but other researchers were struck by his claims – and specifically that they seemed too good to be true. Even as Rosenow built up his dossier, papers appeared that comprehensively torpedoed his key findings, one by one.[18] His streptococci fell at the first hurdle of Koch's Postulates, because identical bacteria were readily found in the brain and tonsils of people and animals who had never had polio. Rosenow clearly had not checked his controls carefully enough. Worse still, the same streptoocci were found contaminating supposedly sterile glassware and the chick-mash medium. Even 'extraordinary care for sterility' – decking researchers out in caps, gowns and gloves and searing the brain surface with a red-hot scalpel before taking samples – could not exclude them. Verdict: Rosenow's streptococci were merely 'airborne contaminants or terminal invaders of sick animals'.

Next, Rosenow's germ could not be persuaded to spawn filterable offspring; put succinctly, 'the streptococci remained streptococci'. Then nobody could replicate Rosenow's finding that his germ caused poliomyelitis. Injecting a heavy culture of the streptococci directly into the brain could produce paralysis, but only because the entire skull contents became infected, just as with dozens of ordinary bacteria. The rabbit turned out to be a hopeless model for polio, as it could be paralysed by injecting virtually anything into its brain. Verdict: 'the cocci have no etiological relation to the lesions of poliomyelitis'.[18]

Finally, others could not detect Rosenow's supposedly protective antibodies. Proper trials – allowing for the variable outcome of polio – showed his immune serum to be no better than letting nature take its course unassisted. Verdict: 'the Rosenow serum is devoid of protective power'.[19]

Rosenow's reaction to his critics was surprising for a senior professor at one of America's flagship research centres. Initially, he fought back on points of detail: their culture conditions were not quite right, the doses they injected were too small, their rabbits were too old. Then he simply turned his back on them and carried on ploughing his increasingly solitary furrow.

And a long furrow it proved to be. In 1930, 14 years after Rosenow's first paper on his poliomyelitic streptococcus, Beatrice Howitt wrote: 'Except for the work of Rosenow, it is generally accepted that poliomyelitis is due to a filterable virus'.[20] But Rosenow stuck to his guns and kept publishing. In 1933, a commentator noted:

> It must be very trying to the pundits, who assert that Rosenow's poliomyelitic streptococcus just doesn't exist, to have it continue as the subject of researches, all of which seem to indicate that the condemned microorganism is very much alive.[21]

In 1944, *Time* magazine was prompted to remark:

> For 27 years, the medical profession has looked politely down its nose at bacteriologist Edward Carl Rosenow ... a stubborn man who has persisted in his obsession that in ways no doctor understands, streptococcus plays a malignant part in infantile paralysis.[22]

And, perhaps with a note of weariness, the *Postgraduate Medical Journal* of January 1949 observed: 'Rosenow persisted in his streptococcal theory'.[23]

Rosenow continued to do so, and somehow could still pull a good crowd, long after his streptococci had been flushed away by evidence. In June 1952, Rosenow was invited to address the assembly of general practitioners at the annual meeting of the American Medical Association (AMA) in San Francisco. Now eight years into retirement, he delivered a one-man festschrift about his own 'forthright bacteriologic studies' which had led him to the bacterium that he still believed to cause polio. He acknowledged that the filterable agent was 'currently considered to be a virus', but made no concessions to his detractors. In fact, he did not even mention them. Over half of the papers that he cited were by himself, and most of the rest were the early (and by now discredited) reports by Hektoen and others of cocci spuriously associated with polio.[4]

This lecture must have been an odd event, as there was no longer any controversy over what caused polio. When the poliovirus came up on the screen of the electron microscope in 1946, it looked nothing like a bacterium and was far tinier than Rosenow's micrococcus. Yet Rosenow spoke as though this revelation, and all the other evidence that had killed his hypothesis, had never happened.

Perhaps Rosenow was invited to the AMA so that his 'poliomyelitic streptococcus' could at last be given the decent burial which it should have had 40 years earlier. But Rosenow remained true to his mission and continued to preach the doctrine of his conviction.

Foreign bodies

Rosenow's streptococcus was not the only non-cause of polio to seize the imagination during the 1920s and 1930s. Serious competition came from 'two of the world's foremost bacteriologists' at the hugely powerful Rockefeller Institute in New York. Their contender had an alluring name: the 'globoid bodies' of Noguchi and Flexner.

The tale of the globoid bodies begins in Japan just before the turn of the twentieth century, with a bright doctor in his early twenties looking to make his name in America. Hideyo Noguchi had already shown determination to succeed against adversity from the age of two, when his left hand was badly burned. Prolonged and painful surgery left him with reasonable dexterity and a longing to be a brilliant doctor. To further this ambition, he dropped his parents' given name, Seisaku, in favour of Hideyo, which means 'excellent' (Figure 4.3).[24]

Figure 4.3 Hideyo Noguchi (1876–1928), who left Japan to work in the United States with Simon Flexner. Noguchi proved that general paralysis of the insane was a late stage of syphilis, but the organisms which he reported as the causes of rabies and yellow fever turned out to be spurious

Figure 4.4 Simon Flexner (1863–1946), founding Director of the Rockefeller Institute for Medical Research, New York, and polymath medical researcher. In addition to polio, Flexner's interests included meningitis and snake venoms. Together with Hideyo Noguchi, he reported the 'globoid bodies' as the causative organism of polio, and was convinced that the infection was spread from the nose to the brain. Reproduced by courtesy of the Rockefeller Archive Center, New York

A couple of years after graduating from medical school in Tokyo in 1897, Noguchi decided that America was ready for him. He bought a one-way ticket to Philadelphia and arrived unannounced one morning in June 1899 at the University of Pennsylvania. There, he told Dr Simon Flexner, world-famous microbiologist, that he had come to work with him (Figure 4.4). At first, Flexner did not know what to do with him, but Noguchi was clever, hard-working and eager to please. He was put to work on one of Flexner's many research interests, snake venom. Noguchi and Flexner produced several joint papers, and Noguchi went on to write a monograph on the topic, which became a classic.[25] Now firmly embedded in Flexner's team, Noguchi was regarded with affection and respect; his nickname, 'the Yellow Peril', was evidently given and received in good humour.

A few years later, Flexner was appointed chief of the Rockefeller Institute, and Noguchi followed his boss to New York to tackle the grandest challenges in microbiology. He churned out dozens of papers in top American journals. An early coup was finding the elusive corkscrew-shaped spirochaete of syphilis in the brains of victims of 'general paralysis of the insane'.[26] Noguchi's discovery settled the long-running controversy about the cause of this pitiful condition, and cemented his reputation. Noguchi went on to claim that another spirochaete was responsible for yellow fever[27] – one of the greatest killers in South America and central Africa – and that a large protozoon 'corpuscle' caused rabies. These reports impressed his peers at the time, but later came back to haunt him.

Noguchi moved into polio in 1910. Flexner had been excited by Landsteiner's findings and, together with Paul Lewis at the Rockefeller, confirmed that polio could be transmitted to monkeys by a filtrate of spinal cord extract.[28] Noguchi and Flexner now set out to hunt down the organism responsible. They used a culture medium invented by Noguchi, which had nourished various bacteria that otherwise seemed impossible to grow. This was based on human ascites fluid (which collects inside the abdomen in advanced heart failure or liver cirrhosis), sealed under paraffin oil to keep out germs and air. For the polio experiments, Noguchi popped in a piece of fresh rabbit kidney. He never explained why, but this macabre ingredient seemed to do the trick. A few days after adding a sample of poliomyelitic spinal cord, a haze appeared around the kidney fragment. Something seemed to be growing, and a high-powered microscope revealed masses of minute round forms, often in chains or clusters.

Crucially, these 'globoid bodies' were small enough to pass through the Berkefelt candle: a chain of a dozen, coiled up, could fit within the outline of a streptococcus. They did not look like any known microbe – hence their non-committal name – but Noguchi and Flexner were convinced that they transmitted polio. According to them, filtered cultures of the globoid bodies injected into the brains of monkeys produced poliomyelitis and paralysis. Globoid bodies could be cultured from the brains of these monkeys and, in turn, induced experimental polio in fresh animals. Koch would have been convinced.

In 1913, Noguchi and Flexner reported that they had cultivated 'the microorganism causing epidemic poliomyelitis'.[29] Their hefty 25-page paper in the *Journal of Experimental Medicine* seemed to have everything: ingenuity, technical brilliance, a clever solution to a thorny problem – and all coming from two of the top names in microbiology. It was also an open invitation for others to repeat the experiments – which they did, but with variable results that soon raised doubts. Harold Amoss was able to replicate the findings;[30] any technical queries would have been easy to resolve, as he worked down the corridor from Flexner, leading one sceptic to remark that the globoid bodies were a phenomenon largely confined to the Rockefeller Institute.[31] Others, however, could not persuade the globoid bodies to grow or to cause polio, and some could not find them at all. Even following Noguchi's method 'with the utmost exactness' sometimes produced nothing but peculiar tiny blobs that could only be artefacts.[31,32]

The globoid bodies wobbled along for over 20 years, generating papers and controversy along the way. They seemed to exist, at least in the depths of Noguchi's idiosyncratic medium, but whether or not they caused polio was a question that divided the polio research community into believers and non-believers. Possibly because of Flexner's stature and intimidating personality, the non-believers were less vociferous than those who attacked Rosenow's streptococcus. Steadily, though, the balance tipped against the globoid bodies. Several researchers challenged the magic of Noguchi's ascites medium, but the killer question was the lack of proper controls. Incredibly for such high-calibre scientists, Flexner and Noguchi had not bothered to see what happened if they used samples of brain or cord from animals that did not have polio.

By the early 1930s, the globoid bodies had been inflated into 'a balloon waiting to be pricked'.[33] The decisive puncture was inflicted in 1936 by Gerald Logrippo, from Flexner's former base at the University of Pennsylvania. Logrippo laid out his stall clearly at the start of his paper:

> Although many investigators have obtained the same results [as Flexner and Noguchi], the general opinion is that the bodies have a doubtful relationship to the etiological agent of the disease.[34]

Following Noguchi's method, Logrippo recreated the tell-tale haze around fragments of spinal cord and found globoid bodies – but in the complete absence of poliovirus. He had simply performed the control experiments that Noguchi should have done two decades earlier, using spinal cord from a healthy rabbit. Logrippo photographed the bodies budding off stringy filaments that were left exposed as the spinal cord fell apart. They were nothing but tiny droplets of fat.

These minute blobs were identical to the typical Noguchi-Flexner globoid bodies that Logrippo saw when he repeated the experiment with poliomyelitic cord. To clinch his case, Logrippo rigged up a Frankensteinesque experiment with a piece of fresh cord from a healthy rabbit impaled on a silver wire, suspended between two electrodes in culture medium. Knowing that fats were attracted towards the positive end of an electrical field, he had dunked the cord in Sudan III dye, which stains fats bright red. Over a few days, a red haze emerged from the cord and spread towards the positive electrode. Under the microscope, the haze consisted of classical globoid bodies, nicely picked out in deep pink by Sudan III.

Unlike Rosenow, who never retracted his claims about his poliomyelitic streptococcus, Flexner eventually admitted that the globoid bodies – whatever they might be – were not the causative agent of polio. In fact, he had already done this in 1928, before Logrippo's experiment.[35] At that time, Noguchi was not available for comment. For reasons that will become clear, he wrote himself out of the story in that same year.

Building the picture

During the 1930s and 1940s, progress was generally slow in the hunt for the true cause of polio. The seeds of confusion sown by the likes of Rosenow, Noguchi and Flexner did not help, but a more intractable problem was the elusiveness of the virus itself. The only method for detecting it was labour-intensive, expensive, unreliable and agonisingly slow to yield results – because the detection system in question was the central nervous system of living monkeys.

As described in the next chapter, the brake on polio research was released when it was discovered in 1949 that the poliovirus could be persuaded to grow outside live animals, in cultures of various tissues. As a result, new knowledge flooded in during the 1950s.

In the meantime, several pieces of the jigsaw were slotted into place, even though the picture of the villain remained tantalisingly incomplete. One seemingly quiet advance in 1931 had extraordinarily wide consequences. This was the discovery by Macfarlane Burnet and Jean Macnamara, working in Melbourne, Australia, that the serum collected from survivors of polio neutralised some polioviruses, but not others.[36] This meant that there must be different varieties of poliovirus. This could explain some peculiarities of the disease, such as why some epidemics were more serious than others. The finding also had huge practical implications, because an effective polio vaccine would have to protect against all the different varieties of poliovirus.

Worryingly, it later turned out that there were hundreds of different 'strains' of polio, but there was also good news: they were all encompassed by just three broad 'types' of poliovirus. The three types have quite different credentials. Type 1 caused most epidemics and 80 per cent of all cases of paralysis. Type 3 was in second place for viciousness, responsible for about 13 per cent of paralytic cases. Type 2 was rather rare and benign, accounting for just 7 per cent of paralytic attacks (at least in humans; unlike the other two types, it was able to infect and paralyse mice). As the 1931 paper by Burnet and Macnamara had anticipated, an infection with one strain produced antibodies that had little or no effect against strains of the other two types. However, antibodies against a particular strain protected against all other strains of that type – meaning that a 'trivalent' polio vaccine containing just one strain from each of the three types could protect against every possible strain of the virus.

Certain poliovirus strains crop up repeatedly through the history of polio, sometimes for unhappy reasons. Type 1 strains are often represented by Mahoney, isolated from the stools of children from a family in Cleveland, Ohio in 1941. The Mahoneys' legacy included a disastrous complication of Jonas Salk's injectable vaccine, which threatened to sabotage the entire polio vaccination programme. Another famous Type 1 strain is Brunhilde, honouring a big-breasted female chimpanzee from Johns Hopkins University in Baltimore.

The Michigan town of Lansing is immortalised in the name of a widely used Type 2 strain, which killed the source patient there in 1938. The Type 3 strain, Leon, is named after a boy who died of polio in Los Angeles in 1930 and who, like Brunhilde, provided the spinal cord which yielded the virus.

Another one to watch for is the Type 1 strain called CHAT, short for 'Charlton', the surname of the little girl from whose stools it was isolated in 1956. CHAT was used in an early oral polio vaccine that was given to hundreds of thousands of African children during the late 1950s. Later, it was catapulted straight into the realms of horror fantasy by the claim that it had been contaminated with a monkey virus which mutated into the human immunodeficiency virus (HIV) and so caused the AIDS pandemic.

All this lay many years in the future when the scientific community first considered the Burnet-Macnamara paper in 1931. To some experts, the notion that there might be different types of poliovirus was of dubious or no significance. Simon Flexner, looking down from the ivory tower of the Rockefeller Institute in New York, had little time for scientists outside America. This paper had emanated from far beyond the pale: an odd-sounding research outfit (the Walter and Eliza Hall Institute) in Melbourne, and written by authors who had not much of a track record in polio research.

Flexner never changed his belief that there was only one type of poliovirus, and even used his last public lecture on polio, in 1937, to sideline Burnet's work – proof that, right to the end, he was not always good at distinguishing winners from losers.

Small, round and beautiful

Science eventually caught up with the poliovirus.[37] It turned out to have a diameter of about 30 nanometres (3 millionths of a centimetre, roughly one-thirtieth of the size of a streptococcus), and weighed in at 10^{-17}g. Unlike bacteria and other more complicated organisms, the poliovirus is invariable in its construction. It can therefore be given a molecular weight (about 6,800,000) and a chemical formula, $C_{332662}H_{492388}N_{98245}O_{131196}P_{7500}S_{2340}$.

The structure of the poliovirus was revealed during the 1950s by X-ray diffraction and electron microscopy. X-ray diffraction had been put on the map during the 1910s by the father and son team of William Henry and William Lawrence Bragg in London, using crystals of simple chemicals such as halite (table salt). Crystals consist of regularly stacks of molecules, which split a narrow beam of X-rays into a fan of sub-beams that show up as an array of dots on photographic film placed behind the crystal. The patterns of dots can be analysed mathematically to reveal the chemical composition and structure of the molecule in question. With heavy-duty mathematics, X-ray diffraction can yield similar information about viruses, which can also be made to pack together to form crystals.

In the early 1950s, this was cutting-edge research that attracted brilliant minds, including two men in Cambridge and a woman in London. The men in Cambridge were an Englishman and an American on a visiting fellowship from the American polio charity, the National Foundation for Infantile Paralysis. Respectively, they were Francis Crick and James D. Watson. The woman was Rosalind Franklin, who had left Cambridge for the world-class X-ray crystallography unit at Birkbeck College after falling out with Crick and Watson during the scramble to elucidate the structure of DNA.

All three helped to unmask the poliovirus. Crick and Watson were working on how small round viruses are assembled from their constituent subunits. They deduced that 60 identical subunits were packed together to form a hollow shell that appears spherical, but is actually a regular geometric solid with 20 faces, somewhat like the polygonal panels on a football.[38] This 'icosahedron' is aesthetically pleasing, and also has a nice resonance with the origins of geometry, as it was one of Plato's 'perfect solids'. It also turned out to be the structure of the poliovirus.

Franklin began work on the poliovirus in 1955. She had already cracked the structure of the tobacco mosaic virus (TMV), a hollow cylinder about 300 nm long and made up of over 2,000 spherical subunits arranged in a tight spiral. The poliovirus was a thornier problem than TMV. Coaxing it to crystallise was a major feat, yielding tiny glassy splinters half a millimetre long which tended to disintegrate under X-ray bombardment.

Even Franklin was defeated; her work had to be finished by her colleagues at Birkbeck, John Finch and Aaron Klug.[39] They used poliovirus crystals grown by Carlton Schwerdt and Frederick Schaffer in Stanford (and smuggled past English Customs by Schwerdt's wife, Patsy, in her handbag).[40] Using X-ray equipment belonging to Lawrence Bragg himself, Finch and Klug showed that the poliovirus consisted of a hollow icosahedral shell of 60 spherical subunits, enclosing a core of RNA. This structure fitted the theoretical model constructed by Crick and Watson, as well as the picture that emerged from the electron microscope: an essentially spherical particle with a finely granular surface[41] (Figure 4.5).

Rosalind Franklin missed the satisfaction of seeing all these pieces of the puzzle click together. In the summer of 1956, at the age of 36, she had been diagnosed with ovarian cancer. She did not survive to see the

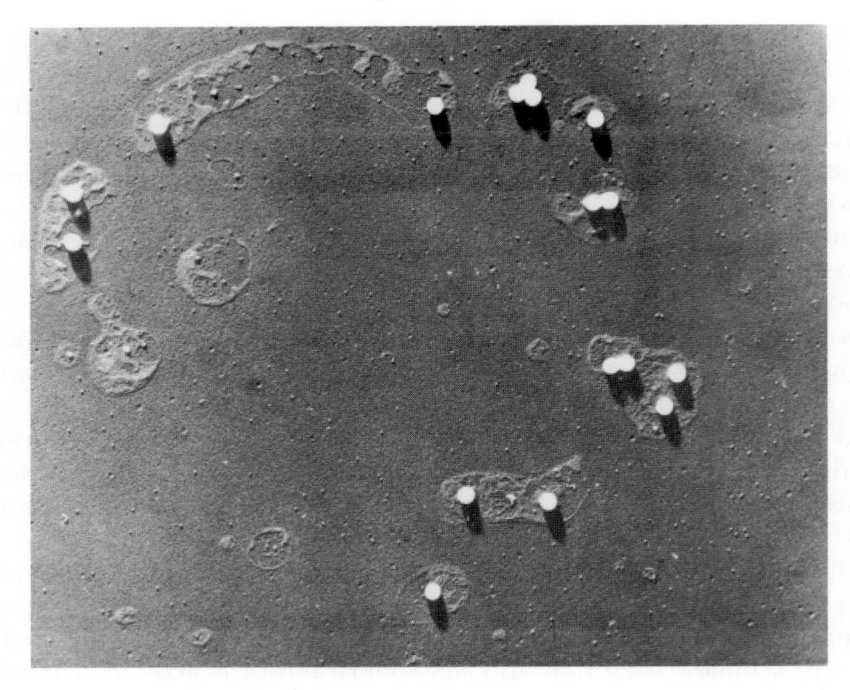

Figure 4.5 An early electron micrograph of poliovirus particles. Their spherical shape is brought out by 'shadowing' with vaporised gold fired obliquely across the sample. Reproduced by kind permission of the Centers for Disease Control, Atlanta, Georgia

sensation created by a huge model of her TMV structure when it was unveiled at the World Fair in Brussels in 1958 – or the papers published a year later by her friends at Birkbeck, on the structure of the poliovirus.

Bare necessities

Viruses are the simplest of all organisms. They are trapped in a biological limbo and only fulfil the most basic criteria for life – growth and reproduction – when they get inside their target cell, be this bacterial, plant or animal. Viruses carry their genetic information in either DNA (like bacteria, plants and animals) or in the alternative nucleic acid, RNA. The poliovirus, like those of influenza, the common cold and severe acute respiratory syndrome (SARS), is an RNA virus.

The polioviruses belong to the wider community of 'enteroviruses', so-called because they inhabit the gut. There are almost 90 different enteroviruses, of which over 60 infect humans.[42] Numerically, the 3 polioviruses are minor players; the others include 29 Coxsackie viruses (named after an Algonquin Indian settlement in New Jersey), several echoviruses and the rhinoviruses that cause the common cold.

Very rarely, other enteroviruses can cause paralysis that, under the microscope and at the bedside, is indistinguishable from classic polio. The main culprits include Coxsackie A7 and B4, and echoviruses 30 and 71. Before vaccination took hold, the poliovirus caused virtually all cases of 'acute flaccid paralysis', the clinical hallmark of polio. Now that vaccination has all but exterminated wild-type polioviruses, the rare cases of acute flaccid paralysis that occur today are caused by these other enteroviruses – or by polioviruses that have mutated from the oral polio vaccine.

Great men and great mistakes

The 'poliomyelitic bacteria' were not the only microbes which ended up in the dustbin of rejected hypotheses. Several others were killed off when 'their' diseases were shown to be vitamin deficiencies, not infections. The bacteria of scurvy were finally declared redundant with the discovery in 1930 of vitamin C (which was later touted as a miracle cure for polio). Beriberi was found to be caused by thiamine deficiency, as had already been deduced by Christiaan Eijkman, working in Djakarta in 1897. Eijkman abandoned his search for the bacterium responsible when he found that experimental beriberi in chickens could be cured simply by giving them rice husks to eat. This would have displeased Robert Koch, who had been Eijkman's mentor in Berlin and who later had microbiologists dispatched to Djakarta to resume the hunt for the beriberi bacterium.

Pellagra also turned out to be another pseudo-infection. The myth was exploded in a series of heroic experiments by the American Joseph Goldberger, who injected himself, his colleagues and family with saliva, blood, urine and faeces from pellagra victims.[43] Goldberger also proved that pellagra was due to dietary deficiency, although the missing ingredient (niacin) was not identified until 1938, nine years after his ashes had been scattered on the Potomac River in Washington, DC.

The two most famous non-causes of polio survived much longer than any of these spurious microbes. Rosenow's poliomyelitic streptococcus was particularly hardy, at least in the mind of its creator. It was the product of artefact and imagination, but still brought Rosenow fame and recognition. He was twice nominated for the Nobel Prize by a friendly Professor of Physiology in Cincinnati, where he eventually retired.[44] The first nomination, in 1938, was for 'fundamental contributions to the theory of disease'; Rosenow's case was evaluated, but went no further. In 1948, Rosenow was put forward again, for having 'established the streptococcal nature of poliomyelitis and the conversion/reversion of this streptococcus into viral form'. By then, anyone with access to the scientific literature would have seen that this was fantasy. The nomination was not even sent out for review.

Rosenow's obsession with his poliomyelitic streptococcus had led him to commit another, graver crime against science. The evidence is buried in a wordy section of his 1918 paper which claimed wondrous effects for his immune serum.[17] Its success was boosted substantially by the deaths of several moribund patients who had not been given the serum. In fact, they had. Rosenow reallocated them posthumously to the untreated group, because they had been so ill that they were going to die anyway. This is statistical cheating at its most blatant, but he got away with it.

This brings Ludvig Hektoen, the magisterial editor-in-chief of the *Journal of Infectious Diseases*, back into the story. Hektoen was a man of legendary seriousness, with only occasional sparks of levity. Once, asked what was new in his field, he replied, 'The skirts are shorter' – but this may just have been a reference to the ongoing debate about the impact of skirt length on the infection risk of women's clothing. Hektoen was also renowned as a tough editor with an eagle eye. Rosenow's statistical sleight of hand cannot have escaped him, and we must presume that Hektoen was content to let it ride. In fact, Hektoen's own science was not perfect, even for that time. His provocative paper of 1918, showing cocci associated with polio, examined dozens of samples from polio victims – but not one from polio-free controls. It is clear from the work of many others that if he had looked, he would have found the same bacteria there too.[45] But then, of course, there would have been no discovery and no publication.

After retiring, Rosenow began a new crusade to prove that streptococci caused neurological and psychiatric diseases, ranging from multiple sclerosis to violent behaviour. This was how he came to inject streptococci from the tonsils of the criminally insane into the brains of mice.[46] It can be difficult to diagnose criminal tendencies in mice, but Rosenow did his best. He also had a simple remedy for all these illnesses: surgical removal of the tonsils, which harboured the brain-damaging streptococci.

Someone else who wobbled off the rails was Royal R. Rife, father of the 'virus microscope', which had revealed Rosenow's 'filter-passing' micrococci in 1932. Like Münchhausen's tales, Rife's inventions became ever more grandiose and fantastic. His 'Universal Microscope' (1933) was a 200-pound beast with 5,682 parts. The 60,000x magnification which he claimed can only be obtained with the electron microscope, but the Universal Microscope enabled Rife to discover the sinister 'Bacillus X' (BX), which caused polio, tuberculosis, cancer and so forth. Luckily, BX could be shaken to death by another Rife invention, his 'Ray Tube', which bombarded diseased tissues with electromagnetic radiation tuned to BW's 'mortal oscillatory rate'.[47]

Rife's career can be summarised in two newspaper headlines: 'San Diegan's cancer work may make cure possible' (*San Diego Union*, 31 July 1949) and 'Scientific genius dies; saw work discredited' (*Daily Californian*, 6 August 1971). By way of epitaph, we have the expert critique of a Rife microscope by a professor of physics at Imperial College, London. He noted that it had lots of knobs and 'had been constructed in such a way as to make the work of microscopy tedious and cumbersome'. It produced a poor image only slightly bigger than an ordinary light microscope, and was good at generating round, coloured artefacts.[48]

What of that other microbial sideshow, the globoid bodies? Long after Flexner disowned them, they continued to generate controversy. G.S. Wilson of the Public Health Laboratory Service in London was a non-believer, and in 1959 stopped short (but only just) of accusing Flexner and Noguchi of fabricating the evidence that the globoid bodies transmitted polio.[33] But 12 years later, Harold Faber from San Francisco was convinced that they had been on to something.[35] Perhaps the poliovirus managed to replicate in a few viable kidney cells, and then stuck to the surface of the fat-rich bodies? Faber even wondered whether Flexner and Noguchi should be credited with having grown the poliovirus in cell culture, 30 years before this was reported by John Enders and his colleagues – who won the Nobel Prize for their discovery.

Flexner had a tough hide, and his reputation easily survived the fall of the globoid bodies. But Hideyo Noguchi, thoughtful and reserved, was more easily bruised by criticism – especially as doubts were growing about some of his greatest discoveries. His rabies 'corpuscle' turned out to be an artefact from his ascites medium, while evidence piled up that yellow fever was caused by a virus, not the spirochaete to which he had devoted 34 papers.

In a last-ditch attempt to prove that he was right about yellow fever, Noguchi set off in late 1927 for Accra on the Gold Coast of Ghana, one of the hot spots for the disease.[24] This was to be Noguchi's last stand. Devastated because his own experiments were now showing that he had been wrong all along, he fell ill with jaundice. It was yellow fever. The virus won the final round, and Noguchi died on 21 May 1928. His last words were, 'I don't understand'.

End of an era

The demise of the bacteria supposed to cause polio and beriberi helped to bring down the curtain on the Golden Age of bacteria and the classic germ hunters such as Robert Koch and Louis Pasteur. New bacteria were waiting to be discovered, but the emphasis was already shifting towards viruses and protozoa.

The new era never quite recaptured the excitement – scientific and public – of the previous age, or the charisma of the old pioneers. The power of scientific celebrity, and its abuse, is nicely shown by Koch's two-month lecture tour to Japan in 1908, two years before his death.[49] He was accompanied by his wife, Hedwig, who had served as guinea pig for her husband's trials of his failed wonder drug, tuberculin.

By then, Koch had built up a worldwide following as the demi-god who had cracked the mystery of tuberculosis, and his trip amounted to a state visit.

He was greeted with adulation wherever he went: standing room only at his lectures, station platforms packed as if to greet the Emperor, hundreds gathering to be photographed with him. And they hung on his every word.

What was his advice about plague? Straightforward: they needed lots of cats to kill the rats that carry plague bacilli; this was so important that the police should check every house for cat ownership. Koch's suggestion of police searches was not followed up, but his other directions were. Koch single-handedly transformed the outlook for Japanese cats. Until then, their employment prospects had been limited to providing skin for the sound box of the *shamisen*, the traditional three-stringed lute. Now, thanks to Koch, they were in demand. The result was rampant feline inflation, with cats changing hands for up to $200.[49]

And what did he think about beriberi? Koch left them in no doubt. Of course beriberi was an infection, and they must send a team immediately to Djakarta to find the bacterium responsible. So they did.

5

Lost in Transmission

What do the following have in common: a cold-shouldered Swedish virologist; a fly-encrusted monkey in a cage; 5,000 snuffly American schoolchildren and 70,000 dead cats in New York City? Answer: they were all victims of bizarre ideas about how polio was spread.

Viruses can enter the human body in many different ways: with food or drink, during sex, through the bite of a dog or a scratch in the skin, or by hitching a lift on a contaminated needle or – as a sad sign of our times – on a spicule of bone from a suicide bomber.[1] During the first third of the twentieth century, the route by which the poliovirus breaks into the human central nervous system was one of the hottest controversies in polio research. This degenerated into a standoff which pitted American science against the rest of the world, and would have left one of the mandarins of polio research looking foolish, if any of his countrymen had had the guts to stand up to him. As a result, tens of thousands of healthy people were exposed to discomfort and risk – and progress towards developing an effective polio vaccine was held up for decades.

To set the scene, we need to trace the journey of the poliovirus through the human body.

Now wash your hands

Like other enteroviruses, the poliovirus inhabits the intestine. It is usually spread by what is delicately termed the 'faecal-oral' route, mostly with the assistance of fingers, water or food. It can also be transmitted in droplets by sneezing and coughing, as the nose and throat are colonised in the early stages of invasion.[2]

Human communities have always provided rich opportunities for oral-oral transmission. Patronising comments were made in 1945 about the Inuit of Hudson Bay who lived on top of their own excreta,[3] but even societies that pride themselves on their sanitation are not above reproach; in present-day Britain, credit cards and banknotes are often heavily contaminated with colonic bacteria.[4] Conversely, basic hygiene can limit the spread of intestinal microbes, including the poliovirus. In 1954, a polio epidemic was in full swing in Western Australia, but the incidence fell off markedly when the Queen arrived on a state visit. The Royal presence was probably catalytic rather than causative, as children were only allowed to join the crowds that greeted Her Majesty if they had thoroughly washed their hands after using the toilet.[5]

The poliovirus can survive in detergents, disinfectants, alcohol and swimming pool-strength chlorine, all at concentrations that kill most bacteria.[6] Powerful chemicals, such as the formalin used to prepare the Salk polio vaccine, are needed to inactivate it so that it can no longer replicate. Unlike bacteria such as cholera, viruses cannot multiply in cell-free media (even when enriched by sewage), but the poliovirus remains viable in water and milk for several days. However, it has weaknesses. It is quickly killed by drying and by the heat of the tropical sun – conditions that do not trouble some other viruses. The smallpox virus could survive in dust for years, and managed to infect mill workers in Bolton several months after dried scabs fell into bales of raw cotton in Madras.[7]

The poliovirus is particularly pervasive and spreads rapidly through families and social networks – and even the pristine household of a professor of virology.[8] In practical terms, if one of your close contacts has polio, you are virtually guaranteed to pick up the virus. However, if you are immune to that particular strain (from having met it previously or from vaccination), then the virus will be beaten back by antibodies secreted into the gut cavity before it can invade the intestinal wall. Even if unprotected by immunity, your chances of becoming paralysed or dying are remote, as explained in Chapter 1.

Once past the teeth, the poliovirus heads south and homes in on the epithelial cells that line the throat, stomach and intestines. Viruses are not truly 'alive', as they cannot replicate without locking into the life-support systems inside intact cells. The poliovirus slips into its target cells and tricks them into shutting down all their normal activities and channelling all their resources into making replicas of it. The whole process is astonishingly quick: within a few hours of entering a cell, a single poliovirus can produce 10,000 progeny.[9] Electron micrographs show the new virus particles lining up under the membrane of the doomed cell, like massed paratroopers preparing to jump.[10] By now, the cell is stuffed with virus and fit to bust – which it soon does, spilling out what is left of its guts and the thousands of polioviruses, ready to enter and infect the next wave of target cells.

The polioviruses march down through the lining of the gut and are carried in the tissue fluid (lymph) to the nearby lymph nodes. These contain immune cells which monitor incoming material, identifying bacterial or viral 'antigens' (key parts of proteins) as 'foreign' and then producing antibodies which will kill the invaders. The immune response may be swift enough to confine the infection to the gut mucosa. Otherwise, the poliovirus can turn the lymphoid tissue to its own advantage, proliferating inside the supposedly defensive cells. From there, a tidal wave of virus sweeps through the lymph nodes and into the bloodstream.

The poliovirus leaves a trail of devastation in the gut which can be followed without a microscope: the mucosa of the throat, stomach and intestine looks angry and red, while the lymph nodes are enlarged and feel hard. These features were first reported in 1888, when John Rissler performed post-mortems on three Norwegian children who died days after developing paralysis.[11] Other Scandinavians, notably Carl Kling from Stockholm, followed the cue and during the 1910s built a convincing case that polio infected the gut.[12] This was backed up in 1915 by William Sawyer, who found the virus in the faeces of polio victims in San Francisco.[13]

The evidence was sound. However, all these findings were studiously ignored by the American polio research community, because it was headed by powerful men who were locked into a radically different impression of polio. They were convinced that polio was an infection of the central nervous system, and that the virus crept into the brain through the nose.

This view prevailed in America until 1938 when John Paul and James Trask – Yale men, and therefore to be taken seriously – again showed that the faeces of polio victims were laden with poliovirus. Moreover, the victims continued to excrete the virus for several weeks. So did apparently healthy people, which neatly confirmed Ivar Wickman's deduction of 30 years earlier that polio was spread by clinically invisible carriers.[14] Old beliefs die hard, and Paul faced a hostile reception when he presented his findings to the American Epidemiologic Society. However, he succeeded in recasting polio as an intestinal infection which occasionally strayed into the central nervous system, rather than one which directly targeted the spinal cord.

It was rapidly confirmed that the poliovirus normally travelled from mouth to anus and so *ad infinitum*. Poliovirus was found to be present in sewage, and in massive quantities during polio outbreaks. Joseph Melnick, applying mathematical wizardry to the concentrations of poliovirus found along the tributaries of the sewage system that fed into one plant in New York City, calculated that there were at least 100 asymptomatic carriers for each case diagnosed during the epidemic of 1941.[15]

From the gut, the poliovirus enters the bloodstream. This 'viraemic' phase lasts only a couple of days and so was harder to nail down. In 1952, David Bodian at Johns Hopkins University in Baltimore found the virus in the blood several days after feeding poliovirus to chimpanzees.[16,17] He then went on to look for the virus during human outbreaks, in daily blood samples from initially unaffected family members of polio cases.

Dorothy Horstmann at Yale University did the same, and both found a transient viraemia which preceded paralysis and coincided with the early phase of minor symptoms and fever.[18,19] From the bloodstream, the poliovirus can breach the final frontier and enter the central nervous system – the 1-in-100 diversion for the poliovirus, which has such devastating consequences for its host.

Roads less travelled

Occasionally, the poliovirus can enter the spinal cord through peripheral nerves, when these happen to end in tissues infected with the virus. This mechanism explains, for example, a strange epidemic of paralysis among the islanders of Western Samoa in 1909. This was particularly striking because paralysis hit the same arms where, a few days earlier, a large dose of an unpleasant arsenic compound had been injected to try to control the tropical infection known as yaws. The victims blamed the injections, and they turned out to be right.[20,21]

Neal Nathanson and David Bodian at Johns Hopkins later showed in experimental animals that the poliovirus settles in damaged muscles, and from there tracks back up the motor nerves to the anterior horn cells, paralysing the same muscles.[22,23] The same phenomenon occurred again in 1955, when dozens of American children developed paralytic polio which started in the arm where Salk's inactivated vaccine had recently been injected. This ominous localisation was the first warning that batches of the Salk vaccine contained 'live' poliovirus.[22]

Tonsillectomy also seemed to precipitate polio, and commonly caused respiratory paralysis and a high mortality. During the Los Angeles outbreak of 1942, polio hit all five children in one family who had recently had their tonsils removed, but spared the two-year-old infant who still had hers – even though she was excreting poliovirus in her stools.[24] Tonsillectomy is an unsubtle procedure which leaves a raw, damaged area that attracts polioviruses; from there, they glide up the nerves and directly into the medulla, close to the respiratory centre.

A strange susceptibility

In the time-honoured folklore of polio, paralysis could be brought on by many factors, such as exercise, cold weather and being dropped on the head. A glance at the official report of the first American epidemic, in Vermont in 1894, will show that much of the mythology came from doctors.[25]

It is true that exercise during the one to two days before the onset of paralysis can precipitate paralysis in the muscles that are overused.[26]

One young man who used both arms as paddles in a canoe race became paralysed in both arms, while the paratrooper who carried on with full combat training despite a bad headache ended up in the iron lung.[27] More subtle but just as cruel was the fate meted out to the violinist who lost the use of his left hand after spending a day practising fingering exercises.[28] Exactly how overactive motoneurones attract the poliovirus and invite disaster remains a mystery. We can, however, rule out the suggestion (offered in all seriousness, and published in a respectable medical journal in 1952) that exercise shakes loose the poliovirus from wherever it is attached in the body and knocks it into the bloodstream.[29]

On the other hand, there is no convincing evidence that heat, cold or a bang on the head predispose to paralytic polio. The same is true of a ragbag of risk factors proposed by people who arguably should have known better.

Claus Jungeblut, Professor at Syracuse Medical College, believed that hormonal imbalances could precipitate to polio.[30] His initial experiments appeared to show that adrenaline and the urine of pregnant women inactivated the poliovirus. Jungeblut followed this up with the claim that there were major hormone deficiencies in polio patients. Unfortunately, his results could not be confirmed by other, more careful studies.[31] The same fate befell his and others' suggestions that deficiencies of vitamins such as B1 and C predisposed to paralytic polio. This idea grew from even odder experiments, including the injection of peach extracts into the brains of rats (which cannot catch polio).[32]

This brings us to Benjamin P. Sandler MD, self-styled conqueror of polio and enemy of ice-cream manufacturers. As a young intern during the New York outbreak of 1931, Sandler became convinced that a low blood sugar level triggered paralysis. In 1941, Sandler injected rabbits with insulin to reduce their blood sugar, while trying to infect them with poliomyelitic monkey cord injected into the brain or poured up the nose. He gave so much insulin that the rabbits nearly died, and a day or two later, some became paralysed.[33]

Encouraged, Sandler extended his low-sugar theory to human polio, but with an imaginative twist. He claimed that the fall in blood sugar which summoned up the poliovirus was actually due to eating *too much* sugar, especially in ice cream, sweets and cakes. The 'strange paradox' of how glucose lowered its own levels in the blood was explained, he suggested, because it formed 'gamma glucose'. This mysterious compound had not been shown to exist (because it does not), but Sandler urged his readers to believe in it, just as trusting in photons and electrons helps to make sense of physics.[34]

During the 1940s, Sandler wrote repeatedly to the American public health authorities, asking them to cut the nation's sugar consumption in order to prevent polio. They ignored him. Sandler's chance to prove his theory came in the summer of 1948, when polio broke out in his town, Asheville, in North Carolina. On 5 August, the front page of the *Asheville Citizen* carried Sandler's magic cure. Grateful for any kind of help, Asheville's 55,000 inhabitants largely complied. Sales of ice cream, cookies and Coca Cola plummeted, with one ice cream company selling 1 million gallons less than expected. And within a couple of weeks, the epidemic stalled and then petered out through the autumn.[34]

Sandler became a local hero, and his book, *Diet prevents polio*, was a bestseller when it was published in 1951.[34] However, the epidemic had to finish some time, and although Sandler claimed that it was cut short by his low-sugar diet, there was no evidence of any link with the declining sales of ice cream or Coca Cola. By then, Sandler's original rabbit experiments had been discredited. Back in 1941, Emerson Kempf of Ann Arbor University, Michigan, had repeated Sandler's experiments with insulin, but more carefully. He found no hint that rabbits could be infected with polio, whatever their blood sugar – and also noted that rabbits were easily paralysed by having foreign material injected into their brains or noses.[35]

The prize for the most bizarre theory about risk factors for polio must go to George W. Draper, America's foremost clinical expert on polio during the 1920s. In his famous 1916 monograph, *Infantile paralysis*, Draper revealed that he could spot children susceptible to polio simply from their 'peculiarities of development': plump and well nourished, with a broad face, freckles and gaps between their teeth. There was no objective evidence to support his clinical impression, but Draper was a man of influence, and gave his claim authority by repeating it in successive editions of his monograph and in papers in high-profile journals.[36,37]

And people believed him. As late as 1930, children with an 'ultra-susceptible' appearance according to Draper's criteria were still given top priority for preventative measures against polio. Even later – and nearly 20 years after his theory had been comprehensively annihilated by careful studies during the New York epidemic of 1931[38] – the 'polio-susceptible' features of the Inuit were blamed in part for the catastrophic outbreak of 1948–49 in the Canadian Arctic.[3]

Up the nose

One of the strangest blind alleys in the story of polio lay in the upper stretches of the nasal cavity. This was a tale of the Emperor's new clothes as applied to medical research, with scientist-courtiers too intimidated by the man at the top to point out that he was wrong.

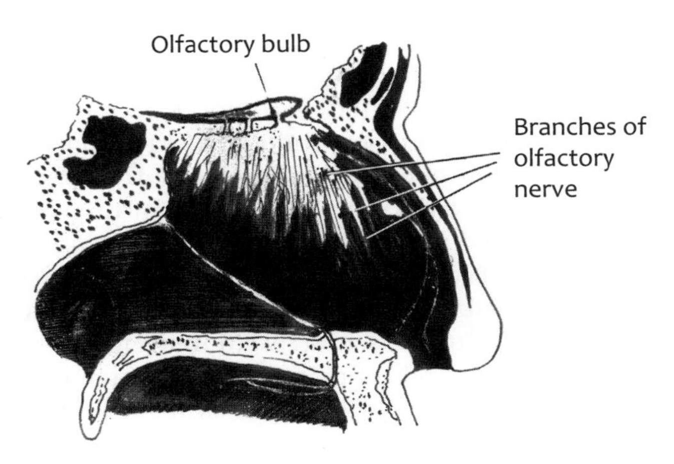

Figure 5.1 The human nose, showing the olfactory nerves running through the 'cribriform plate' into the olfactory bulb. The olfactory bulb is connected to the brain. Illustration by Ray Loadman

The nose enjoys a privileged relationship with the brain, and the hypothesis that this was the poliovirus's way into the central nervous system was not too far-fetched. The top of the nasal cavity lies immediately below the undersurface of the brain, separated only by the meninges and a flimsy wafer of bone known as the cribriform ('sieve-like') plate because it is perforated by dozens of holes. Through these pass the olfactory nerves, serving the sense of smell, formed from hundreds of branches running in from the lining of the nose. On each side, the olfactory nerves feed into the olfactory bulb, which is connected to the brain by a thin ribbon of nervous tissue (Figure 5.1).

The nearness of the brain to the top of the nasal cavity has been exploited in various ways. The Ancient Egyptians prepared bodies for mummification by pulling the brain piecemeal down the nostrils with hooks, while present-day neurosurgeons use the same route to operate on the base of the brain.

The 'olfactory route' for the poliovirus to enter the brain was a neat theory that sprang up at the Rockefeller Institute. Peter Olitsky, one of Simon Flexner's most trusted research captains, had worked on viral infections of the brain. One of these, equine encephalomyelitis, was apparently spread by infected droplets from sneezing horses. Olitsky found a more convenient laboratory model, as the virus readily invaded the brains of mice when it was swabbed into their nostrils. When polio arrived on the Rockefeller scene in 1910, with all the trappings of a 'neurotrophic' virus that went straight for the brain, it seemed natural for it to be put through Olitsky's olfactory transmission protocol – but in monkeys, the species which Landsteiner had succeeded in infecting.

And it worked.[39] Flexner, an authority on the meningococcus which seemed to invade the meninges through the cribriform plate, was immediately excited, and diverted the Rockefeller's resources and researchers into proving the case. Under his direction (and with no degrees of freedom for anyone who wished to continue working at the Institute), supporting facts quickly accumulated.[40] Intranasal inoculation of the poliovirus produced paralytic polio in monkeys just as reliably as when it was injected directly into their brain. The poliovirus could be recovered from the olfactory nerves and bulbs a couple of days after it had been swabbed up the animals' nostrils. Moreover, paralysis could be prevented by 'blocking' the olfactory route before giving monkeys the virus intranasally. Blocking could be done either by surgical removal of the olfactory bulb[41], or – excitingly, because this pointed to a possible preventive treatment for human polio – by applying chemicals to the nasal mucosa.[42]

The olfactory bandwagon rapidly gained momentum. Hypothesis hardened into fact, generating dozens of papers that made the Rockefeller Institute and Simon Flexner famous around the world. And de facto, even though there was not a shred of direct evidence, this peculiar animal model became the template for the human infection.

The experimental protocols for investigating the olfactory route and its blockade drifted steadily away from real life. Swabs soaked in extracts of infected spinal cord were pushed high up the monkeys' nostrils and poked around to paint the mucosa thoroughly – somewhat more invasive than sniffing in tiny droplets sneezed out by a polio patient. The blockade experiments involved spraying various chemicals into the nose, in the hope of interfering with the poliovirus's ability to hitch a lift into the brain via the olfactory nerves. Cobra venom was tried and abandoned, as were over 150 other less exotic compounds.[43] Various astringents, such as alum, silver nitrate and picric acid, seemed to work, although large volumes (1.5 ml into each nostril for a cat-sized monkey) had to be delivered right to the top of the nasal cavity. The lining of the nose is delicate and exquisitely pain sensitive, and these substances are not bland. Alum was used in the styptic pencils which stop bleeding from razor-blade cuts (and sting like fury), while picric acid comes with a warning that it burns flesh.

With the weight of the Rockefeller behind all this, it was inevitable that the human nose would eventually be dragged in – even though Charles Armstrong, one of the pioneers in the monkey experiments, warned that 'this method of prevention is not recommended for human use'.[44]

In Kew, near Melbourne, Australia, specially designed nose clips were tried out during an epidemic in 1937, with no obvious benefit.[45] Around the same time, much more ambitious experiments were being conducted in North American children to test the astringent nasal sprays which Armstrong had advised against using in man. Armstrong had evidently thought again, because it was he who ran the trials.[46]

Solutions of alum, with or without picric acid, were delivered into the apex of the nasal cavity. This was not a child-friendly procedure. Ordinary sprays and drops were not good enough, and indeed were blamed for failure of the early experiments. The state-of-the-art method, developed on the advice of a neurosurgeon, was to push a fine catheter far up each nostril, to within a few millimetres of the cribriform plate (about twice as far as the fingertip of an enthusiastic nose picker). This could only be done by medical personnel; unfortunately, many doctors refused to take part because they were not paid enough.

The first trials, in 4,600 schoolchildren from Alabama, took place during the polio season of 1934. The results were uninterpretable, as there were no untreated controls, but this was clearly not a therapeutic miracle because many treated children went down with polio.[47] Two years later, 5,200 children in Ontario were treated with zinc sulphate given by the catheter technique; this time, untreated controls were included.[48]

Both trials were funded by the National Foundation for Infantile Paralysis. They were viewed as serious science, and in the absence of a safe and effective vaccine, the only glimmer of hope for preventing polio. There were upbeat editorials in medical journals, while newspapers carried advertisements with coupons which parents could cut out and take to a doctor to have their child treated free of charge (Figure 5.2). Polio had broken out in Toronto just before the 1936 trial in Ontario; panic was in the air, and nasal blockade the only straw to clutch at. Apart from the trial, thousands of terrified parents bought picric acid and perfume atomisers to treat their children. One mother whose baby died of polio thanked God that she had used the spray, because it meant that she had done everything possible to save her child.[49]

When the results were analysed, it was obvious that nasal blockade – whether professionally administered or do-it-yourself – was completely useless. The only significant outcomes were severe headache and, in several thousand children, permanent loss of the sense of smell.[50]

Meanwhile, other doubts had been growing about the olfactory route, although few were bold enough to voice them in North America. In his pre-vaccine days, Albert Sabin worked at the Rockefeller, and began as one of Flexner's bright young men who boosted the credibility of the olfactory route. Later, though, Sabin found himself questioning whether this was a valid model for human polio. He looked carefully in the olfactory bulbs of several patients who had died of the disease and found

INFANTILE PARALYSIS
FACILITIES FOR NASAL SPRAYING

Experimental work appears to offer some hope that the proper application of a solution of certain chemicals high up in the nose may afford protection against poliomyelitis. Adequate trial, however, of this method has not yet been made and its value can be learned only by careful observations during the presence of a large number of cases.

As the spray is harmless, with only an occasional child suffering some temporary discomfort, it is believed that many parents may wish to co-operate in a study to determine its possible value. Facilities are being provided through funds made available by the Department of Health of Ontario, and arranged by the Hospital for Sick Children and the School of Hygiene, University of Toronto. The work is being made possible through the hearty co-operation of the nose and throat specialists and the authorities of the various Toronto hospitals. To permit of proper observation the group will be limited to 5,000 children from 3 to 10 years of age resident in Greater Toronto. One treatment will be given to each child by a nose and throat specialist as soon as possible and subsequent treatment at intervals of two weeks if the conditions warrant. Information regarding the effectiveness of the treatment in each child will subsequently be obtained from the parents.

If as a parent you desire to have your children 3 to 10 years of age given this treatment:

(1) Complete the request form printed below and sign it.
(2) MAIL IT IMMEDIATELY, OR BRING IT, to the Hospital for Sick Children (front door, College Street), as only the first 5,000 requests can be accepted.
(3) A card of admission to the clinic will be sent to you immediately, telling you to which hospital to go and the time.

The work is conducted without charge.

NOTE: No children will be treated at any of these clinics without appointment and NO APPOINTMENTS WILL BE MADE BY TELEPHONE. Do not bring your children without the appointment card which will be sent to you when the signed request form is received.

Request Form For Nasal Spraying

Names of Children	Age
....	....
....	....
....	....
....	....
....	....
....	....

I hereby request that the above-mentioned (child) (children) be given the nasal spray treatment as arranged by the Hospital for Sick Children. I understand that the treatment will be given shortly and subsequent treatments at intervals of two weeks if the outbreak continues.

Telephone No. .... Signed .... Parent or Guardian

Date .... Address ....

Figure 5.2 Information and consent form for nasal spraying to try to prevent polio, as printed in newspapers in Toronto and Ontario, spring 1936. Reproduced by kind permission of the *Canadian Public Health Journal*

no evidence of viral damage.[51] This was a significant nail in the coffin of the olfactory theory, and also in Sabin's relationship with Flexner. Later still, David Bodian proved that the nose was irrelevant in chimps given poliovirus orally, as they still developed the infection even after the olfactory nerves had been cut surgically.[52]

All this should have been enough to convince the nose's staunchest supporters that it was time to move gracefully into the gut. However, intellectually attractive ideas often outlive their credibility, even if they are obviously rubbish. Flexner never retracted his many statements that human polio must behave as it did in the Rockefeller's monkeys. And as late as 1946, Haven Emerson, the man who had led the charge against the great New York epidemic of 1916, remained convinced that the nose really was the way in for the poliovirus.[53]

No-fly zone

The nose was not the only scientific cul-de-sac in the long journey to discover how polio was transmitted. Many species of animals – from bedbugs to budgies to cats – were thought to be involved, either by harbouring or spreading the infection. This was another area riven with folklore, anecdote and poor-quality science.

Many animals are co-conspirators with the infectious diseases of mankind. A classic example is plague, caused by a bacterium which is equally at home in the black rat as it is in humans, and which skips between the two hosts in the bite of the flea that is happy to feed on either. Up to the 1920s, polio was also assumed to have other, non-human reservoirs of infection. There were numerous reports of paralytic disease breaking out in domestic animals, sometimes at the same time as human polio. Charles Caverly devoted several pages of his report on the first large American epidemic, in Vermont in 1894, to instances of paralysis in cattle, horses, dogs, cats and poultry which seemed suspiciously close to human cases in time and/or space.[25]

The same theme was picked up by Henry Frauenthal and Jacolyn Manning in their 1914 *Manual of infantile paralysis*.[54] Their selective survey of various outbreaks in the United States and Europe featured many engaging anecdotes but very little science. They focused on dogs that had gone off their legs in Massachusetts; an epidemic paralysis that struck down English sheep every autumn (but was unrecognised in Britain); and a chicken from Iowa 'with a most suggestive history', which at post-mortem had poliomyelitis-like lesions in its spinal cord.

More rigorous investigators were less convinced. In 1914, Carl Ten Broek from Boston examined a menagerie of paralysed animals, including cattle, cats, dogs, rats and chickens in variable proximity to polio cases. Post-mortem revealed several other causes of paralysis, including spinal tumours and bacterial infections, and none of the animals' spinal cords caused paralytic polio when injected into monkeys' brains. His conclusion, 53 monkeys later, was that whatever paralysed the animals was not polio.[55] Similarly, an outbreak of paralysis affecting 20 hunting dogs in Little Rock, Arkansas, in 1917 provided 'no valid reason to consider that the affection had any relation to human poliomyelitis'.[56]

Out on the streets, however, logic did not necessarily prevail. During the great New York City epidemic of 1916, the finger of popular blame was pointed in turn at dirt, flies and immigrants, and then settled on the cat. It was not clear how the rumour started, but 70,000 cats were killed in the city during the month of July – and health officials joined bounty-hunting boys in the massacre.[57]

In a more restrained way, the law of the jungle also applied in the laboratory. Hundreds of thousands of rats, mice, guinea pigs, hamsters, rabbits, dogs and cats perished during experiments to find out how polio found its way into the central nervous system. Poliomyelitic cord extracts were often injected directly into animals' brains, while potential risk factors were simulated in ways that can most charitably be described as crude. Benjamin Sandler gave near-lethal insulin doses to rabbits to mimic the slight fall in blood sugar which he believed (wrongly) was induced by eating ice cream, while others injected fruit extracts directly into the brain to reverse vitamin deficiencies. Unsurprisingly, many studies were negative. Of greater interest are those which reported positive findings – paralysed limbs and destruction of the motoneurones in the spinal cord – in species such as rats, rabbits, cats and dogs, which are not susceptible to polio. As with Ten Broek's domestic animals, whatever caused paralysis was not polio.

Ten Broek's experiments also bring enlightenment to anyone wondering about the title of this section. Some of his experimental monkeys underwent an even stranger test than being injected with the spinal cord of cows or dogs. Nineteen biting stable flies (related to the European horse-fly) were ground up and injected into a monkey's brain; when nothing happened, another twenty-two flies went the same way, just to make sure.[55]

The idea for these experiments was not simply plucked out of the air. At this time – 1914 – there was an extensive literature on the transmission of polio by insects, prompted by the recent discoveries that mosquitoes spread malaria and yellow fever. The biting stable fly, an ugly brute with dagger-like mouthparts, was one of the chief suspects (Figure 5.3). Jacolyn Manning saw 'no reason why the stable fly should not transfer any blood-inhabiting micro-organism from insect to man or from man to man'.[58] And under some inspirationally contrived conditions, its bite could transmit polio – for example, when several thousand starving stable flies enjoyed a feeding frenzy on monkeys with experimental polio and were then left to bite uninfected monkeys.[59] Suspicion also settled on other species of fly, notably the housefly which co-habited with dirt in the polio-infested immigrant slums of New York, and which was easily caught in the act of feeding on faeces.[57]

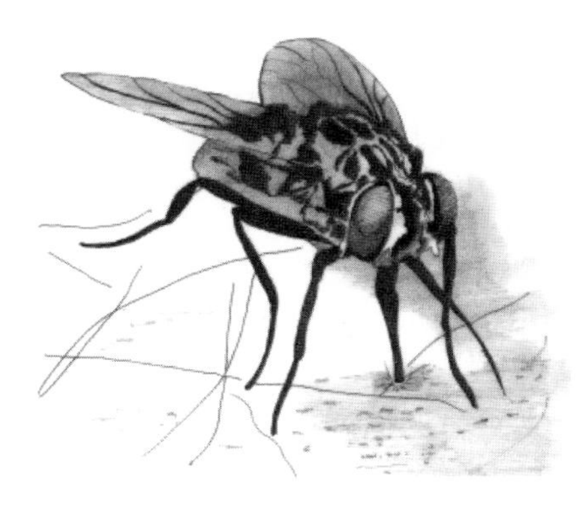
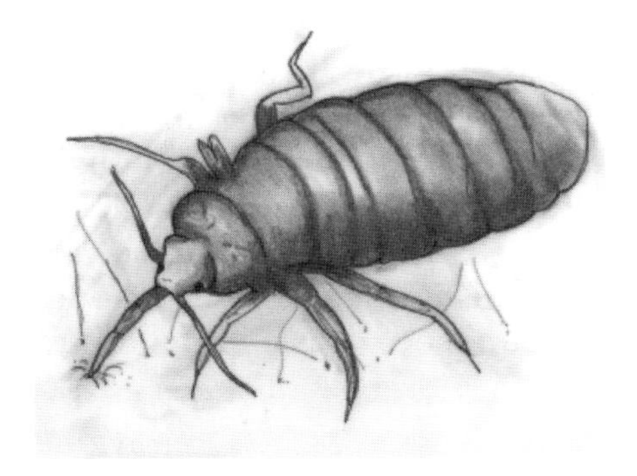

Figure 5.3 Biting stable fly (left) and bedbug (right), two non-carriers of polio. The fly is about the size of a housefly, and the bedbug about 4 millimetres long. Illustrations by Ray Loadman

One hesitates to say that flies created something of a buzz in the polio research community. However, they were still a hot topic in the 1940s. L.O. Howard caught 23,087 flies (98.8 per cent of which were houseflies); Joseph Melnick, one of the pillars of the Yale polio research group, had a mobile insect-processing laboratory which he drove to epidemics; and M.E. Power laid out a feast of bananas and dog faeces to entice the fly population of New Haven, Connecticut, in 1942. Even tiny fruit flies were subjected to intimate body searches for the poliovirus.[60–62]

At ground level, another menace lurked. This was Manning's bête noire which filled several pages of the *Manual of infantile paralysis*: the blood-sucking bedbug (Figure 5.3).[63] Throughout history, bedbugs have 'accompanied man wherever he has gone' and, in Siberia, reportedly grew as big as cockroaches. It had been shown that bedbugs could transmit plague, but there was no evidence of any involvement in spreading polio. This did not deter Manning from her mission to incriminate and destroy this 'pan-American pest'. Her accounts of bedbugs 'boiling' out of an iron bed frame in Cook County Hospital, Chicago, and of the violent end of the bug which crossed her path on the Hudson Tube at Manhattan Transfer, make particularly graphic reading.[63]

The bottom line, after decades of work and hundreds of papers: like fingers, flies can pick up the poliovirus from faeces and carry it for a few days. However, they are not hosts in the way that the yellow fever virus infects and replicates inside mosquitoes.

This conclusion was not reached in time to prevent one of the flimsiest prophylactic exercises in the history of polio. On 8 August 1945, a US military plane appeared in the skies over Rockford, Illinois, where a polio epidemic was in full swing. Its cargo was 1,000 gallons of DDT, and its mission was palpably more humanitarian than that of the B-29 which had overflown Hiroshima two days earlier. As a local newspaper put it: 'Preventative Spraying for Polio as Important to Rockford as the Atomic Bomb'. Unfortunately, although the fly counts dropped off somewhat after the DDT offensive, polio rampaged on unabated.[64]

The same outcome was seen when DDT was sprayed over the village of Luanshya in Northern Rhodesia, except that malaria – which really is insect borne – was eliminated from the locality.[65]

Monkey business

Karl Landsteiner's transmission of polio from a dead boy to a living monkey just before Christmas 1908 was a pivotal event in more ways than one. It transformed polio from a black-box medical enigma into an infection that could be forced to reveal its secrets in the laboratory. However, it also set polio research trundling off down the wrong track for 30 years, to the detriment of science, mankind and monkeys.

Landsteiner's experimental design was eagerly seized by Flexner and others and soon became scientific gospel. Thanks to the monkey, polio could at last be recreated in the laboratory. Originally, Landsteiner had injected ground-up spinal cord into the monkey's abdominal cavity. Soon, on the premise that the virus attacks the central nervous system, Flexner and others were routinely injecting infected material directly into the experimental monkey's brain. The results were spectacular. Manning was bowled over by a tour of Flexner's laboratory (conducted by the great man himself), where she saw dozens of monkeys showing the gamut of paralyses that she recognised from her patients in Wisconsin. She enthused over Flexner's 'epoch-making line of investigation' as 'one of the rare gratifications of life'.[66]

Unfortunately, the 'rare gratification' was a self-fulfilling prophecy that quickly led American researchers on to a false trail. Delivering the virus directly into the brain was logical to Flexner and those who worshipped at the altar of olfactory transmission. If, however, the poliovirus normally had to negotiate defensive barriers in the gut and meninges, this made it far too easy. It was like leading a burglar into the vault and then opening the safe for him. Drawing conclusions about the human disease from what happened when the virus was put straight into the brain could therefore be dangerously misleading.

The experiments were distorted even further by Flexner's favourite poliovirus strain (MV, for 'mixed virus'), which was used in all the Rockefeller's polio research.[39] Flexner's colleague Harold Amoss had originally derived MV by pooling the spinal cords of two patients who had died of polio in 1914, and the product had been passed repeatedly from one monkey's brain to another's, in the belief that this was the poliovirus's preferred habitat.

And as far as MV was concerned, so it came to pass. Mutations arise all the time when a virus replicates, and the alien environment of the monkey's brain selected out those which managed to survive, while progressively killing off the wild type. Years of being locked inside the monkey's brain changed MV into a professional neurotrophic virus, so far removed from its gut-inhabiting ancestor that it had lost any capacity to invade tissues outside the central nervous system. This crucial flaw was unknown to Flexner at the time. All he saw was a virus that behaved exactly as he predicted and that reinforced all his hypotheses. It was only decades later that the extent to which MV had been crippled in Flexner's laboratory became apparent. MV could not be transmitted orally, or even infect non-nervous tissue – a shortcoming that held up the development of a vaccine for 15 years and, as explained later, may have robbed Albert Sabin of a Nobel Prize.

More immediately, the unhappy marriage between the dysfunctional virus and its inappropriate niche produced a steady gush of papers from the Rockefeller. These were published in top journals such as the *Journal of Experimental Medicine* (editor-in-chief, Dr Simon Flexner), and helped to push the Rockefeller to the forefront of polio research worldwide. The success of this theme also provided ammunition for Flexner's quiet but ruthless war of attrition against certain European researchers – Carl Kling in Stockholm, Constantin Levaditi in Paris – who were trying to push their own evidence that polio infected the gut and might be transmitted orally. Flexner was able to contain this heresy where it belonged, on the other side of the Atlantic.

The experimental monkey kept its stranglehold on polio research until the late 1940s, when the poliovirus was successfully grown in tissue culture. During this time, polio research was a lumbering, hugely expensive juggernaut. Monkeys were expensive, difficult to keep and unreliable as experimental subjects. Even straightforward experiments could use hundreds of monkeys, and take many weeks.

The logistical nightmare of using monkeys is nicely portrayed by John Paul's, who was commanded by Flexner to make a field trip to Los Angeles during the epidemic of 1934 and bring back the virus responsible. Paul described the outbreak as 'somewhat odd', thanks to the relatively benign Type 2 strain that was on the loose, and to the idiosyncrasies of the city and its inhabitants. Also, Edward C. Rosenow had beaten them to it and was already there, actively promoting his poliomyelitic streptococcus.[67]

Paul's team hit town like a travelling circus, bringing 75 monkeys with them (they soon had to buy more). Samples of spinal fluid or nasal washings, or extracts of post-mortem spinal cord, were injected into a monkey's brain (0.5 ml) and for good measure into its abdominal cavity (12 ml). The monkey was watched up to 12 days, and killed at any sign of fever or paralysis. The spinal cord was removed for histology and also to check that polio could be transmitted to another monkey. The diagnosis could only be made if poliomyelitis lesions were seen in the original test monkey, and in the one which it infected; this generally took over two weeks. Paul's team succeeded in capturing poliovirus from a few cases, including the first isolation from the human nasopharynx. They also confirmed that the monkey-based poliovirus detection system was woefully insensitive and unreliable, and could only make the diagnosis in 1 in 30 cases of clinically obvious polio. Nonetheless, Paul and his colleagues were given the Freedom of Pasadena, while back in New York, Flexner was delighted.

Even more monkeys were needed to study the (wrong) route of transmission. For example, the experiments that produced a single paper on nasal blockade went through 250 monkeys.[42] The programme which showed that the 200-plus poliovirus strains belonged to just three types used up 17,000 monkeys between 1949 and 1951.[68] Polio research accounted for most of the monkeys used in biomedicine in the United States during the 1950s and 1960s. The body count reached 200,000 in 1957–8. Monkeys were expensive to buy (about $25 each during the 1950s), and ran up big bills during their laboratory lifetime, which averaged a few weeks. Monkey costs mopped up at least a third of most polio research grants awarded during this time.[69,70]

The dependence on monkeys opened up strange new trade routes into the United States. The two species most heavily used were both macaques, the Rhesus monkey from India (and a deliberately colonised island off the coast of Puerto Rico), and the cynomolgus monkey (crab-eating macaque) from the Philippines. The Rhesus monkey was sacred to the Hindus, but this did not save millions of the animals from being trapped and sold into biomedical research, especially in America. By 1978, the Rhesus monkey population was so badly depleted that the Indian government banned further exports, to save it from extinction in the country.[69,70]

Consignments of monkeys flown into the United States converged on a few breeding and distribution centres, notably Okatie Farms, near Pritchardville, South Carolina. This facility was run by the National Foundation for Infantile Paralysis to serve the American polio research community. Monkeys were quarantined at this 'conditioning centre' for long enough to allow nature to take its course with the many that were riddled with tuberculosis and other infections, or terminally weakened by the stresses of capture and transport. The survivors were not guaranteed a clean bill of health, and were prone to ruin experiments by dying at inconvenient moments. Fever, listlessness and even paralysis were often caused by diseases picked up in transit, prompting Sabin to write a detailed article about the pitfalls in using monkeys to study experimental polio.[71]

Unlike the macaques, the great apes – just below us on the evolutionary tree – are naturally susceptible to polio. The gorilla, orangutan and chimpanzee can catch and transmit polio by the faecal-oral route, just like humans.

During the mid-1940s, the chimpanzee carved out a valuable career as the closest animal model to human polio. This began after David Bodian and Howard Howe at Johns Hopkins spotted a ten-year-old report that two chimps in the children's zoo in Cologne, Germany had been struck down with a paralytic illness that appeared identical to that of man. Bodian, and Dorothy Horstmann at Yale, both worked out the transmission of polio in the chimp, from gut to blood to central nervous system.[16–19] Finally, on the threshold of the 1950s, they convinced the doubters that experimental polio in the chimpanzee painted a faithful picture of the human infection. Only then, over 30 years after the first triumphant papers about the olfactory route of polio transmission, was this pervasive and self-propagating myth finally laid to rest.

Meanwhile in the gut

This climate change in American polio research allowed some Scandinavian wisdom, by then 30 years old, to come in from the cold. During the large Swedish epidemic of 1911, Carl Kling, Wilhelm Wernstedt and Alfred Pettersson from the State Microbiology Laboratory in Stockholm, looked for the then-recently described filterable virus in both living and dead victims of polio. They found that samples of saline run through the intestines at post-mortem were able to paralyse monkeys. So could intestinal washings (collected after the bowel had been comprehensively emptied with an enema) from living polio patients – and from their close contacts who had only minor symptoms or even appeared completely healthy.[12]

This was genuinely groundbreaking thinking that should have short-circuited a quarter-century of misdirected and wasted polio research. Kling, then aged only 25, was extremely bright and hard-working, and already on course to become the doyen of the European polio community. When he came to Washington, DC, in late 1916 to present his findings to the 15th International Congress of Hygiene and Demography, he deserved a standing ovation. Instead, that accolade went to Professors Rosenau and Brues of Harvard University, for showing that biting stable flies could transmit polio to a monkey (at least when the monkey's cage was thick with thousands of the insects).[59]

Kling's own reception was cool and dismissive, and the prelude to a prolonged campaign of deliberate neglect. All this was orchestrated by Simon Flexner, so that the theory of olfactory transmission and the primacy of the Rockefeller could not be challenged. Until the mid-1930s, Kling's work was rarely cited by North American researchers. Only a few brave souls, such as the Canadian Maurice Brodie, dared to lift their heads above the parapet and admit that the monkey's nose might be irrelevant and that the Swedes could be right about the pathogenesis of the human disease.[72]

Kling ignored the snub and continued to pursue the poliovirus through the intestines. In 1929, he succeeded in giving polio to cynomolgus monkeys, by feeding them the virus orally. This finding, also revolutionary and a tour de force because lower primates are so difficult to infect by mouth, was similarly ignored in America. It did not help that this article, like his previous papers, was published in French.[73]

Kling was well ahead of the crowd, but left some crucial gaps in the argument that the poliovirus homed in on the intestine and multiplied there. Eventually, it was Sabin who in 1941 found poliovirus in the wall of polio victims' intestines, proving that this was where polio broke into the human body.[74]

And only then was Kling finally vindicated and rehabilitated on both sides of the Atlantic – thanks to American researchers who confirmed what he had found up to 25 years earlier. This pulled polio research back on course and had a hugely important spin-off. Sabin was prompted to start thinking seriously about the rationale and practicalities of an oral polio vaccine.

The world was rapidly moving on. Once again, Flexner was being left behind, but even if he had noticed, he would not have believed it.

A cultured solution

Meanwhile at the laboratory bench, researchers were increasingly desperate for new tools that could make polio experiments quick, easy and affordable.

The first breakthrough came in 1939, from Charles Armstrong of the Hygienic Laboratory in Washington, DC (who had blown hot and cold over olfactory blockade). Armstrong persuaded the Type 2 poliovirus, Lansing, to infect and paralyse the cotton rat when it was injected directly into the animal's brain.[75] The cotton rat (actually a large mouse) is a common wild rodent in North America and, apart from a tendency of adults to fight to the death, it quickly adapted itself to life in the laboratory.

The cotton rat opened up a bright new world of polio research, in which it became feasible to do in vivo experiments that, with monkeys, would have taken months and bust the budget. The cotton rat also contributed to the Allied war effort, being airlifted into England in 1944 for 'Operation Tyburn', a top-secret research programme into methods to control typhus.[76] It was partly superseded when Armstrong further adapted the Lansing poliovirus to infect the most convenient laboratory mammal of all, the white mouse.[77]

This success made up for Armstrong's earlier disappointment with olfactory blockade and won him his place in the Polio Hall of Fame. His reputation was only slightly marred by reports that his colleague, Ardrozny Pakchanian, had actually done the hard work with the cotton rat, but was unlucky enough to be in hospital when Armstrong submitted the paper – with himself as sole author and Pakchanian's name nowhere to be seen.

In 1953, C.P. Li and Morris Schaefer pulled off an even more useful feat, by infecting mice with a Type 1 poliovirus, of the sort responsible for most epidemics and cases of paralysis.[78] But by then, an even greater advance had burst onto the scene – a Nobel Prize-winning breakthrough that was neat in concept and of unimaginable breadth in application. The poliovirus could be grown outside living animals, in cultures of cells kept in an artificial medium.

The notion of tissue culture had been around since the 1910s. Constantin Levaditi claimed to have grown polioviruses in culture in 1913. Strangely, for such a major advance, he only mentioned this in a book chapter 40 years later, and shortly after others had reported their own successes.[79]

In 1936, Sabin and Peter Olitsky tried to grow the MV poliovirus in cultures of various tissues collected from aborted human fetuses. They were partly successful, in that it grew well in brain tissue but not in the other cell types.[80] This was a breakthrough, but one which reinforced the myth that the poliovirus primarily targeted the central nervous system. Sabin's basic error was to have chosen MV, so embedded in its neural niche that it could not survive in other cells.

Years later, Sabin went back and tried to grow MV in non-neural human cells using a recipe that had been good enough to win the Nobel Prize.[81] MV again failed to thrive outside nervous tissue, confirming its peculiar nature – and making Sabin realise that if he had chosen a different strain of virus, he might have been the one to receive a congratulatory telegram from the Nobel Prize Committee.

That accolade went instead to the trio of John Enders, Frederick Robbins and Thomas Weller, working at the Children's Hospital in Boston. They succeeded in growing the Lansing (type 2) virus in cultures of several cell types taken from the limbs and viscera of aborted fetuses – skin, muscle and intestine – as well as nervous tissue.[82]

Their paper, published in *Science* in 1949, opened up vast new horizons. The poliovirus could be grown easily, and its numbers counted accurately for the first time, thanks to its 'cytopathic' (cell-killing) effect which created visible holes in sheets of cultured cells. And huge quantities of poliovirus, enough for industrial production lines, could now be grown up, quickly, safely and cheaply. The most obvious spin-off, although still hypothetical, was a polio vaccine.

Enders, Robbins and Weller duly received their Nobel Prize for Physiology or Medicine in 1954 (Figure 5.4). The expert evaluation of their Prize nomination was particularly effusive, describing their work as 'the most important discovery in the history of virology'.[83]

Figure 5.4 Thomas Weller (1915–2008), Frederick Robbins (1916–2003) and John Enders (1897–1985) at the 1954 Nobel Prize award ceremony in Stockholm. They were awarded the Prize for Physiology or Medicine for 'the greatest discovery in the history of virology' – the culture of the poliovirus in non-nervous human tissues. Reproduced by kind permission of the Division of Medicine and Science, National Museum of American History, Smithsonian Institution, Washington, DC

Birth of a monster

Three intriguing questions about the spread of polio are left dangling. Why was polio a disease of summer and autumn? Why was it rampant in Western countries, and rare in the developing world? And why did epidemics suddenly appear around the turn of the twentieth century?

The first question has no good answer, even though the seasonality of polio in the Western hemisphere has long been recognised. The polio season could begin as early as May, but generally picked up momentum through July and August, heralded by the exodus of the wealthy and mobile from American cities. In Scandinavia, epidemics often ran into the autumn – hence the traditional warnings to Swedish children not to risk the infection by kicking through fallen leaves or eating apples off the ground.[84] In the United States, doctors were reminded that polio 'should be suspected until Thanksgiving, or at least until the snow flies'.[85]

In fact, polio often broke through these seasonal boundaries. Many epidemics grumbled on into the following year, while the vicious outbreak in the Canadian Arctic in 1948–49 reached its peak during the depths of an unusually harsh winter.[3]

The second question – why polio seemed to spare the developing world – has a much clearer answer. Quite simply, polio was as common there as in the Western hemisphere, but nobody had bothered to look properly. When systematic surveys of 'lameness' were carried out during the 1980s, the prevalence of polio turned out to be just as high as it had been in America and Europe during the pre-vaccination era.[86] However, cases cropped up constantly throughout the year, without a seasonal peak or the big-bang epidemics that could cut through the West.

This brings us to the third question: why polio changed its nature so dramatically in the West, from an inconspicuous, fine-print disease of childhood to an epidemic monster that could bring America to its knees.

It seems that humanity must share responsibility with the poliovirus. The transformation of polio into an epidemic infection has paralleled the advance of sanitation, and especially effective sewage disposal that has kept our excreta safely away from our drinking water, food and children. This association is too powerful to be coincidence.[87]

In pre-hygienic communities, polioviruses were everywhere; from birth, children had every opportunity to pick up the infection and acquire immunity. Infants have some built-in protection against common infections, thanks to antibodies from their mother, initially passing across the placenta in utero, and later from breast milk. These antibodies helped to mop up any poliovirus that might stray into the infant's bloodstream, while the child built up its own layer of protection in the gut – the so-called 'intestinal' immunity which is the first and best line of defence against future attacks.

With the general removal of excreta from the scene, children have less chance of meeting the poliovirus before the protection of their mother's antibodies wears off within a few months. This leaves them vulnerable to attack. An epidemic can be triggered when the poliovirus is introduced into a community where the number of susceptible, non-immune people exceeds a critical threshold – with consequences that became all too familiar.

Civilisation brings risks as well as benefits. It would seem that polio, like obesity and cardiovascular disease, should be added to the list of diseases that we have helped to bring upon ourselves.

6

Fear Is the Key

It was obvious from early in the polio season that 1949 was going to be a bad year for several countries. In Washington, DC, the Surgeon General told Americans to expect 'the worst year for infantile paralysis in the history of the USA'. Across the Atlantic, where an English winter had failed to snuff out the tail end of the 1948 epidemic, the *British Medical Journal* warned that the omens were 'sufficiently grave'. Meanwhile, Swedes were preparing themselves for their annual epidemic and wondering if this one might outstrip the biggest on record, in 1936. And in South Africa, an outbreak brewing up in the Transvaal was already filling up the beds in Johannesburg's Fever Hospital.[1–4]

As the American epidemic took hold, the nation slipped into a routine which had become a conditioned reflex ever since the catastrophic outbreak that crippled New York in 1916. This time, Texas was badly hit. One of its hot spots was the city of San Angelo, set in the agricultural lands of Tom Green County some 200 miles south-west of Dallas. Thanks to oil, the mohair industry and nearby US Air Force bases, San Angelo had expanded rapidly since the 1920s and was now an urban sprawl with a chequerboard of good, bad and dire sanitation that was ready to welcome in the poliovirus. The previous year's epidemic had stuttered on, and there were already 50 cases in town by mid-May.[5]

A series of deaths in children in early June made headlines; within a week, several patients each day were being admitted to the city's Shannon Hospital. Emergency medical equipment, including iron lungs, was airlifted in on a C-47 heavy transporter. This was high season for electrical storms. Luckily, thunderclouds could be spotted approaching across the plains in good time to call in relays of volunteers to hand-crank the iron lungs during power cuts – although the levers were so badly designed that even strong men soon tired.[6]

On the streets of San Angelo, visibility was often poor because of the smokescreen of DDT which belched out of 100-gallon tanks on the back of 'fogger' trucks. The foggers were backed up by DDT-spraying teams on foot, who targeted out-buildings and privies. Other measures to kill off the mosquitoes that might just carry polio included draining ponds and cutting down overgrown vegetation (houseowners with weeds over 12 inches high were fined on the orders of the City Health Officer). In early June, the municipal pools were closed, and parents told to keep their children out of swimming holes and the Concho River. Cinemas, theatres, schools and churches were shut down, and indoor meetings banned. Preachers had to resort to local radio for their hotlines to the Almighty to appeal for Divine intervention to stop the epidemic.[7]

The milk of human kindness ran dry in San Angelo that summer. Townsfolk shut their doors on those who had a polio victim in the family. One of the city's most senior doctors diagnosed polio by clinical impression, because he refused to do lumbar punctures for fear of catching it himself. Travellers gave the place a wide berth. Motorists who stopped drove off without having their tyres topped up, to avoid taking polio-contaminated air home with them. As always, some benefited from the misfortunes of others: insurance companies, keen to point out how quickly one crippled child could bankrupt the average family, and chiropractors who claimed to protect children against polio with one simple 'adjustment' of the vertebrae in the neck.[8]

When the outbreak ended late that autumn, San Angelo claimed the record for the community with the highest polio hit rate of America's bad year, with some 400 cases, including 84 paralysed and 28 deaths.[9] However, the entire nation had reeled under the impact of polio that summer, as in every summer during the 1940s and 1950s. Fear stalked the streets of every American city, even where it was not arm in arm with polio. Areas that were spared that year knew that their turn would come sooner or later. As proof that nowhere was safe, 1949 witnessed the final onslaught of the polio outbreak which hacked through the Inuit settlements scattered around Chesterfield Inlet in the Canadian Arctic.[10,11]

Back in the spring, the Surgeon General had made another prediction, that even worse times lay ahead. This was also fulfilled. The American epidemics worked up into a crescendo that reached a climax in 1952, with 58,000 polio cases and 3,000 deaths. Throughout the early 1950s, Americans remained in the grip of fear, despite the best efforts of the public health authorities to deal with the threat in a calm and measured fashion.

An editorial in the *Journal of Pediatrics* in 1951 lamented the irrationality of 'polio hysteria' in the public at large and warned that it was beginning to infect the medical profession too.[12] Unfortunately, doctors had no therapeutic tricks up their sleeve to treat or prevent polio, or to reassure terrified parents. The calm voice of reason could explain that the chances were less than one in a thousand of losing a child to a wheelchair or worse during an outbreak – but this was quickly drowned out by the resurgence of panic when the first cases of the new polio season made banner headlines in the local newspapers. In 1952, a nationwide poll about what kept Americans awake at night found that fear of polio was second only to that of nuclear annihilation.[13]

Compare and contrast with other targets of polio that year. England was still recovering from the rude awakening of the big (for Britain) 1947 epidemic which had paralysed 7,800, killed 700 and exposed wholesale 'ignorance, impotence and insecurity' in the country's capacity to deal with the disease.[14] Some things had been put right, and the nation generally felt calm and better prepared. Doctors and the public had been told all about polio, through articles in medical journals and newspapers, and short films that targeted professional and lay audiences. *His fighting chance*, released by the Crown Film Unit for the Ministry of Health, starred two-year-old toddler Johnny Green, gazing placidly into the camera and unable to lift his head or arms off the mattress. The voice-over was by Michael Redgrave and (a nice touch) America's former First Lady, Eleanor Roosevelt. The tone was upbeat and almost Dunkirkian, and ended with a stirring 'Go ahead, Johnny, and win through!' from Mrs Roosevelt.

As the epidemic got under way, it became obvious that this would match the severity of 1947. But there was no panic in the streets; instead, the mood was one of quiet resignation. In Harrow School, North London, a couple of boys went down with polio in June. A dozen of their peers elected for quarantine at home, mainly because being locked up in school would have disrupted the family holiday. The other 600 remained at Harrow, where business continued much as usual apart from the suspension of cross-country running and other strenuous sports.[15]

However, some of the lessons of 1947 had not been learned. In St. Pancras, London, the coroner grilled a surgeon about what he knew of the connection between tonsillectomy and paralytic polio. Not much, confessed the surgeon, who had taken the tonsils out of David and Alan Luckins, aged four and seven. He clearly felt that their deaths from paralytic polio ten days later were an unfortunate coincidence. In the United States, the surgeon could have been hung out to dry for negligence. In London, the coroner excused himself for not being an expert in polio, and let it pass.[16]

The ancient and Georgian city of Bath (population, 78,000) was one of the foci of the English epidemic. There, the Medical Officer of Health, Dr B.A. Astley Weston, had instructed local surgeons to stop doing tonsillectomies, but otherwise did his best to conserve calm. He took every opportunity to deny that polio was a problem, even though cases were filling up the Isolation Hospital that looked out over the city from the heights of Claverton Down. 'There is no sign of an epidemic of infantile paralysis in Bath', he stated on 13 August.[17] He simply advised people to stay away from crowded places and not to swim in the sewage-polluted River Avon. The chlorinated municipal pools were safe, as were the baths in the Roman Spa where the director had taken unspecified 'additional precautions'. Astley Weston did, however, cancel the Dolphins Swimming Club Gala, because some of the competitors were from Somerset, where there had been a spot of bother with polio.

Astley Weston was no maverick, but was following the party line. Bristol's Deputy Medical Officer of Health, Dr R.C. Wofinden, had declared earlier in August that the fear of polio was out of proportion to the damage it caused; he insisted that 'We have to get this disease into perspective'.[18] On the other side of the country, 'Medico', the agony-column doctor of the *Derby Daily Telegraph*, pointed out that the English epidemic was much smaller than those in the United States or Scandinavia, and quoted the soothing statistic that the risk of paralysis was less than 1 per cent among those who caught polio. He believed that he would do more harm than good by listing the early symptoms of polio, so he did not. His advice to avoid physical exercise because 'doing the wrong thing may be literally suicidal' might have alarmed some, but overall, he was confident that 'the outlook is not as gloomy as one might think'.[19]

The English epidemic of 1949 overran the usual polio season and was still claiming victims at Christmas. Nationally, the final toll was almost as high as in 1947, with 5,900 cases and 650 deaths. In Bath, there were 45 confirmed cases and eight deaths.[20] Bristol, 12 miles to the north-west, was also badly hit, but the phlegmatic Wofinden had kept the disease in perspective and the situation under control. Across the country as a whole, most people simply followed Medico's advice to 'be sensible but don't worry'.

The stiff upper lip (or perhaps *belle indifférence*) also prevailed much further afield. In Sweden, the epidemic was severe, but less so than in 1936. Doctors, hospitals and the public generally got on with their lives, while children faced nothing more draconian than the traditional warnings not to eat windfall apples or kick through wet leaves, as these were thought to harbour the poliovirus.[21] In Johannesburg, there was a brief flurry of anxiety when married nurses at the overfilled Fever Hospital refused to look after children with polio because they might take the virus home to their own families. A radio appeal was broadcast, volunteers poured in and the problem was solved.[22]

At the end of 1949, America had again led the field: 42,000 cases and 2,720 deaths,[23] far ahead of the paltry 5,900 cases and 650 deaths in Britain.[20] But the United States is vast, and its population that year was 152 million, over three times that in Great Britain.

The hit rate of an illness is better judged by its 'incidence', the number of cases per year adjusted for the size of the population. The 1949 league table for the incidence of polio tells a strikingly different story. Sweden was right out in front that year, with an overall incidence of nearly 40 cases per 100,000 people. Sweden's performance was not a one-off, as the incidence was higher than in the United States in most years between 1930 and 1950.[21] In San Angelo, the hottest of America's hot spots in the exceptionally bad summer of 1949, the incidence of paralytic polio had been 80 per 100,000, not much higher than in Bristol and distinctly lower than in the Midlands town of Dudley, near Birmingham (Figure 6.1).[20]

Stepping back further, we can see more oddities about how the dread of polio far outweighed the statistical facts of its morbidity and mortality. While polio was insinuating itself into medical consciousness during the mid-nineteenth century, many other mass murderers were still on the loose. In 1849, cholera killed 55,000 people in England and Wales. Polio was a horrible way to die. Arguably, cholera was even worse, with the victim wallowing in torrential diarrhoea until dying from circulatory collapse. Until the early decades of the twentieth century, British children were wiped out in their hundreds of thousands each year by common infections of infancy, such as measles, whooping cough and scarlet fever. These diseases dragged down average life expectancy, for example to just 15 years in Liverpool in the 1870s.[24]

Returning to Britain in 1949, there were 385,000 cases of measles and 102,000 of whooping cough, which respectively killed 300 and 500

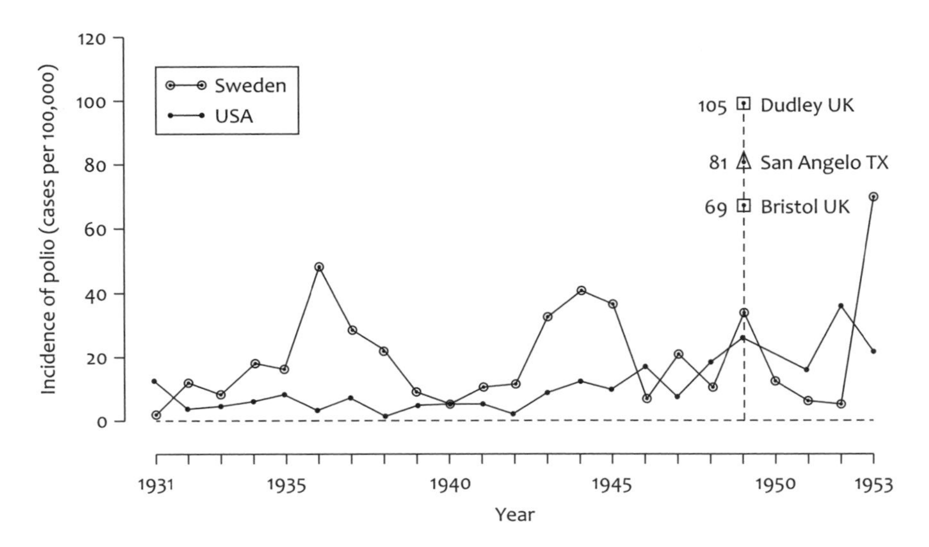

Figure 6.1 Incidence of polio (per 100,000 people) in Sweden and the United States, between 1931 and 1953. The incidences in the Dudley and Bristol (England) and San Angelo, Texas, during the polio epidemics of 1949 are also shown. Adapted from P. Axelsson *Asclepio* 2009;61:23–38

children. The same year saw 52,000 cases of tuberculosis, 3,000 of which were fatal – easily eclipsing the damage done by polio.[25]

All of this raises two obvious questions: how did polio terrify so much, and why was that terror so keenly felt in America?

The answer to the first question includes unfortunate timing, fear of the unknown and poor-quality science. Polio only mutated into a nightmare when the spread of reasonable sanitation transformed its mode of attack into explosive epidemics. Polio outbreaks were smaller than the epidemics of classic scourges such as cholera and plague, but they came into their own while these time-honoured killers were being seen off by the same improvements in basic hygiene which helped along the poliovirus. The contrast between extinct threats and the new curse coming into its ascendancy appeared particularly stark from the 1930s onwards. Tried and tested public health measures were powerless to prevent polio. And this in turn made the sudden and unpredictable strike of polio, deep into the heart of homes that were now safely barricaded against other infections, all the more terrifying.

Polio was also a disease that constantly threw doctors and scientists into confusion. High up in their ivory towers, the experts might argue eloquently about the evidence favouring nasal, oral or fly-borne transmission. But this was all hot air to parents down on the ground, watching their homes being doused in DDT and their children screaming as corrosive chemicals were squirted up their noses. One clever theory after another burst into the headlines, then wobbled, crashed and burned, taking with it fresh hopes and dashed promises. And by 1950, all the might of twentieth-century medicine had signally failed to come up with any treatment that worked. Even the iron lung, the three-ton symbol of salvation that was airlifted into San Angelo in June 1949, could only promise a two in five chance of coming out alive – and the lucky were rescued by spontaneous recovery, which was nothing to do with medical treatments.

There was also the thorny problem of what polio might leave behind. Polio was not like other diseases such as diphtheria or whooping cough, with an all-or-none outcome of the battle for the child's life. If the victims of those diseases survived, then everything was soon back to normal, and the episode could be safely written off as horrific, but history. By contrast, polio only rarely made a clean kill. Parents, family, friends and schools could move on after the death of a child, but it takes a special kind of resilience to cope with the emotional, physical and financial cataclysms of helping a child in a wheelchair or an iron lung to live a full and rewarding life. During the 1930s and 1940s, many parents were torn apart by those stresses, and many more prayed desperately that they would never be put to the test.

All of the above are driven by basic human instincts, which we would take to be broadly similar across the planet. Which brings us to the strange case of the disease which Americans claimed as their own and which, for almost half a century, terrified America like no other nation.

An American story

The great American epidemic of 1916 concentrated its wrath on the East Coast, and especially New York, New Jersey, Philadelphia and Boston. Its impact far outlived the seven months of its attack. Anything that could bring New York City to its knees was bound to leave lasting scars on the American psyche, and the sheer magnitude of the assault was shocking, with 9,000 cases and 2,500 dead between July and December in the Metropolis alone.[26] For a minor illness that many people had known nothing about in spring of that year, this was a commanding debut.

The outbreak also showed the medical authorities at their most impotent. The man in charge was Haven Emerson, Commissioner of Health for New York City, a 42-year-old public health physician whose dedication to his work had earned him the nickname 'the Last of the Great Puritans'. Emerson was highly intelligent and highly regarded, and later became famous as the inventor of an improved iron lung which was so good that the inventor of the original one sued him. However, as the commander of the public health response to the rising tide of polio casualties in 1916, Emerson was soon out of his depth.

It was not all his fault, as there were huge holes in basic knowledge about polio and how it was spread. Flies were still in fashion; other suspects included stray cats and foreigners. Blame was heaped on the Italian community in the squalid and unfortunately named Pigtown, because many of them spoke no English, believed in the 'evil eye' and kissed the lips of the dead. It was irrelevant that they had all been declared healthy a couple of years earlier on passing through the US Immigration Center on Ellis Island in New York Harbour. And it must have been a statistical quirk that the attack rate later turned out to be higher among wealthy Americans living in the smart quarters of New York than among the Italians crowded into the filth of Pigtown.[27]

Emerson responded to the rising body count and public panic in the only way he knew how, with conventional measures to contain the infection by quarantining cases and cutting off likely lines of transmission. Polio victims were confined to their houses – identified for all to see with 'Keep Out' signs in English, Italian and Yiddish – and behind fly screens and closed windows, which added to the ordeal of fever during the stifling hot summer. Cinemas, holiday camps and swimming pools were closed down, and bathing suits ordered to be locked away to remove temptation.[28]

When all that failed to stop the epidemic, Emerson's response was to tighten his isolation measures with ever-greater stringency and desperation. Entire streets were declared no-go areas. From mid-July, children under the age of 16 were barred from leaving New York City unless accompanied by an adult and in possession of a 'health certificate', signed by a doctor, confirming that they were free from any taint of polio. Travel restrictions were enforced by inspectors at railway stations, road checkpoints and ferries and boat landings. Between July and September, an average of 1,100 children were permitted to leave the city each day (Figure 6.2).[29] Cats came into the firing line, literally, and were gunned down in their tens of thousands (even when it was admitted that no bounty was payable after all).[30] Dirt was obviously unhealthy, so Emerson's staff directed the City's streets to be scrubbed clean, to the tune of 4 million gallons of water each day.[28]

Out on the unusually pristine streets, panic and chaos reigned, especially when overburdened hospitals started to close their doors to new polio cases, alive or dead. One distraught father drove his dying son at high speed to Smith Infirmary on Staten Island, but arrived too late. Doctors at Smith refused to accept the child's body, as did the other

Figure 6.2 Women and children fleeing New York, July 1916. Reproduced by kind permission of The Bettmann Archive/CORBIS

medical centres which the father visited during several hours of driving around New York that day.[31] The *New York Times* carried daily counts of new cases and deaths, together with snippets that capture the turmoil of the moment. On Sunday, 20 August, the paper reported that the Brace Fund for Cripples, set up by Emerson, had reached $11,452; that a doctor's son had caught polio, presumably because his father had taken him in the car when he visited a patient; and that Dr Robert Lovett, polio expert from Boston, strongly advised against treating polio victims with strychnine.

When the dust finally settled in December, Emerson had presided over more than a calamity. The outbreak had demonstrated what polio could do, and also that it was beyond the law of even the most draconian public health measures. In just seven months, polio had transformed itself into a nightmare which could not be stopped or prevented.

Dreadful though it was, the American polio epidemic of 1916 could have been allowed to slip into the past, alongside the collective memories of smallpox, the influenza pandemic and the Great War. The Annual Report of the US Public Health Service for 1917 acknowledged that 'poliomyelitis as an epidemic disease seems really to be a development of our time' and that the outbreak in New York had been virulent and carried a high mortality. Nonetheless, polio merited fewer pages in the Report than, for example, maritime quarantine and the control of plague in New Orleans and California.[29] In other countries, such as England and Sweden, the threat of polio was carefully played down to keep it in its rightful place, low in the pecking order of diseases that really mattered.

However, the threat of polio was kept alive for the American public. Everything that was frightening about the disease was deliberately exaggerated to ensure that Americans never forgot the spectre which came back to haunt them every summer. They were bombarded with reminders that polio was an aggressor on the nation's home soil. It struck without warning, picking off healthy children, and killed them or condemned them to a life sentence in leg irons or a wheelchair. Nobody was safe; this was an enemy which was cruel, unstoppable and that spat in the face of medical science. Americans were told that polio must be defeated, and that the solution lay with the parents whose children were being targeted. This solution was money, vast amounts of it, to be hurled into helping polio victims and their families to live with the disease and its aftermath, and into research that would eventually defeat it.

This all-American response to a disease that was portrayed as targeting the United States became the most successful public fund-raising campaign of all time. Arguably, it was also one of the most ambitious exercises ever in social engineering, and generated a climate of fear which enveloped Americans from the mid-1930s until the late 1950s.

The organisation responsible wielded the power of a government department. Its tentacles extended right across the nation and, apart from a few unreachable communities, into every stratum of American society. It quickly became the voice of polio victims throughout the United States, and set the agenda for polio research by dishing out (or withholding) grants which eclipsed the funding for other diseases.

This organisation also made up its own rules, cut out people it did not like and was not accountable to anyone. Yet nobody dared to challenge it, because the man at the top was also the man who led America out of the Great Depression and into the Second World War.

Victim turned victor

American physicians traditionally compared skin lesions with irregular edges to the convoluted coastline of Maine, the state high up the Atlantic seaboard which abuts the Canadian border. Cartographers are kept especially busy by the Bay of Fundy, off the north-eastern tip of Maine and the Canadian province of New Brunswick. The Bay has long been famed for its massive 24-foot tides, among the greatest in the world, and the rich hauls of fish and shellfish to be harvested from its cold, grey-green waters.

Half a mile off the Maine coast at Lubec lies the island of Campobello, an ocean-gnawed sliver of rock measuring just nine miles from top to bottom at its longest and three miles across at its widest. The history of Campobello features the Passamaquoddy Indians, fish, rum-running, more fish and political tussles. In 1866, the American Fenian Fellowship set out in the name of Irish nationalism to capture Campobello and intimidate the English. They failed, but the attempt somehow persuaded Canada to acquire New Brunswick and Campobello. During the 1880s, Campobello faced another kind of invasion: an influx of the wealthy of the East Coast, from Boston to Montreal, seduced by the rocky coastline, the bracing salt-laced air and an exclusive brand of solitude. The invaders included James and Sara Roosevelt of Hyde Park, New York, who visited in 1883 with their one-year-old son, Franklin Delano. They were so taken with the place that they built a 15-roomed 'cottage', where the lad spent his summer holidays.

Campobello remained a pivotal point and a sanctuary for Franklin D. Roosevelt during his adult life, through good times and bad. The good times included a sparkling career at Harvard University and Columbia Law School; his marriage in 1905 to Eleanor Roosevelt, his fifth cousin and a niece of Theodore Roosevelt, who happened to be US President at the time; his election to the New York State Senate in 1910; and his appointment as Assistant Secretary to the US Navy in 1913. The latter post allowed him to prove his leadership in seeing off the threat of German U-boats, and the grateful navy was happy to deliver, by battleship, the materials for the new 'cottage' which Roosevelt built on Campobello after the Great War. This house had just 35 rooms (with the third floor set aside for the servants), and was on the clifftop above Friar Bay, looking west towards the mainland.

Bad times followed in 1920–21, in Roosevelt's thirty-ninth year. A chapter of disasters began with a landslide defeat in Roosevelt's first bid for the White House (as Vice President) and continued with a juicy sex scandal which bubbled up during his watch at the navy. Not for the first time in history, a thriving sex industry had grown up around a naval base, in this instance at Newport on Rhode Island. This area catered for all tastes, heterosexual and otherwise, and it was the latter which the navy was hell-bent on stamping out.

Public outrage erupted when it transpired that sailors had been sent in undercover to entice comrades to reveal their homosexuality *in flagrante delicto*. In July 1921, Roosevelt tried desperately, but failed, to prevent publication of a US Senate report which lambasted him for approving the deception and 'ignoring the rights of every American boy who enlisted in the Navy to fight for his country'. Under the headline, DETAILS ARE UNPRINTABLE, the *New York Times* of 20 July laid into Roosevelt for his 'reprehensible' abuse of enlisted men.

On 7 August, a bruised and battered Roosevelt retreated to his cottage on Campobello to lick his wounds. This time, his beloved island provided diversion but no solace. After two days of intense physical activity – sailing, falling overboard, putting out a wildfire and more swimming – Roosevelt developed a fever and went to bed early on 10 August. The climb to his bedroom was the last time that he walked unaided. The next morning, he had a high fever, with backache, and his right leg was weak; by the afternoon, it was paralysed and the left leg was rapidly losing strength. The weakness ascended steadily, and by 13 August, Roosevelt was paralysed from the chest down, although luckily still able to breathe. He had other symptoms, not typical of polio: exquisite sensitivity of his skin, with the lightest touch triggering waves of agonising pain, and paralysis of his bladder and bowel. He had to be catheterised for six weeks.[32]

Local doctors were first on the scene and, possibly confused by the unusual features, did not perform a lumbar puncture, and diagnosed a chill and a blood clot in the spinal cord. Finally, a fortnight after paralysis set in, a proper expert was summoned: Robert W. Lovett, Boston orthopaedic surgeon and chief of the Harvard Infantile Paralysis Commission. Lovett diagnosed polio, while admitting that aspects of the case were atypical. Having ascended more or less symmetrically, the paralysis was now descending again, but left both legs affected. Roosevelt was still racked with pain below the waist, and his bowels and bladder had not recovered function. Eleanor wrote to Roosevelt's mother with typical understatement – 'Dearest Mama, Franklin has been quite ill' – and then knuckled down to the grim routine of dealing with her husband's catheters, enemas and bedpans, and trying to rebuild his confidence.

Roosevelt's homecoming to American soil was also miserable, thanks to overzealous immigration officials who initially refused him entry.[33] His convalescence was slow and demanded a radical rethink about his ambitions as well as the painful fact that this previously fit, outdoor-obsessed athlete would never walk again without leg irons which locked his knees straight, or an arm or crutch to take his weight.

Rehabilitation

Two events in 1924 proved to be life-changing for millions of others as well as for Roosevelt himself. First, he contacted a friend from his days as a New York lawyer, before he had his head turned by politics; and second, he visited a run-down mineral spa set among the tall trees of Georgia to seek relief from the pain which still tortured him.

The legal friend was Basil O'Connor, 12 years his junior and an alumnus of Dartmouth College and Harvard.[34] O'Connor had no silver spoon – he always claimed to be just two generations away from Irish servitude – but was Roosevelt's equal in intellect, drive and powers of persuasion. The two joined forces to set up a law firm at 120 Wall Street, where Roosevelt remained a partner until he moved into the White House in 1933. O'Connor was an astute and pushy businessman as well as a razor-sharp lawyer. However, he was still capable of being moved by seeing his friend's brave façade crumble when Roosevelt missed a step and crashed to the marble floor of their office. Against his better judgement, O'Connor was also susceptible to Roosevelt's out-of-the-blue suggestion that the two of them should take over the dilapidated health spa in Georgia.

This was Warm Springs, set at an invigorating 1,250 feet up on Pine Mountain, 75 miles south-west of Atlanta. When Roosevelt first visited in October 1924, it was in pre-terminal decline, with its owner desperately trying to drum up trade through newspaper spreads of scantily dressed physical therapists, draped around the poolside like bathing beauties.[35] But the healing waters, or some other magic ingredient of the place, eased Roosevelt's pain and brought him new confidence (Figure 6.3). In 1926, he ploughed one-third of his personal fortune into buying the spa and, with O'Connor on board, set about raising money to turn it into a state-of-the-art centre to rehabilitate polio victims.[36]

The fund-raising campaign picked up momentum while Roosevelt reinvented himself as a politician and tried again for the White House in 1932. This time, he stood against the helpfully unpopular Herbert

Figure 6.3 Franklin D. Roosevelt (1882–1945) at the poolside during his first visit to Warm Springs, October 1924. Reproduced by kind permission of Ed Jackson and the University of Georgia Archives

Hoover, who had no convincing solution for the Great Depression which had opened up after the Wall Street crash of late 1929. And now, thanks at least in part to polio, Roosevelt was a changed man and an appealing presidential prospect. The aloof product of Hyde Park and Harvard, who had taken hereditary wealth and health for granted, had been replaced by a man who understood misfortune and disadvantage at first hand. This transformation laid the foundation for the listening president, whose New Deal to haul the country out of the Depression hinged on resolving the plight of the common man, not the high flyers of his native Wall Street. And so in January 1933, FDR entered the White House as America's thirty-second president – and O'Connor found himself sole partner of their law firm and in charge of the Warm Springs Foundation.

Big bangs from small change

The White House brand proved to be a massive money-spinner. Starting on 30 January 1934 (Roosevelt's fifty-fourth birthday), O'Connor organised the first wave of the annual 'President's Birthday Ball' events held simultaneously across the nation, under the motto 'Dance so that others can walk'.[37] Dollars poured in, over 1 million of them, from the 6,000 balls held that first night, and were channelled into rebuilding Warm Springs. The following year was even more successful. The embarrassment of riches paid off the debts at Warm Springs and left enough to start funding research projects, including a daring bid to produce a vaccine against polio (as explained later, this ended in catastrophe).

Before long, however, the Birthday Balls were losing their novelty, and questions were being asked about the destination of the 'Cripples' Money'.[38] Roosevelt's solution was to scale up the operation, and in 1937 he founded the grandly named National Foundation for Infantile Paralysis (NFIP). The presidential imprimatur gave the NFIP credibility, but by now, buoyed along by America's successful escape from the Depression, Roosevelt was into a hectic second term and busy wooing the nation with his radio broadcast 'fireside chats'. There was only one man who could head the NFIP. O'Connor took the job on, with no salary but unlimited expenses that included a permanently reserved suite at the Waldorf Astoria in New York and, if in need of quiet, an entire Pullman railway carriage to himself.

The NFIP was in good, strong and manipulative hands. O'Connor was a natural networker and string puller and quickly built up a community of hard workers, advisers and A-list celebrities to push the NFIP into the nation's consciousness. They included movie stars such as Humphrey Bogart and Judy Garland, and internationally famous violinist Jascha Heifetz. Best of the lot was Eddie 'Banjo Eyes' Cantor, who had graduated from a blacked-up vaudeville singer during the 1920s to the 'Captain of Comedy' and a seasoned star of the radio and silver screen.

Cantor neatly cashed in on the popularity of 'The March of Time', a newsreel series which had become a national institution during the 1930s, in dreaming up the catchy name for the reborn fund-raising campaign: 'The March of Dimes'. This was launched in January 1938 on Cantor's evening radio show, which was hugely popular across America (with the possible exception of his five daughters, whom he teased mercilessly on air). The unit of currency was also an inspired choice. The dime was small enough to give without feeling the loss, yet big enough en masse to make a difference. It also had a special resonance for post-Depression America. The plaintive ballad "Brother, can you spare a dime?" had come out in 1932, notably in a melodramatic recording by Al Jolson.

Cantor simply told his listeners to put their dimes in an envelope and send it to the White House, and they did. The White House had to take on extra staff to deal with up to 150,000 donations each day. The campaign quickly seized Americans' imagination, and other radio celebrities followed Cantor's cue. Film stars ran five-minute appeals that were shown before the main features in cinemas across the nation. Then the lights would come up long enough for the ushers to comb the captive audience with plastic collection buckets that generally ended up heavy with silver.

The money poured in, far exceeding the expectations of FDR, O'Connor and Cantor (Figure 6.4). During that first year, over $1.8 million was raised, including 2,680,000 dimes.[39] Placed side-by-side, the dimes alone were enough to map the 40-mile boundary of the District of Columbia; the total amount, in dimes, would have stretched most of the way from the White House to the Empire State Building in New York.

This was just the start. The March of Dimes became an annual fixture each January, with house-to-house collections and fund-raising events. Neighbourhood takings were sometimes enough to require the protection of armed guards.[40] When America entered the Second World War in December 1941, the dimes were marching in to the tune of $3 million each year.[41] The War itself had little impact on the campaign, other than to provide photo opportunities for O'Connor with General Patton.[34] When the War ended in 1945, annual income had reached $20 million, and continued to rise steadily to $55 million in 1954, the frantic year of the field trials of Jonas Salk's vaccine. That amount corresponded

Figure 6.4 President Franklin D. Roosevelt and Basil O'Connor (1892–1972), counting dimes in the White House. Reproduced by kind permission of the March of Dimes

to a 6,200-mile trail of dimes, roughly the distance from Washington to Tokyo, or London to Cape Town.

The phenomenal success of the March of Dimes quickly catapulted the NFIP into top place among the United States' medical charities and grant-giving bodies. It outstripped the long-established American Heart Association and the Arthritis Foundation, even though these laid claim to 100-fold more patients nationwide than the NFIP's polio victims. Throughout the 1940s and 1950s, the NFIP kept a virtual monopoly (or stranglehold, in view of its critics) over the entire polio community, embracing patients, families, doctors, researchers and therapists. During this time, the National Institutes of Health (NIH), the US government's medical research funding agency, typically spent just a few per cent of the NFIP's budget on polio – but then it did not need to.

The NFIP spent its vast fortune on a mixture of philanthropy, propaganda and speculation. Roughly half the budget went on practical help for polio victims and their families, with everything from crutches and information leaflets to building dedicated polio units in hospitals, and buying iron lungs and shipping these into areas struck by outbreaks. The NFIP contributed to the medical costs of polio patients in hospital and during rehabilitation. Families were means-tested, and costs were covered in full for those who could not pay – a godsend against the mercenary background of American medicine.

Without that help, tens of thousands of American families would have been crippled financially. In Dallas in 1949, the daily cost for a child with polio began at $11, rising steeply to $55 if ventilation and intensive nursing care were needed.[42] A typical six-week stay for a lucky case often cost over $500; even when the NFIP covered half of the expenses, this could still leave a huge dent, as the average family income was just $350 per month. [43] Of the 58,000 polio cases in the 1949 epidemic, about 40,000 received financial help from the NFIP.[44]

The NFIP did its best for polio sufferers, but this was only as good as the 'experts' whose advice they sought. Early on, it paid for the Bradford frames onto which paralysed children were strapped (and were made worse), and trials of treatments that should never have been tested. Later, O'Connor and some of his advisers were seduced by the controversial therapy brought against resistance to America by the flamboyant Australian nurse Sister Elizabeth Kenny. Like many NFIP ventures, this episode ended unhappily when O'Connor lost patience and severed all connections with Kenny.

In the end, the misjudgments and wasted millions of the NFIP were largely vindicated by the product of all the support which they pumped into Salk's laboratory in Pittsburgh. The NFIP brand was all over Salk's polio vaccine, from its embryonic stages to the massive media frenzy which greeted the public announcement that the vaccine worked and was safe. But even in its greatest hour, the NFIP would lay itself open to accusations of clumsiness, favouritism and cover-up.

Once revived by regular transfusions of cash, Warm Springs thrived as a centre for the rehabilitation of polio patients and a showcase for the NFIP's commitment to them.

FDR's aura was evident, with his 'Little White House' residence and regular sightings of the man in person – apparently able-bodied in the swimming pool, until he hauled himself across the poolside on his belly, trailing his spindly and lifeless legs. In many ways, Warm Springs was ahead of its time; in some aspects, it was a prisoner of that time. 'Togetherness' was a recurrent theme in O'Connor's speeches, but it did not quite happen at Warm Springs. When Eleanor Roosevelt asked if a cabin should be built for Black polio victims (who were notable by their complete absence), she was told that this was 'not desirable' in Georgia.[45]

However, the First Lady's hint was followed up with a solution that would not threaten the flow of dimes from White America: a parallel, all-Black polio rehabilitation centre, not quite so well appointed and 80 miles away across the State boundary in Tuskegee, Alabama. By unfortunate coincidence, this facility was placed alongside the unit where the American Public Health Service was meticulously charting the natural history of syphilis. The research programme had been running for ten years when the NFIP polio centre opened in 1941. At that time, penicillin was not yet widely available. It was not until several years later that the Public Health Service doctors decided to withhold it from 400 Black patients so that their long-term follow-up of syphilis would not be spoiled by an effective treatment.[46]

Pushing the boundaries

The thousands of scientific papers which gratefully acknowledged the financial support of the President's Birthday Ball Commission or the National Foundation testify to the breadth and variety of the NFIP's research portfolio. This ranged from immunology to epidemiology; from single molecules to entire populations; and from the scientifically sublime to the seriously dodgy.

On the one hand, all those dimes paid for John Enders' Nobel Prize-winning breakthrough in tissue culture; for James D. Watson's fellowship to Cambridge, where he teamed up with Francis Crick to work out how the code of life is engraved on the structure of the DNA molecule; and for the massive poliovirus-typing programme which drew Jonas Salk into the limelight and lined him up to develop his vaccine.

On the other hand, the NFIP also funded George Draper's attempts to predict susceptibility to polio from a person's facial appearance; a jumble of peculiar experiments by John Toomey on various combinations of vegetables, vitamin deficiencies and polio risk; and numerous scientifically threadbare studies on the transmission of polio which achieved nothing beyond the slaughter of thousands of monkeys.[47]

Knowledge transfer was another crucial catalytic role played by the NFIP. Its series of International Poliomyelitis Conferences brought together researchers and clinicians of the worldwide polio community, giving the up-and-coming the chance to present their research and rub shoulders with the stars whose names and work they knew from the journals. The first Conference was held in New York City in November 1948, followed by Copenhagen (1951), Rome (1954), Geneva (1957) and a return to Copenhagen in 1960.

These events were carefully choreographed by O'Connor and his advisers, and followed their vision of inclusivity. One of the extras at the Rome Conference was Pope Pius XII (a good catch for the Catholic O'Connor); carefully excluded was Sister Kenny, who had to wangle a press pass to get in and was then barred from speaking or asking questions.[48] The proceedings of these Conferences are a treasure trove of snapshots of polio research in their times. The birth, life and death of many scientific ideas and treatments can be followed through these volumes, and there are some gems – such as Enders' preliminary reports on growing poliovirus in tissue culture and updates by Salk and Albert Sabin on their vaccines – which shine out in the history of medicine as a whole.

Early on, extracting research funding from the President's Birthday Ball Commission and the NFIP hinged more on whom the applicant knew rather than what he or she knew. Knowing the Research Director was particularly helpful, and this explained some of the odder funding decisions. At the helm during the early years was Paul de Kruif, perhaps the most volatile of the many mavericks attracted by the idiosyncratic power base of the NFIP.

De Kruif began his career as a microbiologist at the Rockefeller Institute during Simon Flexner's reign. He worked at the laboratory bench long enough to assimilate the language and personalities of science and showed promise as a researcher.[49] However, de Kruif is best known as a writer. *Microbe Hunters,* his account of the great microbiologists from Antonie van Leeuwenhoek to Paul Ehrlich, was an instant bestseller when published in 1926, and fired up a generation of budding clinical researchers, including Sabin.[50] De Kruif was also notorious as a serial offender of bosses and a dedicated bearer of grudges. His free time for writing increased smartly in 1921 when Flexner sacked him for publishing 'Our medicine men', an article about the research culture at the Rockefeller which was either gently ironic (de Kruif's view) or suicidally critical (Flexner's).[51] By way of revenge, de Kruif fed inside knowledge to novelist Sinclair Lewis for his book *Arrowsmith*, another bestseller, in which the publicity-hungry McGurk Institute and its mercenary, hairy-handed director Dr A. DeWitt Tubbs are easily recognisable as the Rockefeller and Flexner.[52]

At first sight, de Kruif seemed a good prospect when O'Connor appointed him as the Research Director of the President's Birthday Ball Commission in 1934. One of de Kruif's early decisions was to fund the development of an experimental polio vaccine by William Park and Maurice Brodie.[53] This undoubtedly had promise at the time, and it was not de Kruif's fault that the vaccine failed catastrophically, taking its inventors with it. However, it soon became clear that de Kruif was a loose cannon. The combination of de Kruif's prickly personality ('constitutionally incapable of moderation') and O'Connor's low boiling point ('close to room temperature') was unlikely to end happily.[54–56] In 1936, O'Connor gave up trying to bring de Kruif into line and sacked him.

De Kruif's revenge, nearly 20 years in incubation, was pure malice: spreading the rumour that Salk's vaccine, O'Connor's pièce de résistance, was so dangerous that the US government had secretly stockpiled thousands of child-sized coffins for the expected fatalities. As described later, de Kruif used all his powers of persuasion to ensure that the smear, broadcast on prime-time radio, would inflict the greatest possible damage on his former boss and his organisation.[57]

O'Connor powered his way through that storm and many others. He was a thick-skinned, abrasive 'pirate' and 'iconoclast', who was happy to be quoted as saying that 'science is too important to be left to the scientists'.[58,59] In Salk's view, though, O'Connor was also someone 'who had it within his power to cause almost anything to happen'.[60] If fund-raising was the raison d'être of the NFIP, then he was the right man at the top. The NFIP's structure was a vast pyramid, grounded on 90,000 year-round volunteers (who pulled in another 2 million each January for the March of Dimes offensive), organised into 3,000 local chapters that covered the nation and were divided among five regional directors. They in turn reported to O'Connor in his ever-expanding headquarters at 120 Broadway.[61]

O'Connor was a people person, as long as they were his people. His personal favourites included Salk, 'a man who could see beyond the microscope', but not Salk's arch-rival Sabin, whom O'Connor accused publicly of 'envy and ignorance'.[62,63] O'Connor also believed that committees were there to do his bidding, and were to be fired and replaced if they failed to do that.[64] He occasionally met his match, although Tom Rivers, dictatorial virologist and chair of the NFIP research committee, maintained that they would tell him to 'go to hell' if he tried to interfere.[65] Possibly because both were outspoken and bullies, O'Connor and Rivers had a sneaking respect for each other. O'Connor's view that Rivers was 'learned but not particularly bright' was warmly reciprocated.[66,67]

On the positive side, O'Connor was a powerful orator, who revelled in having the first and last word at the International Conferences. His speeches can still stir today, even though his views on togetherness and scientists might not ring entirely true[68]:

> Together we will conquer poliomyelitis... Mankind will ascend to the heights of true civilisation under the leadership of men of science, men of faith and men of peace.

Interestingly, the Pope's address to the International Conference in Rome in 1954 reads almost as though it had been ghosted by O'Connor[69] – which is possible. FDR's legendary silver tongue owed some of its fluency to O'Connor, who wrote many speeches for him, including the address when he accepted the presidency in 1934.[70]

Perhaps surprisingly, the NFIP channelled only 10–12 per cent of its total expenditure into research – less than for 'fund-raising expenses'.[71] O'Connor repeatedly fought off accusations by 'pressure groups [which] have gone to extremes of irresponsibility and distortion' that the NFIP was bloated and top-heavy, pointing out that he drew no salary.[72] However, he never volunteered information about how many dimes were mopped up by his luxury hotels, fine dining and first-class travel.

Nation shall speak unto nation

Organisations similar to the NFIP sprang up in many other countries. Some carved their own path, while others adopted the NFIP's fund-raising tactics and adapted them to their own settings; none could compare with the sheer mass and momentum of the American model.

The first was born while plans for the inaugural President's Birthday Ball were still in gestation. In April 1933, the Association pour les Paralysés de France was founded by André Trannoy, whose attack of polio in 1925 – four years after Roosevelt's – left him tetraplegic. Trannoy built up a network of helpers across France to channel information, moral support and practical help to the tens of thousands of *handicappés* who felt abandoned by the State and its doctors.

Next off the mark, pre-empting the formation of the NFIP, was the Greek Society for the Protection of Crippled Children, established by the Rotary Club of Athens in 1936.[73] Two years later, the Swedish National Association against Poliomyelitis began operating, followed in 1939 by the Infantile Paralysis Fellowship in Britain. The Canadian Foundation for Poliomyelitis appeared in 1948 with its bold FIGHT POLIO! on a distinctive red maple leaf, and the Spanish Associación de Lucha contra la Poliomielitis in 1957.

March of Pennies

Britain's Infantile Paralysis Fellowship was dismissed by Dr W.H. Bradley at the First International Conference as a 'very small body indeed' with an income (just £4000 [$10,000] in 1947) that was 'mere chickenfeed'.[74] In strictly numerical terms, this was true, if unpatriotic. The Fellowship's limited size and money-grabbing power were due as much to the absence of polio hysteria, fanned so effectively in the United States by the NFIP, as to the smallness of Britain. On 6 May 1949, as the English polio endemic was picking up momentum, the *Derby Evening Telegraph* carried a brief piece entitled, 'Happy in spite of their suffering'. This marked the tenth birthday of the Fellowship and its fight against 'one of the most cruel of all diseases ... more often than not, its victims never walk properly again'. But this appeal was barely heard over the background of reassurance that polio was a minor threat which must not cause concern.

Like the British Diabetic Association five years earlier, the Fellowship came into being following a letter to the *Times*. The writer was Frederic Morena, formerly a Shakespearean actor and now, thanks to polio, a still-theatrical language teacher confined to a wheelchair. Morena and Patricia Carey, who fell victim to polio at the age of eight, launched their organisation to 'help the sufferers help themselves and afford each other mutual aid and sympathy' (Figure 6.5).[75]

The Fellowship did great things, in a muted, British kind of way that did not draw inspiration from the massed glitz of the President's Birthday Balls. There were meetings at the Lyons Corner House on the Strand and the Tea House in Kensington Gardens; the annual Christmas party in London's Guildhall eventually pulled in Vera Lynn and entertainer Michael Flanders, who sang from his wheelchair and later teamed up with pianist Donald Swann. Before it went glossy in the late 1940s, the Fellowship's *Bulletin* comprised two sheets of cyclostyled foolscap.[76]

British versions of the March of Dimes included the 'March of Bobs' (shillings) with which 'Wee' Georgie Wood raised £500, and during the Blitz, a stall in a bombed-out shop in Tottenham Court Road. More ambitious was the 'March of Pennies', which eventually raised £1 million ($2.5 million), but took a decade.[77]

In Britain, the rehabilitation facilities of Warm Springs were eventually emulated by the hydrotherapy pools and gyms in many National Health Service hospitals. The Fellowship provided a different kind of support,

Figure 6.5 Patricia Carey (born 1916) and Frederic Morena (1893–1960), with Mimi Morena, on the set of *This Is Your Life*, 18 January 1960. Reproduced by kind permission of the British Polio Fellowship

with dedicated hotels where polio-disabled people and their friends and families could take a break without facing embarrassment or prejudice. A pilot experiment with John Groom's Crippleage in Clacton-on-Sea fell through, but seaside 'Lantern Hotels' were opened in Worthing on the South Coast in 1950 and in Lytham St. Anne's on Morecambe Bay a few years later. Demand was brisk, even though a shortage of funds meant that no concessions were made initially for people in wheelchairs. Later developments included caravans fitted with cardboard partition walls and furniture that could be easily shifted around to accommodate the new breed of 'responauts' determined to travel despite being chained to their cuirass respirators or iron lungs.[78]

The Fellowship did much to raise polio victims' morale as well as fight their corner. Regular events included outings to West End theatres or the sands at Lytham; the annual taxi cavalcade from Glasgow to Troon on the Ayrshire Coast, with the taxi drivers in fancy dress; and rallies for the 'trikes', the three-wheeled cars issued to the 'handicapped' by the Ministry of Health after the War.[79]

The National Games organised by the Fellowship in 1938 at Stoke Mandeville Hospital, featuring archery and a wheelchair relay race, fed directly into the movement which led ultimately to the Paralympics. Thanks to the Fellowship, wheelchair users could at last clock in easily at work, while the paralysed could communicate through the 'Possum' writing machine that could be operated by anything from a toe to a stick held in the teeth. In the early 1950s, Morena himself won an important psychological victory – without help from politicians or doctors – over the Association of Baths Superintendents. He persuaded them that paralysed polio victims could not infect other swimmers and, for the first time, British polio patients could use any public pool in the country.[80]

There were also many defeats and frustrations. After the War, the Fellowship campaigned hard to raise public and government money for a Warm Springs-style national rehabilitation centre. Not surprisingly, with Britain in the grip of post-War austerity, they failed.

Compare this disappointment with the stirring event in London's Grosvenor Square on 12 April 1948, attended by the Royal Family and Winston Churchill.[81] This was the unveiling of a statue which had been funded by a public appeal; 160,000 donations had immediately flooded in, covering the full cost within a couple of days. The statue is of a strong, cloaked man, standing firm and showing no sign of needing the walking stick which protrudes from under his cloak: the thirty-second president of the United States, Franklin D. Roosevelt.

Wages of fear

So how was the NFIP so singularly successful at extracting millions of dimes from the pockets of Americans? Tom Coleman, head of public relations at the University of Pittsburgh during the 1950s, had no illusions. The NFIP, he claimed, were 'slick, arrogant fear-mongers, raising money through a campaign of terror'.[82] His view may have been jaundiced, but Coleman knew the territory, as he had worked close to Roosevelt in the White House.

The 'campaign of terror' was a classic stick-and-carrot strategy, but with both elements carved from fear. The stick was obvious. Polio, this peculiarly American nightmare, could carry off anyone's child and would remain a threat forever unless enough dollars were raised to banish it. There was no escape from March of Dimes propaganda, as the NFIP's well-funded and well-connected press office exploited all the media to hammer home the horrors of polio. Millions of posters and leaflets were pinned up in public places or pushed through letterboxes by up to 2 million volunteers across the land.[83]

Figure 6.6 The first March of Dimes poster (1946), showing Donald Anderson from Prineville, Oregon. Reproduced by kind permission of the March of Dimes

March of Dimes posters, beginning in 1944 with the picture of five-year-old Donald Anderson from Prineville, Oregon, soon became an art form (Figure 6.6). The backdrop ranged from laboratories and hospitals to battle-weary GIs in the Korean War, but the focus was always a crippled child: smiling bravely or gazing apprehensively into the future, cute but disfigured by leg braces or crutches. Articles were fed to newspapers and popular magazines such as *Readers' Digest*, often with powerful images: close-ups of infants in their tiny child-sized respirators, and serried ranks of iron lungs during polio outbreaks (Figure 6.7).

From 1940, American cinemas were invaded by *The Crippler*, a 20-minute melodrama which makes a fascinating contrast with *His fighting chance*, produced in England nine years later. *The Crippler* depicts polio as a black cloud hanging low over the rooftops, which condenses into a demonic figure with talons that stretch out towards sleeping children. Luckily, the monster is beaten back by an ordinary American heroine, played by fresh-faced Nancy Davis.[84] The film was all pervasive; the 1945 *Film Daily Year Book* mentions that it had been shown in another 2,000 cinemas that year.[85]

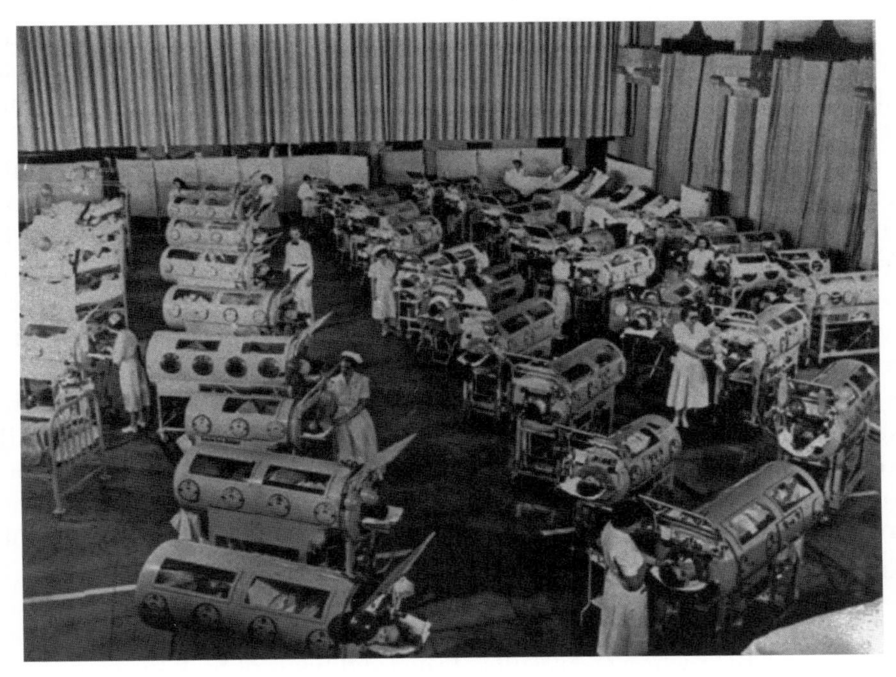

Figure 6.7 Ward of iron lungs in Los Angeles County Hospital during the polio outbreak of 1952. Original credit, the March of Dimes. Reproduced by kind permission of the World Health Organisation

The carrot was more subtle: warm thanks for earlier generosity, with a glimpse of the rewards already reaped by the NFIP. Poster boy Donald Anderson strides away from his former self, left clinging to his cot sides, and proclaims, 'Your dimes did this for me!' The trademark of the NFIP was everywhere, from the obligatory footer on the first page of scientific papers to MARCH OF DIMES plastered across iron lungs, uncomfortably reminiscent of the messages stencilled on American blockbuster bombs. And implicit in all this, whether stick or carrot, was the universality of the threat. Those children could be anybody's, and yours could be next.

Men on top

FDR was not alone in being scarred by The Crippler's talons. In spring 1950, O'Connor's daughter Bettyann, then aged 30 and the mother of three, fell ill with paralytic polio. Seeing his vivacious and self-assured daughter transformed into just another patient at Warm Springs hit O'Connor hard, even though Bettyann's ordeal was the currency in which he had dealt for over 15 years.[86]

Falling prey to the monster which he had helped to create exposed a surprising weakness in this tough-talking, short-fused bully. One consequence was the special relationship which sprang up between O'Connor and Salk (see Figure 12.8, p. 289).

This association began in November 1951 on board the *Queen Mary*, bringing home the American contingent from the Second International Conference in Copenhagen. The proceedings had been dominated by Enders' continuing revelations about growing the poliovirus in human tissue culture, but it was Salk who had O'Connor's ear on the return voyage. Bettyann had accompanied her father on the trip and, somewhat against expectation, found Salk to be a listening doctor, warm and caring.[87]

O'Connor became a powerful fan who sided with Salk throughout his many battles with Sabin and others. O'Connor's loyalty to Salk repeatedly overrode sound judgment, such as when he pushed State medical associations into running the trial of Salk's vaccine as Salk wanted (without proper controls), and tried to bulldoze the NFIP research committee into rejecting Sabin's rival vaccine.[88]

O'Connor's devotion to Salk was ultimately rewarded, on the momentous day in April 1955 when the results of the field trials of Salk's vaccine were trumpeted to the world. On that occasion, however, the other architect of the NFIP's victory over polio was notably absent.

Roosevelt's presidency was the longest in American history, and only the strongest of men could have guided the nation safely through the catastrophes of the Depression and the Second World War. In photos and newsreels, FDR appeared entirely able-bodied. He was generally flanked by burly Secret Service men. A would-be assassin had shot at Roosevelt during his White House campaign in 1933, killing one of his close friends, but the Secret Service men were not there to stop bullets for the president. They were there to hold him up.

Roosevelt went to great extremes to hide his disability. These 'splendid deceptions' – enough to write a book about[89] – included tailored leg braces concealed under his trouser legs; his giving interviews while sitting in the presidential limousine; and impeccable stage management to cover up the moments when he needed assistance from others. Photographs of him were carefully vetted, and most pictured him only from the waist up; just two are thought to exist which show him in a wheelchair. Roosevelt's polio attack in 1921 was carefully hushed up to avoid damaging his political career, but was impossible to hide for someone so active and in the public eye – especially after the *Atlanta Journal* revealed all in 1927, with a scoop from Warm Springs under the headline 'Franklin Roosevelt will swim to health'.[90] But this was not really news, as FDR had already been hailed as 'a real hero' at the 1924 Democratic Convention for climbing onto the stage on crutches.[91] Roosevelt himself made no secret of his having 'recovered from polio'. So these 'splendid deceptions' were a folie à deux in which the American public actively colluded, because the president they needed to believe in had to be nothing short of Superman.

In the end, it was a mundane assassin, nothing to do with polio, which killed the President. In April 1945, with Allied victory in Europe almost in the bag, Roosevelt returned to Warm Springs to rest and to try to bring down his dangerously high blood pressure. Just after 1 p.m. on the 14 April, while sitting to have his portrait painted, Roosevelt collapsed with a 'terrible headache'. He died from a cerebral haemorrhage a couple of hours later. FDR was just 63 years old, three months into his fourth term as president, and within three weeks of the German surrender in Europe.

There are many enduring images of Franklin Delano Roosevelt, but three deserve special mention. The first is the American dime, which has carried FDR's head since 30 January 1946, which would have been his sixty-fourth birthday. The second is of Roosevelt sitting between Winston Churchill and Joseph Stalin at Yalta, his disability given away only by a slight sagging of the trouser legs over his wasted thighs. Here, he is among equals, and one of the three most powerful men on the planet.

Finally, Roosevelt's statue in Grosvenor Square, London, shows an able-bodied man, the only concession to his encounter with The Crippler being the slender cane which is barely taking his weight. Around the base of the plinth are carved the Four Freedoms which Roosevelt promised his nation in his State of the Union address in January 1941. These are Freedom from Want, Freedom of Speech, Freedom of Worship, and Freedom from Fear. It is ironic that Roosevelt himself was instrumental in the campaign of terror which deliberately denied Americans this last Freedom, and that he did not live to reassure himself that the end justified the means.

7

First Do No Harm

When polio first entered the scene in the late eighteenth century, patients were given the same time-honoured 'cures' which were dished out for everything from period pains to smallpox. During the following two centuries, dozens of treatments were proposed, ranging from the rational to the extraordinary. Some were obviously the work of quacks – useless and/or dangerous and aiming simply to extract money from the naive. Unfortunately, the same is true of many 'treatments' conjured up by men of science and medicine, some of whom skilfully sidestepped the checkpoints of peer review and clinical trials.

The outlook for polio victims has improved dramatically, but this is mainly thanks to palliative measures which buy time to allow Nature – if it is so inclined – to heal. These include the iron lung, better nursing care, and antibiotics to treat pneumonia and other bacterial infections that complicate paralysis. Essentially nothing can be done to control the infection itself; Simon Flexner's claim in 1911 that his laboratory would produce a 'specific remedy' within six months[1] proved to be hot air. Even today, there is no therapeutic miracle that can stop the poliovirus in its tracks.

To be judged fairly, treatments for polio must be seen in their proper context of time and place, and calibrated against the practices of the day. Up to the 1950s, therapeutics advanced in fits and starts, and often in an ethical vacuum. The history of another infection – syphilis – gives us some insights into the bumpy road to drug discovery during the twentieth century.

In 1900, as during the previous four centuries, the mainstay of treatment was mercury salts, which had dreadful side effects that far outweighed their feeble activity against the spirochaete bacterium responsible for syphilis. The early antimicrobial, neosalvarsan (1910), was better, but still more of a dud than a magic bullet.[2] While everyone waited for an effective antibiotic to come along, fever therapy ('pyretotherapy') became popular during the 1920s. Syphilitic patients were injected with blood from malaria victims in the belief that a high temperature would kill the spirochaete.[3] In the 1930s, Dr Orval Cunningham of Kansas City promised to cure syphilis in his high-pressure oxygen tank, at just $3 for an afternoon's treatment. Cunningham's Tank Therapy was repeatedly denounced as quackery by the American Medical Association (AMA).[4] However, the stance of the medical establishment was not entirely consistent. In July 1938, the AMA granted a licence to the Burdick Corporation of Milton, Wisconsin, for their Fever Therapy Cabinet which could cook a patient at up to 106° F for as long as was necessary.[5]

All these treatments were swept aside when penicillin became widely available in the late 1940s to treat syphilis and other bacterial infections – except for hundreds of Black patients in Tuskegee, Alabama. The wonder drug was deliberately withheld from them for another 30 years, so that the American Public Health Service could complete the study of the natural history of untreated syphilis, which they had begun in 1932.[6]

And so to polio.

Back to basics

Early treatments for polio were millennia older than the disease itself, and originated with physicians of Ancient Greece, Rome, Arabia and China. Bleeding, either by scalpel cuts or peckish leeches, was the usual opening gambit for European and American doctors during the seventeenth through the nineteenth centuries, irrespective of the patient's complaint. Edward Jenner, physician-scientist and vaccination pioneer, was a dedicated bleeder, and during the 1780s collected enough human blood to conduct extensive experiments on its use as a fertiliser (disappointingly, manure turned out to be better). Leeches were prescribed in Jenner's day to treat smallpox – and still were in 1884, by a Belgian anti-vaccinationist doctor who believed that smallpox was caused by emotional shocks.[7]

Other archaic treatments included cupping, blistering and cautery. Cupping probably originated in Ancient China. A hot glass cup a few centimetres in diameter was placed on the skin; as the air inside cooled, a vacuum was created that left a circular sucker mark of ruptured capillaries. Alternatively, the skin could be blistered with extracts of 'Spanish fly' which contain the caustic toxin, cantharidin. (This can also cause erection when taken orally; in the pre-Viagra era, Spanish fly was a rather poor aphrodisiac with the performance-limiting complication of sudden death).[8] Less subtle still was cautery, a full-thickness burn usually inflicted by a red-hot poker.

All these measures were 'counter-irritants', supposed to draw the illness away from the affected part. In polio, they were later rationalised as relieving high pressure and swelling in the spinal cord. They were generally applied to the skin over the spine, at the level corresponding to the paralytic lesion.[9]

Medicinal drugs included compounds that induced vomiting (e.g. nux vomica) or profuse diarrhoea (e.g. mercury salts such as calomel). They were similarly worthless and sometimes harmful, but they all enabled the doctor to charge the patient for services rendered.

The therapeutic assault on polio began in 1789 in Michael Underwood's first article about 'debility of the lower extremities', when he stated that 'brisk' camomile purges and blisters were effective. Like many after him, Underwood assumed that his opinion was enough to prove that a treatment worked. When Underwood's *Diseases of children* reached America in 1793, the only New World amendment was to recommend bleeding as well.[10] Purges and blisters remained in fashion for over a century. Samuel Gross, a distinguished surgeon at Jefferson Medical College in Philadelphia during the 1870s, threw the therapeutic book at his young polio patients.[11] He began with leeches, blisters, cupping, calomel ointment rubbed over the spine, and smacking the limbs and back with a wet towel to stimulate skin blood flow. A 'valuable adjunct' was the red-hot poker, applied over the spine for long enough to leave 'a good issue' (i.e. a nicely weeping raw area) when the scar was peeled away after a few days.

The red-hot poker was omitted from Mary Putnam Jacobi's hefty chapter on infantile paralysis in the *American system of medicine* (1886).[9] Mary Jacobi was a rare bird: one of America's first female medical graduates, trained at the Salpetrière in Paris, and an astonishingly young professor of therapeutics in New York. But old habits die hard, and Jacobi still recommended calomel ointment rubbed along the spine, followed by blisters. A few years later, William Osler's voice of reason and authority reverberated across America. In *Principles and practice of medicine* (1892), he made his views typically clear. 'No drugs have the slightest influence ... the application of blisters and other forms of counter-irritation is irrational and only cruel to the child'. However, even Osler remained wedded to the idea of cupping.[12]

In their *Manual of infantile paralysis, with modern methods of treatment* (1914), Henry Frauenthal and Jacolyn Manning reminded their readers of the 'notorious faker in the West' whose spinal blistering therapy had killed a child.[13] They also noted that many drugs had crept into use because of doctors' 'desperate helplessness', and felt strongly that they did more harm than good. By contrast, 'elimination', or emptying the bowels, was invaluable. Calomel did the trick, as did large volumes of water given orally or rectally. Frauenthal and Manning also recommended lumbar puncture for therapy as well as diagnosis, believing that it reduced pressure within the central nervous system. Helpfully, they described a neat way of strapping up the patient during a lumbar puncture to prevent the back from arching, which could snap the needle off inside the patient's back.[14]

To summarise: various 'treatments' had been parroted from one supposed authority to another, and most could ultimately be traced back to obviously peculiar ideas that had been recycled uncritically for a couple of thousand years. When Frauenthal and Manning published their *Manual* in 1914, research into the nature of polio was well under way. Unfortunately, it took many years for new knowledge to be translated into new treatments. And if a therapeutic lumbar puncture sounds unpleasant, be warned that there is worse to come.

And so to bed

Isolation and strict bed rest were the traditional cornerstones for treating paralytic polio, and indeed for anyone who developed suspicious symptoms – fever, a stiff neck or a sore back – during an outbreak. As Frauenthal and Manning put it, isolation resulted in 'the exhaustion of the material on which [the epidemic] feeds'.[15] Cases were generally isolated for two weeks, even after it was shown during the 1940s that patients' stools could remain loaded with the virus for several weeks. Bed rest also seemed entirely logical. It is a basic human instinct to lie down when ill, and it was widely believed (and eventually confirmed) that exercise while incubating polio could precipitate paralysis.

However, both measures heaped further misery on the polio victim. Apart from nursing and medical staff, who were often gowned, gloved and masked, confinement was initially solitary. In many 'fever' hospitals, parents could visit just once a month, and then only behind a glass partition – all very character-building for a frightened child, in pain and desperate for a kiss and a cuddle (Figure 7.1). 'Strict' bed rest meant lying flat all the time, often for weeks or months. If back pain was intolerable, then the patient was nursed face down. The toilet and commode were out of bounds; imagine using a bedpan after a colonic washout, while keeping the shoulders and heels firmly on the mattress.

Some polio experts pushed their own variations on the theme of therapeutic bed rest. Aladar Farkas, an orthopaedic surgeon in New York, kept his patients tilted head down for several months to prevent 'decomposed nervous tissue', supposedly crawling with polioviruses, from entering the spinal fluid. He claimed that this 'positional-nutritional' treatment also helped to nourish and oxygenate the brain, and had restored muscle function in 24 hopeless cases. This report was published in 1952 in a reputable journal (*Archives of Pediatrics*), but included no data or untreated controls, making it impossible to know whether he had done good, harm or nothing at all.[16]

Until the 1950s, other forms of torture awaited victims of polio. These were plaster casts, splints and frames, tightly applied to keep paralysed

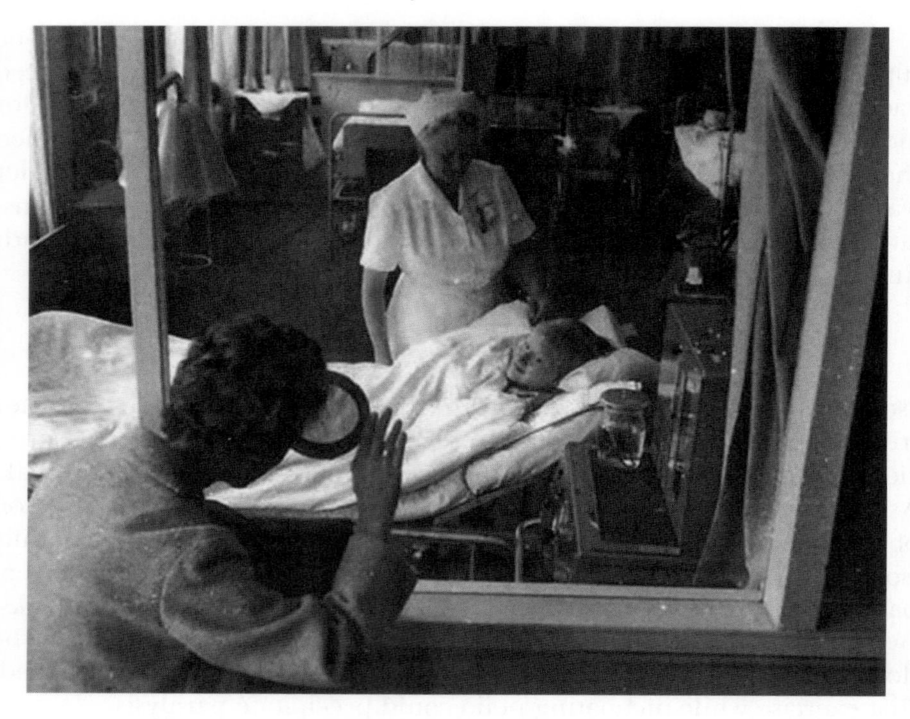

Figure 7.1 Visiting time in the polio isolation ward, Blegdams Hospital, Copenhagen, 1952. Many isolation hospitals allowed parents to visit their children only once a month. At best, visitors wore full gowns, caps and face masks; at worst, they were kept behind plate glass partitions. Image from the Blegdams Hospital Archives, Copenhagen

limbs in the normal position while waiting (hopefully) for strength to recover. The aim was to prevent permanent deformities, which could be as disabling as paralysis itself. Children's limbs were encased in plaster of Paris to maintain the 'anatomical' position and stop joints from being pulled out of line. Infants, who tended to wriggle and undo the benefits of immobilisation, were strapped into a moulded plaster carapace, like the top of the tortoise's shell. Patients with weakness of the muscles that bend the vertebral column to either side were trussed up in a tightly laced celluloid jacket to prevent curvature of the spine. One famous deformity 'preventer' was the Bradford frame, on which acute cases were pinned out with their legs and back straight and arms raised (Figure 7.2). The frames were made in the workshop of the Sick Children's Hospital in Toronto during the late 1930s, and proudly distributed free of charge to all polio victims in Ontario.

All these devices were well intentioned, but did more harm than good. Muscles that no longer do the everyday work of contraction and relaxation quickly lose strength and waste away, making the chances of eventual

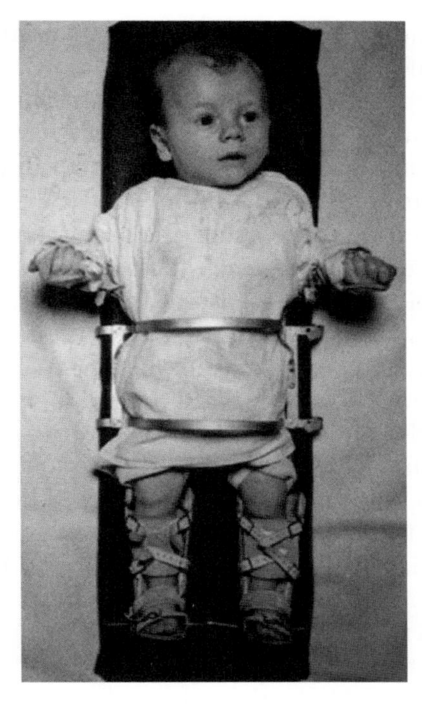

Figure 7.2 Baby strapped into a Bradford Frame during the epidemic in Ontario, Canada, 1937. The frame was supposed to prevent deformities in polio victims, but actually made them worse. Reproduced by kind permission of the Hospital for Sick Children Archives, Toronto

rehabilitation even slimmer. The 'atrophy of disuse' was recognised in the nineteenth century, in limbs immobilised in plaster after a bad fracture. Some doctors, such as the Bostonian orthopaedic surgeon Robert Lovett, were well aware of the risk. From the 1910s, Lovett recommended early mobilisation with gentle passive movements (done by a physical therapist) and active exercises (using the patient's own muscles) as soon as paralysis stopped advancing, to prevent deformities and build up strength.[17,18] To most orthopaedic surgeons, however, the key was immobilisation, ideally enforced by a special apparatus named after themselves.

The management of acute polio and rehabilitation, which could take years, encouraged the growth and development of physical therapy (physiotherapy) as a profession. In turn, polio helped physiotherapy to survive during the lean years between the Wars, when the supply of battle casualties needing intensive rehabilitation dried up. One early pioneer in the late nineteenth century was Dr Metzger of Amsterdam, who brought in new methods of massage. These included the exotic-sounding *effleurage* (firm stroking, aiming to stimulate blood and lymph flow), *pétrissage* (kneading knotted muscles into submission) and 'frictions', which are as uncomfortable as they sound.[19]

Robert Lovett, who would later look after America's most famous polio victim, brought physical therapists in as key members of his multidisciplinary orthopaedic team. As well as providing routine care, they developed new treatments and monitored the outcome using devices that Lovett had invented to measure muscle strength. This was a rare application of science to a discipline that was still essentially an art.[17,18]

The availability of physiotherapists was patchy, as was the quality of their training. Well-staffed units in the United States boasted production lines of therapists and could brag about 'the truly wonderful results obtained by massage'[20] – which must have stuck in the throat of all those elsewhere in the United States and Europe, let alone less affluent countries, who had little or nothing to offer.

Hydrotherapy became particularly popular. The buoyancy provided by water cancels out the effects of gravity and gives weak muscles a better chance to start working again. For most patients, hydrotherapy was more bearable than doing similar exercises in bed or the treatment gymnasium, and valiant attempts were made to make it fun. Henry Frauenthal recommended diverting the child with the 'expandable, floating paper toys made by the Japanese'.[21] Hydrotherapy facilities ranged from the child-sized Hubbard tank to large pools such as the classically styled one at the Orthopaedic Hospital in Bath, England.

Properly controlled trials of physical therapy were few and far between, making it impossible to measure or even prove the benefits of particular regimens. Nevertheless, intensive physical therapy seemed to sustain, improve and even restore strength in many cases. Frauenthal maintained that it induced the 'vital force' of electricity and the 'interchange of cellular contents'.[22] Even in 1916, this was gobbledygook. Successful ingredients probably included the psychological benefits of personal attention and empathy, as well as physiological processes such as enhanced protein synthesis in exercising muscle and the rewiring of mildly damaged motoneurones in the cord.

Aftermath

Once the acute paralytic phase was over, numerous mechanical contraptions – the descendants of the handiwork of Jacob von Heine and many others (Figure 7.3) – came into their own to prevent or deal with the long-term damage left by polio.

Some devices were devilishly ingenious and devilishly expensive: custom-built in lightweight tubular steel, with straps and cushions of the finest leather. At the other end of the spectrum were clunky pieces

Figure 7.3 Deformities of polio and a solution – an exercise machine to help rebuild strength in paralysed limbs, designed and built by Jacob von Heine. From Heine's book on 'lameness of the lower extremities' and its management, 1840. Reproduced by kind permission of the Wellcome Library, London

of 'lumbersome and complicated machinery with which surgeons have loaded their patients', such as leg irons that weighed one-eighth as much as the child. Frauenthal felt that these were fit only for the scrapheap, or for 'one of the working class'.[23]

The iron calipers which stood in for wasted thigh muscles and prevented the knee from buckling were something that many polio victims had to get used to at an early age (Figure 7.4). The leg braces became a publicity icon for the National Foundation for Infantile Paralysis (NFIP), to highlight the cruelty of a life sentence with the consequences of polio. Other appliances included 'artificial muscles' – in fact, heavy rubber bands or spiral steel springs – for example, to pull up the toe of the shoe to correct foot-drop. There was also a range of walking machines to take the weight of the body while the muscles of the legs and trunk re-learned their trade.[24]

The measure of last resort (or an earlier one, for gung-ho practitioners) was surgery to correct deformities or try to restore function. Operative surgery in general leaped ahead in the second half of the nineteenth century, thanks to the introduction of anaesthetics (from the 1850s), effective pain relief (1860s) and antisepsis (1870s). These essential enabling technologies greatly reduced the chances of the patient expiring on the

Figure 7.4 Callipers (leg irons), to keep a paralysed leg straight and prevent the knee from buckling. Modern versions are lightweight and have a hinge at the knee, allowing the knee to bend for sitting; the original devices could weigh one-eighth as much as a child. Reproduced by kind permission of the World Health Organisation

operating table or shortly afterwards. However, their absence had not deterred orthopaedic pioneers such as Georg Stromeyer from pushing ahead with inventive operations that were heroic for both surgeon and patient, such as shortening or lengthening tendons (1831).[25]

Some orthopaedic operations were straightforward, such as cutting the Achilles tendon at the back of the heel to correct foot-drop.[26,27] A sharp-edged 'tenotome' was pushed through a skin incision under the tendon, about an inch above the heel. The foot was pulled up smartly, and the tendon severed 'with a gradual sawing motion'. Plaster-casting the foot in the correct position allowed the cut ends of the tendon to reunite, resulting in a foot that looked normal and could be walked on. More ambitious were tendon transfer operations, in which tendons were cut and re-attached to another bone, using a U-shaped nail and taking care not to 'strangulate the circulation'.[28] A transferred tendon could substitute for a paralysed muscle. For example, the tendon of a muscle from the back of the forearm was passed around to the palmar surface of the hand to enable a paralysed thumb to meet the little finger – which restores at least some functionality to a previously useless hand.

Bones and joints often needed surgical attention in long-standing cases, for example to correct the floppy 'flail' ankle joint of an equinus foot. The joint could be fixed in the correct position by drilling a hole straight through the key bones and banging in a peg of bone, usually appropriated from someone else's recently amputated leg.

Technically less demanding was joint fusion, which can be simulated using a pork knuckle (uncooked) and a sharp kitchen knife. Cut open the joint and then shave away the smooth, shiny cartilage until the underlying bone is exposed. When the human version is put back together with the foot in the right position, the bones fuse. The good news: the joint is now stable and can bear weight. The bad news: it cannot be moved at all and may be excruciatingly painful. Frauenthal reserved this 'mutilating' operation in cases where other operations had failed, and for the 'very poor of the working class' who could not afford anything better.[29] Even more invasive was the insertion of long iron rods into the vertebral column to correct curvature of the spine.

By the 1920s, polio had largely become an orthopaedic condition. Some leading orthopaedic surgeons such as Robert Lovett in Boston and Sir Robert Jones in Liverpool were genuine innovators and took their responsibilities seriously. It was, however, easy to contradict Jones' sweeping statement that 'by a proper appreciation of the available therapeutic and mechanical agencies, we need rarely, if ever, encounter any paralytic deformity'.[30] Nearly 60 per cent of Philip Lewin's 330-page monograph on polio (1942) was devoted to orthopaedics.[31] The British neurologist F.M.R. Walshe might have complained in 1945 that 'in this country there is growing up the illusion that from beginning to end poliomyelitis is a matter for the orthopaedic surgeon',[32] but in 1962, Hugh Paul (a public health physician) conceded that this was indeed the most important person to manage polio.[33]

For many years, orthopaedic surgeons ruled supreme. They belonged to a powerful, exclusively male cartel, renowned (with some justification) for their brute strength, machismo and the unshakeable conviction that they were always right. Predictably, they were thrown into confusion and fury when someone challenged the dogma of immobilisation – and even worse, when that someone saw polio through completely untutored eyes, had a funny accent and was not even a bloke.

A woman's touch

The untutored eyes in question belonged to Elizabeth Kenny, Australian iconoclast, self-publicist and thorn in the side of the surgical establishment.[34] Even though she had never been through nursing school, she called herself 'Sister Kenny'. To her supporters, the title carried gravitas and near-religious mystique. To all those who wanted to send her back to the Australian bush, this was a shameful fraud, and typical of someone who had once paid a local tailor to run off a nurse's uniform for her.

Kenny was embarrassingly ignorant of the spinal cord and what polio did to it. On the other hand, she had decades of hands-on experience in tackling the physical legacy of polio.

Elizabeth Kenny was born in 1880 and grew up in Nobby, Queensland. During her twenties, she worked in the outback as an unqualified nursing assistant and saw occasional cases of polio. Unencumbered by any knowledge about the disease, she reached her own conclusions about what was wrong.[34] The quiet, paralysed muscles failed to impress her; instead, her attention was seized by others that stood out, hard and knotted, and often intensely painful. These non-paralysed muscles were locked in spasm because the spinal cord was being driven to fire on all cylinders, and they were the only ones able to respond.

Kenny focused her efforts on what she saw as the main abnormality, using intensive massage and 'hot packs', heavy wads of wool soaked in near-boiling water, to relieve pain and spasm in the bunched-up muscles. She also waged psychological warfare against the depression and demotivation that gripped patients and their families after being told that nothing more could be done for them. For the first time, children were cajoled into moving limbs that had been dead weights for months or years. And it seemed to work, at least according to Kenny and an expanding circle of believers.[35,36] Kenny also saw the damage done by splints and immobilisation frames, and set out on a crusade to get rid of them.

The First World War gave Kenny the chance to fill in some gaps in her curriculum vitae on the troop ships which carried munitions and fresh volunteers from Australia to England, and then returned with the injured. These 'dark ships' were a soft target for German battleships and U-boats during each seven-week passage, made only slightly less dangerous by sailing without lights. Two years of military service strengthened her conviction that she knew how to treat paralysis, and earned her the British Commonwealth's title for a senior nurse in the Armed Forces: 'Sister'.[35,36]

After the War, Kenny picked up where she had left off. Rumours, with Biblical overtones, that she had made the lame walk again spread rapidly through Queensland via the bush telegraph. In 1928, she set up a clinic in a hotel in Townsville, funded by supporters and the Queensland government. This soon pulled in polio patients from across the province. Claims of miracle cures by the 'Kenny Method', and her lack of formal qualifications, set her on collision course with the medical establishment, who demanded a Royal Commission to investigate the claims. The Commission considered only 47 patients, but deliberated for two years. Their report, published in 1938, concluded that *Miss* Kenny had 'inadequate knowledge' and that her claims were 'fanciful'.[37]

This was precisely the hatchet job that Kenny's critics had longed for. To their disgust, the Queensland government threw out the Commission's findings and pumped more money into the Kenny franchise. Kenny Clinics sprang up, staffed by thoroughly indoctrinated and boldly uniformed Kenny Nurses. Other Australian provinces followed Queensland's example.[35,36]

In 1940, the 60-year-old Kenny decided that the United States was ready to receive her wisdom.[38] Her American crusade was funded by the Queensland government, a further poke in the eye for her medical enemies. She armed herself with letters of introduction to two men who could make America see things her way: Tom Rivers, the grand master of virology, and Basil O'Connor of the NFIP. She got off to a bad start, with apathy in her first ports of call in California, and a calculated snub from Rivers, who had been briefed by friends in Australia and refused to meet her.[39] Angry and despondent, Kenny was on the point of heading for home when she called in to Minneapolis. There, she found kindred spirits who rallied around, found her space and instant credibility in the medical school, and rustled up money and a growing stream of polio patients, many rescued from the clutches of the orthopaedic mafia.[35,36] O'Connor came to visit and was impressed; funding began to flow in from the NFIP, together with the blessing, kudos and publicity of that organisation.

The Kenny brand flourished in Minneapolis. The first Sister Kenny Institute opened for trade in time for Christmas 1942. Professor of Orthopedics J.F. Pohl co-authored an adulatory book about Kenny's concepts and methods, the other author being Sister Elizabeth Kenny.[40] Over in Boston, another professor of orthopedics declared himself a convert, and others soon found themselves on their own road to Damascus. The Kenny boulder had begun to roll.[35,36]

As in Australia, Kenny set doctor against doctor. Plain-speaking Tom Rivers, appalled by her scientific ignorance, took every opportunity to put her down.[39] He was, however, impressed by some peripheral aspects of her campaign, describing Kenny's 'niece' (actually a trainee whom she had taken under her wing in Australia) as a 'good physical therapist and a doggone good-looking girl'.[39] Kenny and O'Connor fell out over a debacle in South America. In 1943, the Argentine government contacted Washington, DC, and implored them to send Sister Kenny to rescue their nation from a polio epidemic. The NFIP paid to airlift a Kenny assault team into the front line – when many of them deserted to become personal therapists to the Argentine super-rich.[39]

But by then, Kenny had momentum and money of her own and, with Kenny Clinics proliferating across America, she no longer needed the NFIP (Figure 7.5). In 1943, she established the Sister Kenny Foundation,

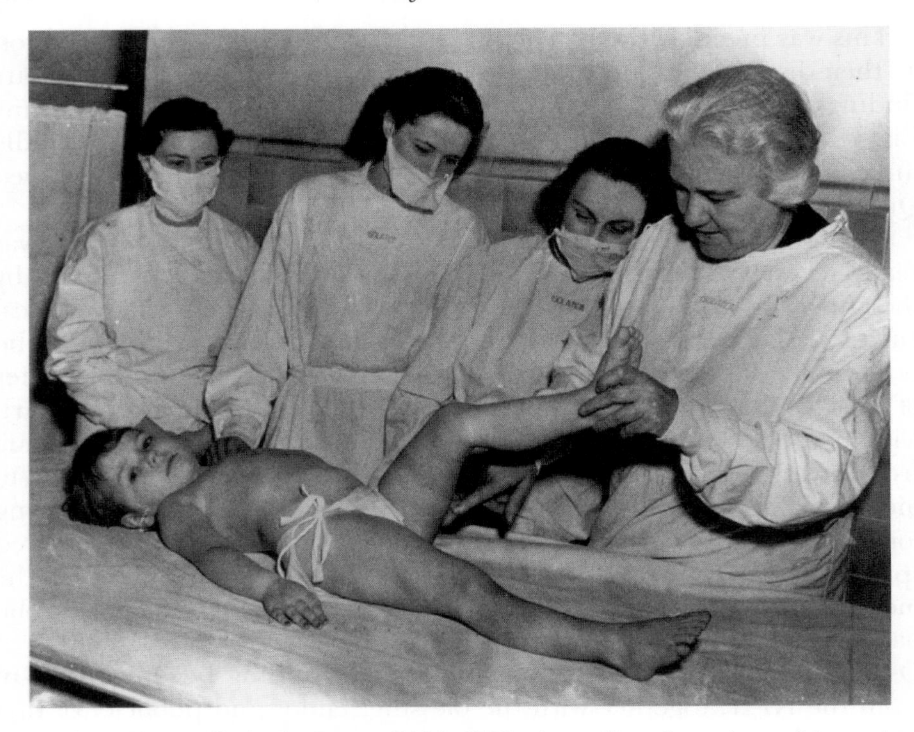

Figure 7.5 Sister Elizabeth Kenny (1880–1952), Australian therapist and iconoclast, teaching with the help of polio patient Nona Bibeau, c. 1950. Kenny's views on the management of polio divided the medical profession around the world. Reproduced by kind permission of the Sister Kenny Rehabilitation Institute

which unashamedly muscled in on the NFIP's territory, providing practical help for patients and their families, funding polio research and even sponsoring scientific conferences about the disease.

From America, Kenny's reputation and notoriety spread around the world. In England, the War years were enlivened by a spirited debate sparked off by a rose-tinted paper in the *British Medical Journal* (BMJ) from Dr F.H. Mills. The Queensland Commission described it as 'confusing and unsubstantiated', and noted sniffily that it was 'remarkable that such should be published in a reputable medical journal'.[37] The *BMJ* responded with a string of anti-Kenny articles, with titles such as *The fact and fantasy in poliomyelitis* (1943) and *The Kenny Method criticised* (1944).

Kenny's influence later percolated to the highest levels. On 22 July 1949, Aneurin Bevan, Minister of Health and architect of the National Health Service, faced pointed questioning in the House of Commons. John Viant MP, dismayed by 'the failure of medical science to discover a cure for poliomyelitis', wanted the Minister's guarantee that Kenny therapy would be available to all English polio sufferers. Bevan's reply politely echoed the Queensland Commission: there was no evidence that Kenny's techniques were better than any other, and their use was 'a matter of medical discretion'.

Like it or not, Kenny left her mark. Her methods eventually infiltrated the treatment and rehabilitation of polio victims around the world. Despite Kenny's contempt for the conventional, this was a two-way street, and her methods quietly assimilated some aspects of 'orthodox' therapy. The hybrid that had evolved by the early 1950s was founded on common sense, which was fortunate, because Kenny's 'scientific' rationale was still rubbish.[35,36] The basic elements have stood the tests of time and medical hostility and are still in use today, with warm packs and massage to relieve painful muscle spasm, positive encouragement and early active mobilisation. And Kenny deserves praise for having helped to rid the world of the Bradford frame and the other instruments of immobilisation.

Just like the NFIP, the Kenny phenomenon was a triumph of clever marketing and social engineering. Nobody seemed to remember that her 'innovative' hot packs had been used many years earlier to relieve painful muscle spasm. Her success was partly fuelled by her portrayal as an evangelist who thrived despite victimisation by medical men in suits. She was easy to mock, with her heavy, almost masculine features, frumpy dresses and ridiculous hats. But she was also impossible to ignore, even though Rivers and many others did their best. To them, the 1951 Gallup Poll that named her the Most Admired Woman in America, and the breathless books about her life[35,36,40] – not to mention *Sister Kenny*, the Hollywood movie starring Rosalind Russell that followed in 1946 – were all further proof of human gullibility.

Who was the real Elizabeth Kenny? Pictured cuddling sick children, the doctor-bashing battleaxe is transformed into the epitome of the warm, caring nurse. But like everything pumped out by the Kenny propaganda machine, the photographs were carefully staged. There were no images of the children who began crying at the sound of the trolley carrying the hot packs being wheeled onto the ward.[41]

Room to breathe

Elizabeth Kenny called it 'an instrument of torture'; to others, it was one of the iconic technologies of twentieth-century medicine. This was the iron lung, an imperfect solution to the complication which killed more polio victims than any other. At its best, it was a life-saver and a sanctuary from the panic of suffocation. At its worst, it was terrible tomb in which to die.

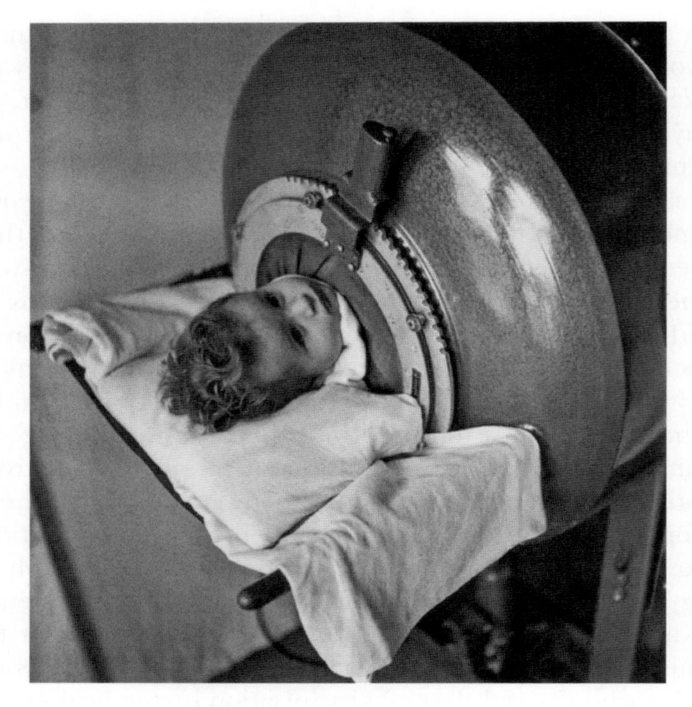

Figure 7.6 Girl in an iron lung during the polio epidemic in Israel, 1958. Reproduced by kind permission of the World Health Organisation

The iron lung is at the heart of images that could have come from science fiction or the Brothers Grimm (Figure 7.6). The little girl whose invisible body is encased in a massive steel cylinder, her neck gripped in a tight rubber diaphragm, her hair neatly combed out on the pillow and watching the world back-to-front in the mirror angled above her face. The bustle of a polio ward during a big American epidemic, with nurses tending dozens of iron lungs pulled into lines, each machine like a gigantic pupa with a human head protruding (see Figure 6.7). The 'two-up, two-down' room respirator at the Boston Children's Hospital, with four patients' heads sticking out of a wall.[42] For many, this was the stuff of nightmares, not medicine.

Designs for mechanical ventilators went back a surprisingly long way, and the first devices used in polio were borrowed from other areas of medicine. The 'pulmotor', designed in Germany in the early 1900s to resuscitate victims of mining disasters, alternately forced oxygen under high pressure into a tightly fitting face mask and then sucked it out again. It was introduced into the United States in 1911, and used mainly to treat coal-gas poisoning. Frauenthal and Manning suggested its use in polio,[43] but the idea did not take off. The pulmotor killed more people than it saved, partly because its high pressures damaged the lungs. Yandell Henderson of Yale University fought a long battle against the 'life-loser' pulmotor, culminating in a two-page rant in *Science* in 1943 entitled, '*The return of the pulmotor as a 'resuscitator': a back step towards the deaths of thousands*'.[44] Henderson had an interest to declare: together with Howard W. Haggard, he had developed a rival device, the 'H&H inhalator'. This killed fewer people but, like the pulmotor, could not keep paralysed patients alive for days or weeks.[45]

The first respirator to treat polio victims is credited to Philip Drinker, a professor in Harvard University's School of Public Health.[46,47] An expert in mercury poisoning, Drinker was also an inspired technical innovator. He dreamed up the helium-oxygen mixture that prevents the 'bends' and has saved the lives of countless submariners, beginning in 1939 with the spectacular rescue from 200 feet down of the crew of the crippled *Squalus*. In the 1920s, Drinker built cat-sized metal chambers with a large syringe and pressure gauge attached, so that controlled volumes of air could be withdrawn and replaced. The cat's head stuck out from an airtight seal at one end of the chamber. This was a 'negative pressure' ventilator system. Pulling on the syringe reduced the pressure in the chamber, causing the cat's chest to expand and draw air into the lungs; pushing air back into the chamber then emptied the lungs. With Drinker's apparatus, cats that had been paralysed with curare could be kept alive for as long as the operator continued to fill and empty the syringe.

With Louis Agassiz Shaw, Drinker developed a scaled-up, motorised version of his respirator suitable for long-term use in humans.[48] Thanks to a steep rise in accidental and deliberate deaths from domestic gas, their research was co-funded by the Consolidated Gas Company of New York. The final product was a formidable beast: a steel cylinder three feet in diameter by seven feet long, and weighing a third of a ton. The life-preserving motor and bellows were housed on a trolley below the chamber. The patient was pushed in like a loaf into an oven, and the head forced through a tight hole in a heavy rubber diaphragm at the centre of the end plate, which was then closed and screwed into place. There were two small access ports for nurses to attend to the routine needs of the body inside the chamber.

Drinker unveiled his invention in 1929, in one of the all-time landmark papers of the *Journal of the American Medical Association*.[49] The respirator was first used on 12 October, in an eight-year-old girl with paralytic polio. About to breathe her last, she was loaded into the adult-sized machine, still experimental and powered by two vacuum-cleaner motors. Drinker had the honour of switching it on, mainly because everyone else refused to do it. The result was miraculous. Within minutes, the patient was pink, breathing comfortably in the grip of the machine, and peacefully asleep. Unfortunately, her story did not end happily. Three days later, she died of pneumonia – a common end for patients in the respirator until the vital importance of keeping secretions out of the lungs was appreciated. Drinker's second patient did much better, regaining the ability to breathe alone after the machine had kept him alive for two weeks.[48] The Drinker respirator quickly acquired its grim nickname of 'iron lung' and became a heavy-duty warhorse in the battle against polio.

However, Drinker's respirator and its results were far from perfect. Even in centres of excellence, 60–80 per cent of patients entrusted to the iron lung died inside it.[50] Many problems lay in nursing care, and the attempts to sort these out helped to lay the foundations for the speciality of intensive care nursing. Secretions had to be sucked out of the patient's pharynx to prevent aspiration pneumonia, while the bowels and bladder had to be carefully tended, and bedsores prevented. Also, it was soon realised that patients with bulbar paralysis fared especially badly in the iron lung. Their uncoordinated respiratory efforts often fought against the machine, while loss of swallowing and coughing reflexes made them more prone to pneumonia.

Drinker's no-frills design was also an invitation for improvement. John Haven Emerson rose to the challenge in 1932, with a user- and patient-friendly respirator that was nicknamed 'The Alligator' because the entire top was hinged and lifted up to engulf the patient (Figure 7.7). It also featured larger ports to make the nurses' tasks easier. The Emerson respirator was lighter, cheaper and obviously better than the original.[51] Drinker promptly sued Emerson, but lost the case because his own design was not that original either. The lineage of mechanical-negative pressure respirators was astonishingly long,[50] and included Thunberg's motor-driven 'Barospirator' tank (Germany, 1920); Dr W. Steuart's wooden box, sealed with clay at the shoulder and waist (South Africa, 1918); Dr E.J. Woillez's 'Spirophore', which won the Silver Medal at the Le Havre Exhibition of Life-saving Equipment in 1876; and an 'apparatus for promoting artificial respiration' which the Scottish physician John Dalziel presented to the British Association for the Advancement of Science in Newcastle, 1838.[52]

Emerson's respirator became the standard model in America, and was produced in its thousands. Its manufacturing costs were covered by the NFIP, which also paid for machines to be airlifted into the centres of emerging outbreaks.

Elsewhere, cheaper and more utilitarian alternatives were soon being produced, including one machine which presaged the era of

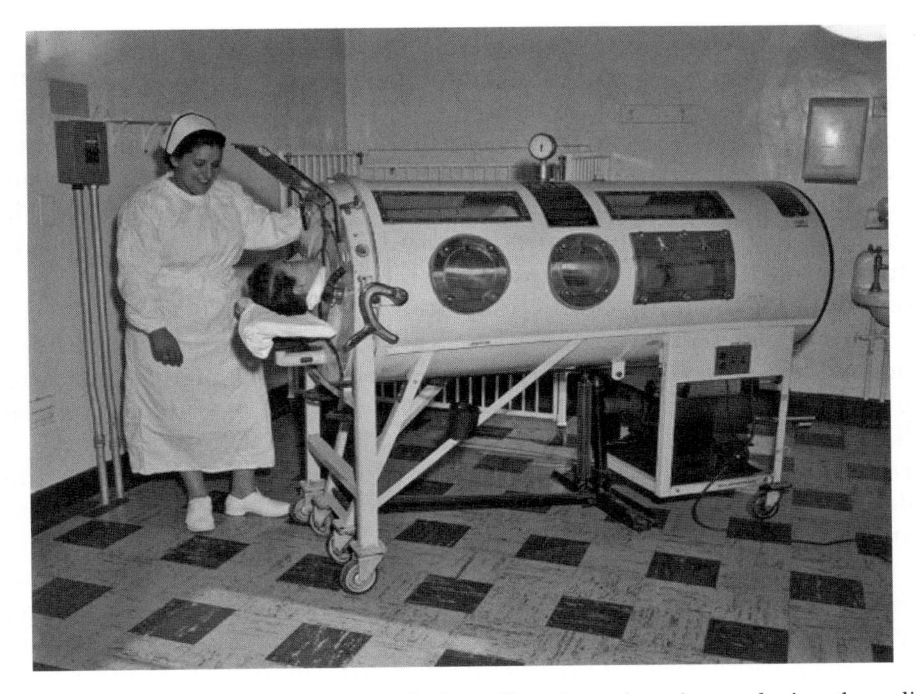

Figure 7.7 The 'Emerson respirator' design of iron lung, here in use during the polio outbreak on Rhode Island, United States in 1953. Reproduced by kind permission of the Centers for Disease Control

flat-pack furniture. Like many good innovations, this was a response to a crisis, specifically the large polio outbreak in Australia in 1937. American respirators were unaffordable, prohibitively heavy to ship and, unhelpfully, had to be returned to the United States for repairs (which were frequently needed). Enter Edward and Arnold Both, inventors of medical equipment, who retired to their workshop in Adelaide with sundry machine parts and lots of plywood. They emerged a fortnight later with the Both Portable Cabinet Respirator, which cost barely £100 and could be built by hospital technicians. It bore an unfortunate resemblance to a coffin, but did the job as effectively as Drinker's machine.[53]

The Both respirator soon infiltrated Britain. Edward Both visited England in 1938 to promote the brothers' new electrocardiography apparatus, and happened to hear an appeal broadcast on the BBC Home Service about a young woman with worsening respiratory paralysis in a small rural hospital that had no iron lung. Both promptly hired a workshop, ran off one of his respirators and donated it to the hospital in time to save the patient's life. A short film about Both's respirator was seen by Robert McIntosh, the canny professor of anaesthetics in Oxford. He showed it to William Morris, founder of the Morris Motor Car Company and about to be ennobled as Lord Nuffield. Morris was so taken with Both's idea that he turned over part of his Morris Minor factory at Cowley to making the respirators, promising to supply one free of charge to every hospital in Britain and the Commonwealth. The Both respirators were mass produced at one-thirtieth of the cost of the Drinker machine. Over 1,700 were sent out from Cowley, and many more were built locally using Both's template.

Not everyone approved of Nuffield's *grand geste*. Sir Frederick Menzies, medical officer of health for London County Council, moaned in a letter to the *BMJ* that Nuffield's gift was a 'wanton waste of benevolence' and the 'height of folly', as hospitals had no idea how to use it.[54] This was true initially, but the Both-Nuffield enterprise succeeded in spreading life-saving technology, and the incentive to make the most of it, across a large part of the world.

Many designs of respirator were developed, generally driven by technological advances or commercial competition, but sometimes precipitated by a brush with fate. The Smith-Clarke respirator, widely used in Britain, was created in a disused air-raid shelter by Captain G.T. Smith-Clarke in 1951, soon after he retired from running the Alvis Motor Company. Another 'alligator' design, this had internal strip lighting providing warmth as well as illumination, a split rubber collar that hurt much less than its predecessor, and an alarm that the patient could activate inside the cabinet. A tasteful touch was the cover over the port for intravenous drip lines, which looked uncannily like the petrol cap from an Alvis racing car – because that is exactly what it was.[55,56]

Further afield, respirators for use in the tropics had the perishable rubber in the bellows replaced with the fine leather used in pipe organs. During the polio outbreak on the island of St. Helena in 1949, a polio victim who was pregnant proved that the ports on the machine flown in from South Africa were large enough to deliver a baby safely.[57]

Improvisation was sometimes required. Royal Air Force (RAF) engineers built a respirator into an aeroplane for Paul Bates, the 'Horizontal Man', so that he could take part in the London-Paris Air Race of 1961.[58] In the Canadian Arctic, a polio patient was airlifted to hospital in a hastily constructed respirator that could be operated with one hand by the pilot while flying his two-seater aircraft. And a letter to the *BMJ* in 1956 included a sketch of a respirator built on the high seas from sheets of marine plywood, sealed with glue and paper, with tubing and pumps from smoke masks (Figure 7.8). The device kept a passenger who had been paralysed en route to New Zealand alive for several days.[59]

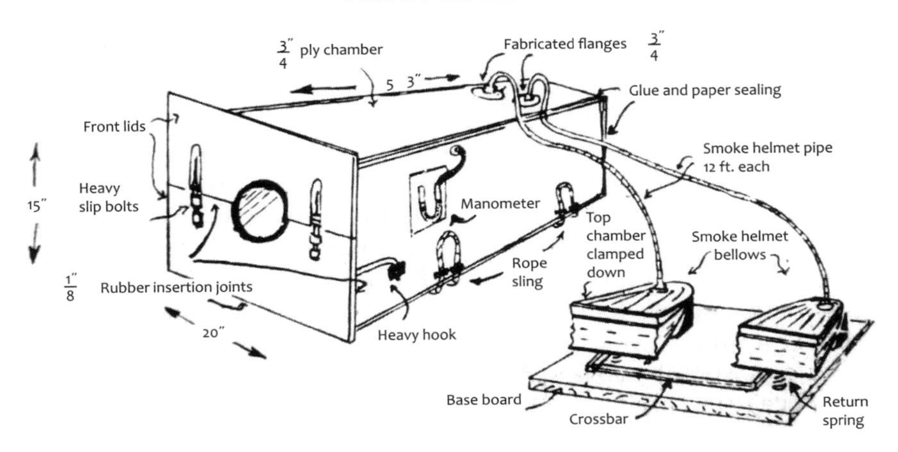

Figure 7.8 Improvised 'iron lung', built from plywood, plasticine and paper in 12 hours by Chief Officer S.G. Robinson and Second Engineer J.D.A. Bright of the New Zealand Shipping Company, in mid-Pacific in spring 1956.[56] Reproduced by kind permission of the *British Medical Journal*

Staying alive

Life inside the iron lung was difficult and sometimes horrific. Some patients could muster no respiratory effort at all and, whenever the airtight seal on the iron lung was broken, they had to be ventilated using a face mask connected to an oxygen cylinder and a rubber bag squeezed by an attendant. The more fortunate could breathe on their own for short periods, and some only used the respirator at night. Inside the chamber, the patient wore only a light gown; until internal heating came along, it could get very cold at night. Bodily functions were a grim and ever-present concern, made worse for everyone by having to grapple with a bedpan through one of the handling ports. Patients with paralysis of the bladder had to be catheterised, sometimes for weeks. The greatest curse was constipation, the result of immobility and a hospital diet. This was often so obstinate that it defeated laxatives and enemas – which left only manual disimpaction by the nurse's gloved finger.

Mental adjustment could also be a huge challenge, which ultimately crushed many patients. There were few windows into normal life. Without the angled mirror or the glass shelf on which books were placed face down above the patient's face, the only view was the end of the respirator and the ceiling above. Luckily, most children seemed remarkably resilient. Lewin recommended placing a 'treasured article' inside the chamber beside the child's hand, noting that 'a small doll or stuffed animal serves nicely'.[60] Photographs from the polio epidemics show other 'treasured articles' – a teddy bear, a toy revolver in a holster, the Stars and Stripes – carefully displayed within gaze.

The hum of the iron lung's motor and the regular sighing of the bellows provided the background to the soundscape of the polio ward. Some found the sounds soothing and comforting, but to others these were a constant reminder of their frailty and the fact that their life depended on the machine. When 14-year-old Mary Berry awoke one morning in December 1949 in the polio ward in Bath, it took her a few moments to realise why the room sounded so strange. The iron lung across the room was now silent and empty; her friend Buffy had died during the night.[61]

Many patients were haunted by the fear that the machine would break down or the power would fail. Paul Bates gives a humorous but chilling account of panicking while watching a hospital electrician, called to the polio ward to change light bulbs, switch off the mains. Even when all the respirators fell silent, the man was unaware of what he had done. Fortunately, a nurse arrived in time to save the day.[62] A similar phobia, but from an unexpected source, stalked 19-year-old Hugh Gallagher, who was completely paralysed and in an iron lung during the Philadelphia epidemic of 1952. Gallagher used to be woken in the middle of the night by the terror of imminent suffocation and unable to say a word, because an apparently sadistic physiotherapist had deliberately switched off the motor to see how long he could last.[63]

Not surprisingly, some patients became psychotic while incarcerated in the iron lung. Many had vivid 'iron lung hallucinations', usually imagining that they were travelling by aeroplane or train. These were apparently caused by prolonged immobilisation and were commoner at night, when sound and visual cues were lost. There were strong parallels with the 'sensory deprivation' hallucinations experienced by American prisoners held in solitary confinement during the Korean War. Later, the iron lung was used experimentally to induce sensory deprivation in normal subjects.[64,65]

Most polio patients spent only a few weeks in the respirator before Nature took its course, one way or the other. Thereafter, some needed occasional support overnight or if they became tired (some iron lungs were kept on standby in the back bedroom), while a small minority of patients needed constant ventilation for the rest of their lives. They invented many ingenious gadgets and tricks to allow them to leave the house, attend sporting events, or go on holiday, all in mobile respirators. Some polio patient support organisations ran hotels with rooms specially modified to take iron lungs, such as the British Polio Fellowship's hotel in Worthing, on the South Coast of England.[66]

Many polio survivors went on to prove that it was possible to survive a normal lifespan in the iron lung. According to the *Guinness Book of Records*, the longest was over 61 years, achieved by Julie Middleton of Melbourne, who died in October 2009 at the age of 83. Not far behind her was Diane Odell of Jackson, Tennessee, whose 60 years in the iron lung were tragically terminated in 2008 by a power cut.[67]

In their heyday (1959), iron lungs kept alive 1,200 polio survivors across America. By 2008, there were fewer than 50. Iron lungs still reside in the basements of hospitals around the world, awaiting the call to help in weaning non-polio patients off the slicker, less threatening ventilators of today. At any time, 30 iron lungs are reckoned to be in use in the United States.

Alternative therapies

The iron lung was not the only invention that kept paralysed polio victims breathing and alive. In early 1938, Aubrey Burstall, Professor of Engineering at the University of Melbourne in Australia, invented the 'cuirass', named after the chest protector in a suit of armour.[50] Burstall's prototype, a six-pound carapace of hammered aluminium, certainly looked the part. It was sealed with rubber flanges around the upper chest and mid-abdomen and, like the iron lung, operated on the negative-pressure principle. Every few seconds, a pump sucked air out of the cuirass to inflate the lungs. Burstall's cuirass came into its own towards the end of the Australian epidemic of 1937–38, especially as several units could be plugged simultaneously into one pump. Later designs used lightweight fibreglass and compact pumps, and could be taken on to public transport and on holiday (Figure 7.9).

A more visceral device was the rocking bed designed in 1940 by Jesse Wright, medical director of the D.T. Watson Home for Crippled Children outside Pittsburgh (who later collaborated with Jonas Salk in the early clinical trials of his polio vaccine). The bed was a fearsome contraption that relied on the shifting weight of the guts, pressing against the diaphragm like a plunger, to shunt air in and out of the lungs. This must have been a very rough ride, with the bed swinging every few seconds from 45° head-down tilt (exhalation) to 45° head-up (inhalation). But the bed was much cheaper than the iron lung, and it worked in milder cases.

During the 1950s, technology literally took a more positive turn, as the negative-pressure respirators were superseded by machines which pushed pulses of pressurised air directly into the trachea. These were descendants of the pulmotor, but without its murderous tendencies because they used lower pressures. Modern 'intermittent positive pressure ventilation' (IPPV) was put on the map in Blegdams Hospital in Copenhagen on the morning of 27 August 1952 with Bjorn Ibsen's last-ditch attempt to save little Vivi Ebert (p. 46).[68]

The technology of IPPV is simpler and cheaper than the iron lung, and soon slashed the mortality rate among ventilated polio patients from 60–80 per cent to just 20 per cent. Various pumping systems were invented, including the Bang Respirator of 1952 in Denmark, and James McCrae's portable Clevedon Respirator (1953), named after a seaside town near Bristol.

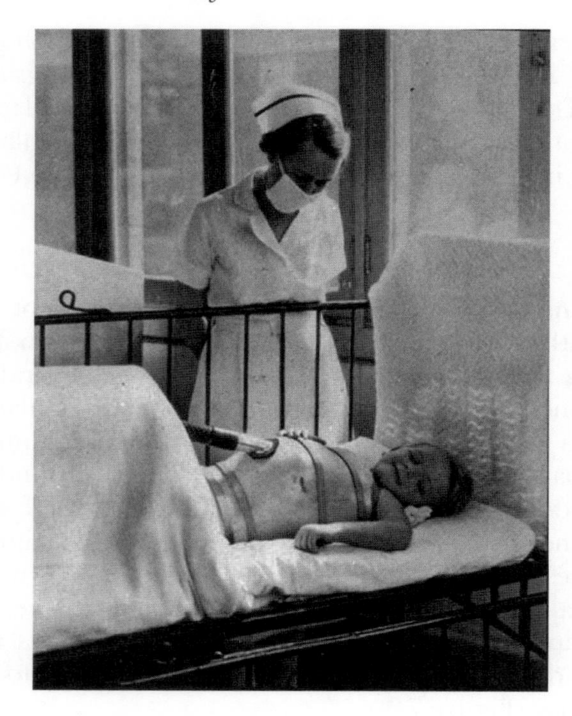

Figure 7.9 Child being ventilated in a 'cuirass', Toronto, 1937. Reproduced by kind permission of the Hospital for Sick Children Archives, Toronto

The business end of IPPV is the cuffed endotracheal tube which is pushed down into the trachea ('intubation'). The cuff carries an inflatable cuff which is blown up to block off the lumen of the trachea, creating an airtight seal and preventing secretions from running down into the lungs. For short-term ventilation, the trachea can be intubated through the mouth, but the tube damages the vocal cords if in place for more than a few days. For the weeks or months of IPPV which most polio patients required, the tube was inserted through a small oval window cut in the front of the trachea. This tracheostomy usually healed remarkably well after the tube was taken out in patients who had recovered their breathing.

IPPV is now the routine method for artificial respiration on the intensive care unit – a satisfying outcome for something that began as a desperate response to the inadequacies of iron lungs.

Marvels of modern medicine

Polio was cruel, terrifying and unpredictable, and before the advent of polio vaccines, completely unstoppable. This was a disease that cried out for a miracle cure, and many candidates were put forward, ranging from electric shocks to snake venom.

Electricity found its way into medicine soon after the Leyden Jar (the first battery) was invented in the mid-1740s. An electrical current passed between electrodes placed over a motor nerve could make muscles spring into action, which made polio an obvious condition to target. Electricity was tried right at the start, in one of Michael Underwood's five cases in 1789, and became respectable a century later when Mary Putnam Jacobi strongly recommended its use.[9] This was within a few years of the foundation of the American Electrotherapeutic Association and the introduction of the electric chair and its 'cruel and unusual' mode of execution.

By 1916, electrotherapy was all the rage in America and Europe, with miraculous cures claimed for everything from muscular dystrophy to shell shock. According to leading lights such as Robert Lovett and Henry Frauenthal, electricity was crucial in treating polio. Practitioners developed complicated treatment protocols using galvanic (direct current) or faradic (rapid pulses) stimulation, sometimes shocking the patient in time to his or her heartbeat. Those on the receiving end might have wondered whether 'cruel and unusual' was unique to the electric chair. Electrodes were placed over the spine in the same position as blisters and the red-hot iron, and the current turned up to (and sometimes beyond) the 'point of toleration'. No wonder that children could end up in a 'hysterical condition'.[69]

Did it work? Of course it did, according to the enthusiasts. Frauenthal believed that this 'hidden force that actuates cell vitality' accounted for most of the benefits of rehabilitation, and that it sterilised the tissues by electrocuting the poliovirus. Frauenthal presented several cases of electrical 'cure' to the American Electrotherapeutic Association in 1914, noting proudly, 'I feel positive that I have relieved the compression of the cord and hastened recovery'.[69]

Others were unpersuaded. Dr Henry L. Taylor, Professor of Orthopaedics at New York's Hospital for the Ruptured and Crippled, was one of many who were blunt in their condemnation: 'Conventional therapy by electricity and massage is completely ineffectual'. With the benefit of hindsight, it is clear that Dr Taylor was right, but electrical stimulation therapy was still widely used in Britain during the 1950s.

Washout

Electricity was not the only physical therapy claimed to have magical effects. In its June 1935 issue, the influential *Literary Digest* carried a punchy article entitled 'Paralysis Stalks Carolina', which predicted (correctly) a bad polio season that summer. Fortunately, the *Digest* reported, there were new weapons in the battle against the disease: prototype polio vaccines, the iron lung and a 'highly experimental' but promising treatment devised by George M. Retan, Professor of Pediatrics at Syracuse University in New York.[70] This was 'forced spinal drainage', known colloquially as 'brain-bath therapy', because it was supposed to stimulate the production of cerebrospinal fluid (CSF) in the hope of washing infectious toxins out of the brain. The patient was given large volumes of hypotonic (dilute) saline solution intravenously. Simultaneously, CSF was drained off through a lumbar puncture needle left in situ; alternatively, a longer needle could be pushed inside the skull through the foramen magnum and into a large CSF space between the cerebellum and the medulla (a glance at Figure 1.6 will confirm that this was a brave thing to do).[71]

Retan had developed his treatment in animals, and claimed that it saved the lives of monkeys given supralethal doses of poliovirus, whereas untreated controls all died. In 1931, he tried his method on a three-year-old girl with pre-paralytic polio, who drank large volumes of water while hypotonic saline went into her arm and CSF dripped out of her back, all for 36 hours. She recovered, with no paralysis. Over the next few years, Retan experimented in patients with polio, encephalitis and syphilis, all apparently with good results, and refined his therapy during polio epidemics in New York. Pre-paralytic patients or those with early weakness were ideal; mild cases were not usually selected, although one was treated because the mother insisted. The patient was strapped into a Bradford frame with a window in the canvas stretcher so that a lumbar puncture needle (blocked with a removable stylet) could be left in place for several hours. Large volumes of dilute saline (nearly a gallon in an adult) were infused intravenously over 3–5 hours, and the stylet pulled out every 30 minutes to allow CSF to drain. Typically, the session was repeated daily for five or six days.[71]

The results reported by Retan were spectacular and sometimes seemed miraculous, such as the three-year-old boy with early paralysis in both arms who could lift heavy weights after just one session. Patients with established paralysis were generally too far gone, but otherwise 'complete recovery' was usual. Even patients with respiratory paralysis could breathe again, and Retan predicted that his therapy would make the iron lung redundant. He insisted that the benefits of his treatment 'cannot be questioned' – and for many years, they were not.

By the summer of 1935, Retan's forced drainage therapy was one of the brave new hopes in a field that was desperate for a breakthrough. At the American Medical Association meeting in Atlantic City in May 1935,

Retan gave a keynote lecture and a demonstration about his technique. That summer, he was summoned to help with the polio outbreak in Louisville, Kentucky, and soon had 18 patients being dripped and drained in the isolation ward at City Hospital. This brought him to the attention of *Collier's*, the magazine famous for aggressive investigative journalism. *Collier's* had exposed the 'Great American Fraud' of useless patent medicines, which eventually led to the formation of the Food and Drugs Administration (FDA). Now, it had nothing but praise for Retan and his visionary therapy.[72]

By now, Retan's dossier included scores of patients and results that were beyond expectation – and to those who believed that they could detect the odour of a rat, beyond credibility. Retan's explanation of 'forced spinal drainage' did not add up, even when he changed the name to 'forced perivascular drainage'. Pushing dilute saline into the bloodstream might encourage fluid to enter the brain, like a dried prune soaking up water – but as the brain and spinal cord were already swollen in polio, was this a good thing? And where did all the washed-out toxins go, especially when Retan decided to abandon CSF drainage and just left the dilute saline running into the veins?[73] And what about matched, untreated controls, and statistics? Answering the last question was easy, because there were not any, but the others remained imponderables.

Nonetheless, Retan cruised on for years, managing to avoid tricky explanations or a decisive test of his treatment – and his charmed run even continued for a year or two after after 1 July 1942. On that day, the *Journal of Immunology* carried an article by Sidney Kramer and colleagues from the Michigan Department of Health in Lansing, describing meticulous experiments in monkeys which showed that Retan's basic ideas had been fatally flawed from the start.[74] Retan fought back[75], and even though few now believed in him, he was never held to account or made to retract his publications.

Panacea

Few wonder drugs have been as wondrous as ascorbic acid, better known as vitamin C. It arrived with a bang in 1932, identified as the elusive 'anti-scorbutic factor', whose absence from the sea-faring diet caused scurvy. It also netted a Nobel Prize for its discoverer, Albert Szent-Gyorgyi, who isolated it from paprika, which gives fire to the goulash in his native Hungary. Within two years, vitamin C seemed to be far more than the cure for scurvy. In the test tube at least (and at unfeasibly high concentrations), it could wipe out the virus of rabies, the bacteria of scarlet fever and tuberculosis, and the lethal toxins of diphtheria and tetanus.[76] It was only a matter of time before someone tried it in polio.

That person was Claus W. Jungeblut, Professor of Bacteriology at Columbia University, New York. Jungeblut had already flirted briefly with various 'poliocidal' substances including the serum of pregnant mares.[77] Now, he showed that 'very small amounts' of vitamin C inactivated the poliovirus in vitro. He then studied monkeys infected with poliomyelitic cord extract, injected into the brain or intranasally, and found that large subcutaneous doses of vitamin C reduced the risk of paralysis by an astounding 90 per cent.[78] Flexner published the paper, but instructed one of his bright young men to check out Jungeblut's claim.

The bright young man was Albert Sabin, then aged 29 and not yet thinking about polio vaccines. They had already met, as Sabin had shot down Jungeblut's claim to have invented a simple test that identified subjects susceptible to polio (an episode described later). On that occasion, Sabin had written a paper with Jungeblut, which tactfully put the record straight. This time, there was no diplomatic solution. Sabin published what he had found, namely that vitamin C had no protective effect whatsoever in monkeys infected with polio intranasally.[79] In response, Jungeblut produced a further paper, still insisting that vitamin C protected against paralysis – and then quietly moved into researching leukaemia.

Meanwhile, others had been seduced by vitamin C, which was now billed as the cure for rattlesnake bites and scorpion stings as well. During the 1940s, Frederick R. Klenner, a family doctor in Reidsville, North Carolina, began experimenting with vitamin C in polio – but in people, not monkeys. In Klenner's view, Jungeblut had been far too timid. Arguing that 'when proper amounts are used, it will destroy all virus organisms', Klenner injected up to 30 grams each day, several thousand times the daily requirement.

During the epidemic of 1948–49, Klenner treated 60 polio victims with stratospheric doses of vitamin C.[80] His claim to have cured all of them, including some 'especially incredible cases', was greeted with suspicion by the medical establishment – which was all the more reason for Klenner to become an instant hero to the growing clan of vitamin therapists.

If Klenner was to be believed, vitamin C was the answer to polio: 100 per cent effective, 100 per cent safe and, by now, cheap. It was crying out to be proved effective in a proper clinical trial, which would have been extremely easy to do compared with some other polio treatments. Strangely, nobody – either pro or con – rose to the challenge. Perhaps Klenner had inflated or invented his results and feared being exposed, and perhaps the medical establishment thought it was too daft to pursue.

Klenner continued to make extravagant claims, complaining that 'the spectrum of vitamin C is so broad that many doctors and nurses have found it difficult to believe the reports'. Indeed. But even his own supporters eventually lost interest. By 1952, high-dose vitamin C therapy was still heavily promoted in medical journals for serious bacterial infections including scarlet fever and tuberculosis.[81] However, polio had dropped off the list.

Meanwhile, vitamin C was well into its Golden Age. Linus Pauling, Nobel Laureate in Chemistry (1934), was later swept up on a messianic and possibly unhinged crusade to use vitamin C to treat the common cold, cancer and many other diseases. Pauling gave credibility to the vitamin gurus, who went on to invent 'Orthomolecular Medicine' (mission statement: 'therapeutic nutrition based upon biochemical individuality'). In 2004, Pauling was admitted to the Orthomolecular Medicine Hall of Fame, followed a year later by Klenner.[82]

Poisons and antidotes

Many other 'miracle cures' blinked in and out of favour, as spin-offs of pharmacology, immunology, toxicology and even chemical warfare.

One short-lived candidate was the antidote to Lewisite, an especially nasty chlorine-arsenic compound nicknamed the 'Dew of Death' because a couple of dozen drops on the skin can kill a man. The antidote, developed during the Second World War in a top-secret programme at the Strangeways Laboratory in Cambridge, England was dimercaprol. This compound, later given the stirring name 'British Anti-Lewisite' (BAL), was good at removing toxic metals from the brain, notably copper in the inherited disorder called Wilson's disease.[83] For unknown reasons, BAL was tried in a four-year-old girl with acute paralytic polio. She made a rapid recovery – dramatic stuff, but well within the frame for spontaneous resolution.[84] This promising lead was never followed up, but BAL was touted as a miracle cure for polio by Ralph Scobey, the poliovirus-denialist doctor of the 1950s,[85] and is still cited today as evidence that polio is caused by toxins, not a virus.

Another intriguing line of research, also initially conducted in high secrecy, centred on cobra venom. For once, this had a reasonable scientific rationale. Cobra venom toxins kill motoneurones in primates, an effect first observed in 1904. The research programme, funded by the NFIP, aimed to modify the toxin so that it would bind safely to the motoneurones and stop the poliovirus from gaining access. Progress was made in emasculating venom toxins with hydrogen peroxide, but no compound suitable for use in polio emerged.[86]

Nonetheless, Murray Sanders, who led the research team, was nominated for a Nobel Prize in 1966. He was unsuccessful, but US Patent No. 3,888,977, filed in June 1975 to Sanders, covers the use of modified snake toxins to treat motoneurone disease, another devastating condition for which a therapeutic breakthrough is long overdue.

Curtain-raiser

This brings us to the threshold of vaccination against polio. To us now, this seems an obvious goal. However, for reasons soon to be explained, the medical establishment in America did its best to discourage this development and took delight in humiliating the perpetrators when the first experimental vaccines fell short of the mark.

Meanwhile, another kind of immunity against polio had been explored – repeatedly, badly and inconclusively. This was 'passive' immunity, conferred by antibodies against the poliovirus which other people had made while under attack by the infection. ('Active' immunity, by which individuals produce their own antibodies, is what vaccinations seeks to achieve.) The rationale of passive immunity was straightforward. Those who survived polio only very rarely caught it again, thanks to the protective antibodies generated by the first attack. Therefore, antibodies from polio survivors should be able to protect healthy people against the infection. Antibodies are 'immunoglobulin' proteins, tailor-made by the immune cells to recognise and destroy the infectious agent which provoked their formation. Immunoglobulins are concentrated in the 'gamma globulin' fraction of serum, the liquid left after blood has clotted. Serum collected from convalescent cases, recovering from polio, was assumed to have the greatest protective effect.

As so often in medicine, the first report of this new treatment for infantile paralysis – from A. Netter in Paris in 1916 – spoke of a miracle. Thirty patients had been rescued from certain paralysis and death by serum injected into the spinal fluid through the lumbar puncture needle that had been inserted to make the diagnosis. However, less excitable experts looked at the sketchy clinical descriptions and decided that these cases might have got better anyway, without the magical serum.

John Paul made it clear in his *History of Poliomyelitis* that the whole business should have been dropped then and there.[87] But interest had been excited, especially as another immune serum – developed by Dr Simon Flexner at the Rockefeller Institute in New York – really did work miracles in meningitis, cutting the death-rate to just one-third of that in untreated patients. Over the next 35 years, immune serum was tried in various doses and injected under the skin, into muscles, into the spinal fluid or even directly into the spinal cord.

The story of immune serum can be summarised as (in chronological order): Yes! – No – Maybe – Yes! – Probably not. Some notably bad trials were performed, for example during the New York epidemic of 1931.[87] Attempts were made to randomise polio victims to be given the serum or not – which was essential to tease out any genuine effect. Randomisation looked fine on paper, but was often impossible in the front line. Doctors found it difficult to explain to parents why their child, with polio just diagnosed thanks to the lumbar puncture needle conveniently stuck in the child's back, was not going to be given the life-saving serum. So few cases were untreated that the results were uninterpretable.

Immune globulin was finally put to the test in 1951, in a massive, well-designed and impeccably conducted trial run by William Hammond of the University of Pittsburgh. From trials on 54,000 children during outbreaks of polio in Utah, Texas and Iowa, Hammond showed conclusively that gamma globulin could protect individuals against polio if it was given before the first encounter with the poliovirus, but that protection lasted only two to five weeks.[88]

A couple of years later, the role of immune serum was assessed by a group of experts whose collective title – the National Advisory Committee for the Evaluation of Gamma Globulin in the Prophylaxis of Poliomyelitis – made it clear what they had been hired to do. Their verdict: 'a preventative effect has not been demonstrated ... Nevertheless, the Committee cannot say that the use of gamma globulin by mass inoculation produced no effect'.[89]

Basil O'Connor would have known how to deal with a committee like that; Tom Rivers said that if the report managed to prove anything, he would eat it for breakfast.[90] Hammond protested, diplomatically but firmly, arguing that the evaluation, not the serum, was at fault. But this marked the end of the trail for immune serum.

The abandoning of another bright hope deepened the sense of despondency and frustration at the failure to get to grips with polio. This was neatly captured by an English medical student, M.H.F. Johnson of Hatch End, Middlesex. Johnson must have been delighted to have his curriculum vitae enhanced when the *British Medical Journal* published his letter on 1 October 1949:

> Sir – As a mere medical student I hesitate to write this letter and hope that in doing so I am not simply revealing a lack of fundamental knowledge.
>
> I am at a loss to understand why the lead given by Jenner and Pasteur for vaccination has not been followed in the case of other virus diseases, notably poliomyelitis.[91]

It was indeed time to move on.

8

Dead or Alive

The first half of the twentieth century was a period of profound mood swings for those hoping that polio would be conquered. Polio had become one of the hottest topics in medical research, but its treatment was lagging far behind the cutting edge of scientific discovery. Numerous 'treatments' – convalescent sera, electrical stimulation, Retan's brain wash-out therapy – had appeared, but all had gone through the depressing cycle of hope, hype, reality check and oblivion. There was no wonder drug capable of dragging those in the grip of an acute attack back from the jaws of death or paralysis.

Meanwhile, polio seemed to be tightening its grip. Anxiety ran especially high in America, where one epidemic after another scythed across the nation and the brutality of the disease was skilfully played up by the National Foundation for Infantile Paralysis (NFIP) in its eagerness to liberate dimes from pockets. The failure to find a cure must have seemed all the more cruel and frustrating, because this was the dawn of the era of antibiotics. A new front was opening up in the war against bacterial infections, including some of the greatest killers of mankind.

To the rescue eventually came vaccination, the neatest solution to the problem of infectious diseases, and ultimate proof of the principle that prevention is better than cure. The advent of effective polio vaccines marked the turning point in our relationship with the disease. The names that intuitively come to mind in this context are Jonas Salk and Albert Sabin, whose photographs grace the pages of every book about the iconic figures of twentieth-century medicine. Salk and Sabin are rightly celebrated for having created the two vaccines that, in combination, have rescued millions from death or paralysis and have pushed polio itself to the brink of extinction.

In fact, the story of polio vaccines began almost 40 years before the era of Salk and Sabin, in the first decade of the twentieth century. It was a lively saga, in which scientific endeavour and the thrill of discovery were often eclipsed by the baser aspects of human nature. This was not a dignified unfolding of the scientific process, but an unseemly race which often descended into backstabbing and bitchiness.

Sir Isaac Newton claimed to have seen further by standing on the shoulders of giants. The pioneers of polio vaccines did that too, but also had no qualms about trampling their competitors underfoot in their haste to get ahead. And the losers in this race included honesty, ethical principles and the safety of the very people whom the vaccines were supposed to protect.

In the footsteps of giants

The idea of developing a vaccine against polio was not particularly inspirational, even in the early 1900s. It was already clear that polio, like other infections, conferred immunity against itself in those who survived: the paradox of an attacker turning into a guardian angel which would fight off its own kind in the future. Second attacks of paralytic polio were vanishingly rare; even in 1935, a painstaking trawl through the literature revealed only a dozen cases.[1] These were probably caused by meeting a poliovirus of a different type, or perhaps another enterovirus altogether that caused a polio-like syndrome.

Various researchers tried to exploit 'passive' immunity, giving antibodies already made by someone else who had survived an attack of polio. This was the rationale that underpinned convalescent serum therapy. As already described, these prefabricated antibodies provided only limited, short-term protection, because, like all proteins in the bloodstream, they are eventually broken down.

A much greater prize – and a far greater challenge – was to induce 'active' immunity. Healthy people are given a virus or bacterium that has been subtly altered from the 'wild type' original, so that it is incapable of causing a dangerous infection yet is able to stimulate an immune response that will also kill the wild type. The key lies in the patchwork of proteins that cover the surface of the infectious agent. In a successful vaccine, the proteins that enable the wild-type virus or bacterium to invade the body are disabled. At the same time, enough of the original 'antigens' (the protein components that trigger the host's immune reaction) are preserved so that the antibodies produced against the vaccine will also target ('cross-react' with) the wild-type microorganism. These antibodies will kill the microorganism before it can invade the body, should it ever try to attack.

By the turn of the twentieth century, several vaccines had been developed against bacteria, namely typhoid and cholera (both in 1896) and plague (1897). There were also two time-honoured vaccines against viral infections, namely Edward Jenner's cowpox vaccine against smallpox (1798) and Louis Pasteur's 'attenuated' (weakened) preparation of the rabies virus (1885).[2]

The archetype vaccine was Jenner's ingenious exploitation of cowpox. The word 'vaccine' comes from Jenner's contrived Latin name for cowpox,

Variolae vaccinae ('smallpox of cattle'), and was later applied at Pasteur's suggestion to all immunisations that protect against infections.

Jenner owed his success to the accidental generosity of Nature. The cowpox virus (vaccinia) and the smallpox virus (variola) are so closely related that the immune response raised against vaccinia also picks off variola. This translated into a generous deal for any humans who caught cowpox off the blistered teats of infected cows. The infection was unpleasant – a nasty weeping sore where the virus had crept into the skin – but was not fatal and cleared up within a couple of weeks (Figure 8.1). The trade-off was lasting immunity against smallpox, a mutilating brute of a disease which, before vaccination, killed one person in twelve worldwide and left millions hideously scarred or blinded.

Jenner's stroke of genius was inspired by a chance meeting in his mid-teens, while apprenticed to an apothecary in rural Gloucestershire.[3] He was told by a milkmaid that she could never catch smallpox because she had previously suffered from cowpox. This was common folklore in the milking parlour, but was news to Jenner, and indeed unknown in the medical literature. As well as being observant and clever, Jenner was

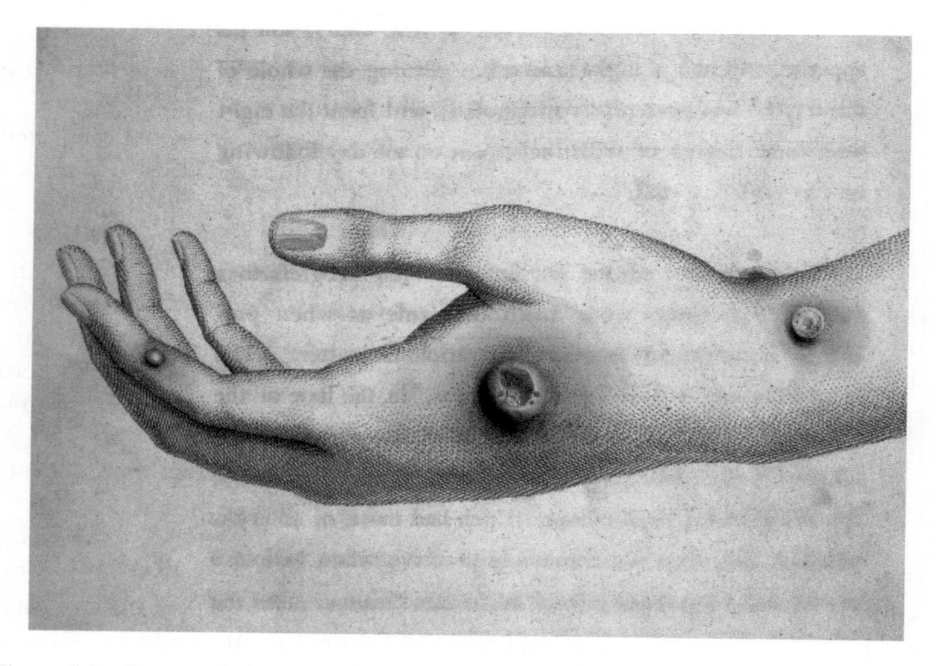

Figure 8.1 Cowpox lesions on the hand of Sarah Nelmes, who provided the blister fluid for Jenner's first vaccination on 14 May 1796. From Jenner's *Inquiry*, the landmark monograph which reported the effects of vaccination against smallpox. Reproduced by kind permission of Dr Jenner's House, Berkeley, Gloucestershire

disorganised and easily distracted, and it took him another 30 years to crystallise his hypothesis and do the experiments that went on to change the course of history. His idea (which also occurred to others) was to inoculate healthy subjects with cowpox, hoping that they too would become immune to smallpox.

The pilot experiment famously involved Blossom (a roan and white cow infected with cowpox), Sarah Nelmes (Blossom's milkmaid, who had a cowpox blister on her left hand), and James Phipps, the eight-year-old son of the Jenners' gardener, who had never had smallpox. On 17 May 1796, Jenner scratched cowpox pus from Sarah Nelmes' blister into James Phipps' arm. There were no ill effects. Jenner then had to confirm that the cowpox inoculation had protected his human guinea pig against smallpox. So, a few weeks later, he deliberately infected the lad with smallpox. To his delight, this failed to take hold.

Jenner repeated the experiment many times in children and adults who had never had smallpox, with the same results. He wrote up his findings in a 60-page paper which he sent to the Royal Society, of which he was already a Fellow for having discovered how the cuckoo chick clears other residents out of its adopted nest. The Royal Society rejected the paper on the grounds that his experiments were unconvincing and could damage his scientific reputation.

Suspecting that the Society's referee intended to steal his secret (which was possibly true), Jenner abandoned the niceties of peer review and paid to have his paper published by a printer in Soho. The result, *An Inquiry into the Causes and Effects of the Variolae Vaccinae*,[4] was an instant sensation. Within a few years, vaccination had entered mainstream medical practice around the world, laying the foundations of immunology as well as vaccines against many other infections (Figure 8.2). Jenner's vaccine not only protected the inoculated individual, but eventually went on to fulfil his prediction that it would rid the world of smallpox. Not a bad result for what was arguably a piece of vanity publishing.

The observant reader may be worried to learn that Jenner, saviour of mankind, had deliberately infected children with the deadly smallpox. In fact, this drew no comment whatsoever at the time because therapeutic inoculation with smallpox was a well-established medical procedure. This was called 'variolation', from *variola*, meaning 'spotted', the Latin name for smallpox.

Variolation apparently sprang up during the seventeenth century in places as widely separated as Africa, Turkey and West Wales – and even earlier in ancient China. Smallpox pus, ideally collected from a mild case, was scratched into the skin of children at risk of smallpox, with the aim of producing a survivable attack that would protect against the natural infection. Astonishingly, variolation worked, with most subjects

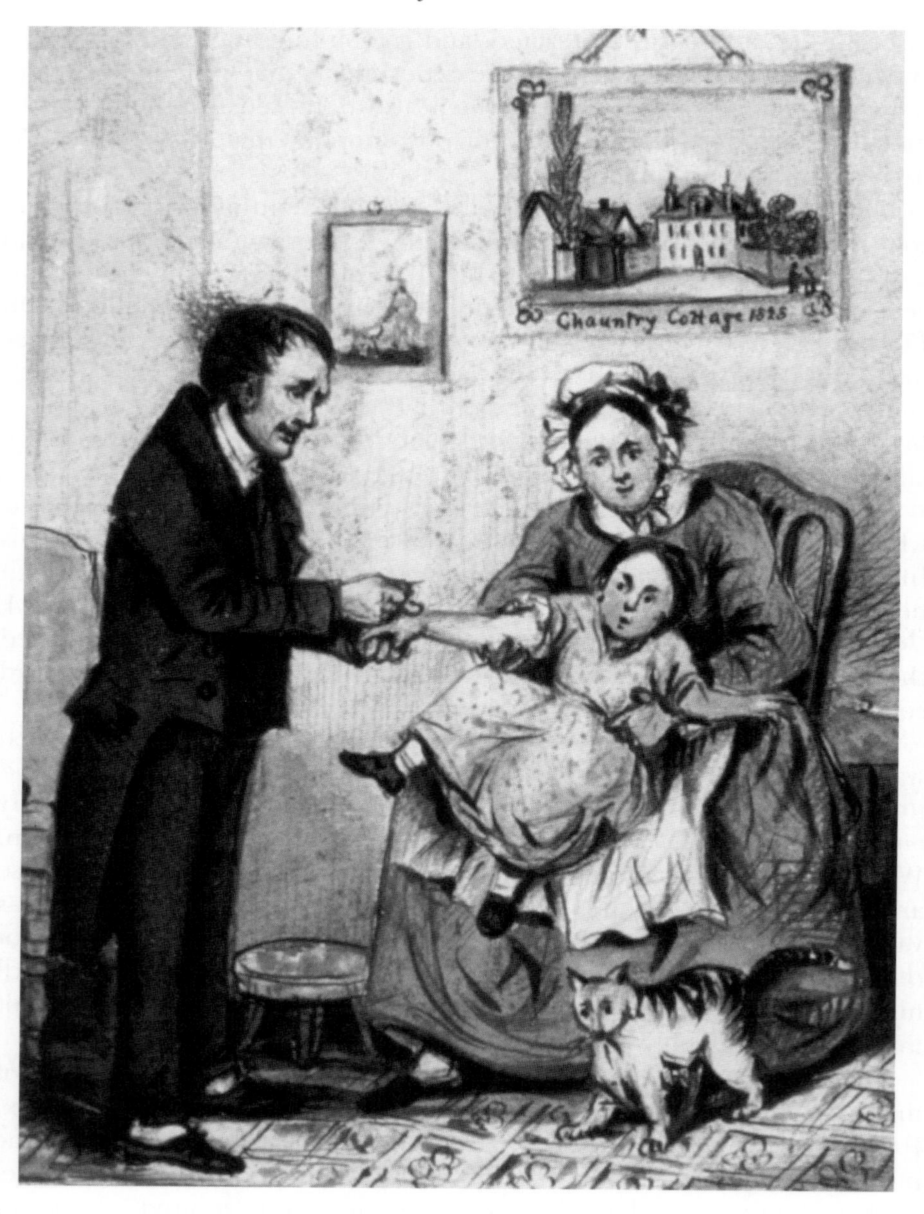

Figure 8.2 Edward Jenner vaccinating a child in his home, The Chantry, in Berkeley. Watercolour by his nephew, Stephen Jenner. Reproduced by kind permission of Dr Jenner's House, Berkeley

enjoying lifelong immunity to smallpox. In careful hands, this was also one of the safest procedures in eighteenth-century medicine: the risk of dying was only 2 per cent, as compared with 30 per cent or more with naturally acquired smallpox.[5]

Unnatural solutions

Unfortunately, Nature has failed to provide other benign doppelgängers which, like cowpox, induce immunity that can also stop a killer infection in its tracks. Human ingenuity has had to fill the gap, by altering virulent organisms in the laboratory so that they become safe to give to people, yet retain enough of their antigenic identity to raise antibodies that will protect against the original agents.

This is a difficult tightrope to walk. On one side is a changeling that no longer stimulates a useful immune response; on the other, something so close to the original that it still causes a dangerous infection.

Two different strategies emerged for modifying infectious agents so that they could be used safely in vaccines: killing and 'attenuation'.

Killing (inactivation) is the safer option, producing organisms that can no longer replicate or infect the host. Many bacteria are easily killed, for example by cooking at 56° C for an hour, or exposure to toxic chemicals like phenol (the active ingredient of Joseph Lister's antiseptic spray) or glycerine. The typhoid, cholera and plague vaccines were all based on dead bacteria, killed by heat and/or phenol. Inactivation wrecks the proteins on which the microorganism depends to replicate and invade, yielding an inert husk that has been robbed of the capacity to do harm. However, enough recognisable protein fragments remain in the debris to stimulate the formation of antibodies that will also attack the original, virulent organism. Even though viruses are not strictly alive to begin with, some are notoriously hard to 'kill' or inactivate, and require prolonged marination in toxins such as formalin.

'Attenuation' was a term invented by Pasteur while experimenting with his first vaccines (against chicken cholera and anthrax in farm animals) during the early 1880s.[2] He forced these bacteria to survive in an alien environment, far removed from the comfort of their host's tissues, and made even more hostile by high temperature, drying and poisons such as chromium salts and ox bile.

The aim was not to kill them, but to pick out potentially useful strains from among the hundreds of thousands of mutations that occur spontaneously when microorganisms replicate. It was hoped that these mutants, selected to survive in such unusual conditions, would be unable to replicate in their natural environment and would therefore have lost their virulence. At the same time, they would have to keep enough of their antigenicity so that the immune response against them would also recognise the virulent ancestor.

Attenuation was a process of trial, error and disappointment that often tested the endurance of the experimenter as well as the infectious agent. Pasteur's protocol for weakening the rabies virus was relatively simple: spinal cord from rabbits infected with rabies was cut into strips and hung inside special desiccation bottles to dry out for several days.[2] The recipe could almost have been inspired by the air-cured charcuterie of Pasteur's native Jura.

By contrast, the production of the Bacille Calmette-Guérin (BCG) to immunise against tuberculosis was much more complicated. The bovine tuberculosis bacterium, isolated from the milk of an infected heifer, was pushed through 230 changes of a medium featuring potato, glycerine and ox bile. The vaccine that emerged 13 years later has remained in use since 1927.[6]

Protection racket

The race to develop a polio vaccine eventually polarised into two competing strategies, with Salk pursuing inactivation and Sabin leading the charge for a live, attenuated vaccine.

However, the earliest attempts to produce active immunity – in monkeys – used unmodified, wild-type poliovirus in the hope that it would trigger a protective immune response before it could invade the central nervous system. The first experiments were conducted in New York, by the ever-energetic Simon Flexner and his colleague, Paul Lewis. This was in 1910, just a year after Flexner set off on the trail of Karl Landsteiner's filterable virus and before Hideyo Noguchi and the globoid bodies entered his life.

Flexner and Lewis simply trusted in nature. They injected diluted filtrates of polio-infected monkey spinal cord under the skin or into the muscles of healthy monkeys.[7] The unusual route of entry was supposed to nobble the virus, which (according to Flexner) was normally programmed to invade the brain via the nose and olfactory nerves. By the time the poliovirus had broken out of its injection site, it would hopefully have triggered an immune response which would mop it up before it could home in on the central nervous system.

This was how variolation worked, but the strategy failed in polio. Flexner and other teams in America and Europe ended up paralysing large numbers of monkeys, either because the 'vaccine' had behaved just like wild-type poliovirus, or because 'vaccination' had failed to protect the animals when poliomyelitic spinal cord was injected into their brains.

Flexner was discouraged, but others went on to try various manoeuvres that had successfully killed or attenuated various bacteria – heat, or mixing with aluminium hydroxide, phenol or formalin.[8] The poliovirus proved difficult to deal with. It was hardy and survived prolonged contact with chlorine and glycerol, which were both good at killing bacteria. Indeed, Flexner had shown that the poliovirus could still paralyse monkeys after being pickled in glycerol for over six years.[9]

Unfortunately, the antigenicity of the poliovirus seemed to be poised on the same knife edge as its viability. When inactivated by heat, phenol or formalin, it also lost its ability to stimulate the production of protective antibodies. Other ingenious approaches were tried, such as giving wild-type poliovirus to monkeys together with immune serum, relying on passive immunity to keep the virus under control while the host's own immune system geared up to produce home-grown antibodies against it.[10] This strategy had been used successfully with the lethal toxin secreted by diphtheria bacteria, but like all the other efforts to immunise using wild-type polio, the results did not inspire confidence enough to risk experiments in humans.

By 1930, the prospect of developing a usable polio vaccine had defeated some of the brightest minds in polio research. There was another, even more powerful deterrent: Flexner, who was convinced that this was a therapeutic avenue which led nowhere. Given his personality, he probably believed that if he could not make it work, then nobody could. Flexner was so put off by the idea that he went further, pronouncing that any attempt to vaccinate humans against polio was not only doomed to failure but was also irresponsible and dangerous. In 1910, he had declared: 'At present, the experimental basis is entirely inadequate to justify the attempt to induce active immunity as a protective measure in human beings'.[7] And 20 years later, he was even more set in his view.

For the American polio research community, Flexner reigned supreme, even when the scientific foundations of his views appeared shaky. As demonstrated by the globoid bodies and the olfactory route of polio infection, his diktats tended to go unchallenged by those who wanted to continue to do polio research in the United States. And so, as the 1930s began, any notion of a polio vaccine was officially dead in the water and untouchable – which made it all the more tempting for outsiders and risk-takers to try to make their mark.

Where angels fear to tread

The long hot summer of 1935 laid down an important milestone in the history of polio, although this was heavily defaced at the time and was completely overshadowed by later events.

This was a busy time for medicine. The 17 August issue of the *Journal of the American Medical Association* (*JAMA*) provides some fascinating snapshots of the moment. Light relief came from a report that deep-freezing could preserve mammalian life, the proof being a newly thawed monkey presented before an amazed audience in San Diego.[11] The editors of *JAMA* had smelled a rat, noting that the inventor had previously called himself Prince Raphael Napoleonovich Lubomirski and was closely associated with the Cunningham Tank Therapy scam that promised to cure everything with hyperbaric oxygen. Those present in San Diego were less cynical, and 300 of them signed up on the spot to follow the monkey into the freezer. As the editors remarked, it could only happen in California. Regrettably, there was no follow-up.

Elsewhere, polio was – as always – the preoccupation of the summer months. A mid-weight epidemic was sweeping down the East Coast and closing in on Virginia, where the International Jamboree was due to bring together tens of thousands of Boy Scouts from around the world. *JAMA* printed the announcement from President Roosevelt himself that the decision had been taken, with great regret, to cancel the Jamboree.[12]

Within a couple of months, *JAMA* published a pair of papers that at first sight heralded a thrilling new era in the battle against polio. In a remarkable example of parallel evolution, two different polio vaccines had come from nowhere, raced through animal testing and had made the bold leap into humans more or less simultaneously. The results of clinical trials in tens of thousands of children had recently been announced to great public excitement, just two days apart.

The two vaccines had been developed by John Kolmer of Philadelphia and Maurice Brodie and William H. Park of New York. Both vaccines were steps in the evolutionary process that led 20 years later to the Salk and Sabin vaccines which are still used today. Salk's vaccine, like Brodie's, used inactivated poliovirus, while Kolmer's and Sabin's were both based on attenuated viruses.

The vaccines may have looked promising, but both were already doomed when the papers were published. By Christmas, both vaccines had been killed off with much publicity and brutality. The so-called 'Brodie–Kolmer fiasco' went on to cast a long shadow, stalling research into polio vaccines for a dozen crucial years while the United States was swept by ever-worsening polio epidemics.

And the vaccines and their inventors rehearsed disasters that later tripped up both Salk and Sabin.

Neck and neck

Maurice Brodie was a clever, pleasant Canadian whose unassuming personality did not equip him well for the rough-and-tumble of American polio politics.

Born in Liverpool in 1903, he was brought up in Ottawa and went to Canada's premier medical school, McGill in Montreal. After carrying off the Gold Medal for the top graduate of 1927, he trained with Wilder Penfield, the great neurosurgeon who worked out how the image of the human body is 'drawn' onto the surface of the brain.[13] Brodie stayed on at McGill and moved into polio research. He began with the perennial puzzle of how the virus is transmitted, and his initial experiments in monkeys (using intranasal administration) agreed with the pervasive Flexnerian view that the nose was the portal of entry.

But Brodie was ahead of his time, at least for North America. He suspected that the human infection might pass through the bloodstream, where it could be attacked by circulating antibodies; viruses safely dug into the brain were assumed to be beyond the reach of the immune system.[14] This idea opened the door to the possibility of vaccination, and Brodie set out to translate theory into reality.

He began experimenting with 'killed' poliovirus, injected into monkeys, looking to see if protective antibodies were raised. Like Flexner, Brodie chose formalin as his killing agent. This powerful denaturing compound destroys the chemical structure of proteins in everything from human tissues right down to the capsule of viruses. It has long been used to embalm cadavers, and its pungent smell propels many doctors straight back to their first days as medical students in the dissection room. Formalin also kills viruses, even the toughest. Up to the 1960s, the corpses of smallpox victims in Britain were buried in putty-sealed coffins filled to the brim with sawdust saturated in formalin.[15]

According to Flexner, formalin held no promise in polio vaccines. Formalin-treated viruses either continued to cling to their version of life and caused paralysis, or were killed so dead that they could not induce antibodies which cross-reacted with the wild-type virus.[7] But Brodie persevered and after several months came up with a recipe that seemed to work: an emulsion of polio-infected monkey spinal cord, steeped in 0.1 per cent formalin for two days. Hefty doses (5 ml) injected under the skin could protect normal monkeys from being paralysed when large amounts of infected cord were administered directly into their brain.[16] Brodie published his findings in a series of solid papers in top-flight journals, including *Science*.[17] In 1933, his growing reputation brought the 30-year-old to the attention of a powerful old man across the border.

This was William H. Park, Director of the Pathology, Bacteriology and Disinfection Laboratories of the New York City Board of Health (Figure 8.3). Park was a tough product of that no-nonsense city; according to John Paul, the gentlemanly chronicler of polio, Park was 'never one to let the grass grow under his feet'.[18] Park was famous for his massive textbook, *Pathogenic bacteria*, and for solving the problem of 'Typhoid Mary', the woman whose unfortunate combination of unusual medical condition (chronic carrier of typhoid) and job (cook) had left a trail of devastation across New York.[19] Park was also an authority on diphtheria;

Figure 8.3 William H. Park (1863–1939), Director of the Pathology, Bacteriology and Disinfection Laboratories in New York from 1893–1936. Park was 70 years old when he brought in Maurice Brodie from McGill University in Montreal to develop his inactivated polio vaccine for use in humans

it was he who had devised the stealth vaccine that combined diphtheria toxin with its antibody. The previous year, Park had been given the Public Welfare Medal, one of America's most prestigious career achievement awards, for pioneering the 'application of scientific discovery to the prevention of disease. (The 1939 Medal, celebrating the prevention of crime, was awarded to J. Edgar Hoover, Chief of the Federal Bureau of Investigation [FBI]).

Park been appointed Director at just 30 years of age. That was back in 1893. Now, the 70-year-old Park was desperate to end his long academic career on a high note, and was also keen to score points off an old adversary across town. This was none other than Flexner at the Rockefeller, whose intense dislike for Park was warmly reciprocated.[20]

Park applied to the Birthday Ball Commission for a grant to bring Brodie to New York to refine his vaccine for human use, under Park's direction. In early 1934, Park was blessed with extreme good fortune, when the Commission awarded him a massive $65,000. Over in the Rockefeller, Flexner's anger over the breaking of his embargo on polio vaccines would have intensified when he learned who had handed out the grant, the biggest awarded by the Commission that year. This was Paul de Kruif, another of Flexner's vendetta victims who had moved to the Commission after Flexner sacked him from the Rockefeller in 1932.[21]

Brodie, loosely supervised by Park, quickly got to work – and with accelerating haste when they realised that they were no longer alone in their unfashionable pursuit. John Kolmer, a fearsomely hard-working microbiologist from Philadelphia, had also started to develop his own polio vaccine. Brodie had always been a methodical worker, but his vaccine project now degenerated into a race against the clock and Kolmer. Brodie's latest formulation, published in *Science*,[17] had only been tested in 20 monkeys, but he decided to short-circuit further animal studies and went straight into man. Brodie and Park followed the example of many pioneers in therapeutics, and in turn set the trend for Salk and Sabin:

> Before giving it to children, it was deemed advisable to try it upon ourselves, not that we had any misgivings about the possibility of infection [*i.e. of causing paralytic polio*], but rather to determine whether the vaccine produced any disagreeable local or general reactions.[22]

They judged the results in themselves and four colleagues as 'perfect safety'. The only problem was minor soreness at the injection site, probably due to the formalin. Reassured, they quickly went on to vaccinate 12 children aged 1–6 years, with one or two 5-millilitre doses. Apart from local discomfort, there were no problems, and the vaccinated children developed significant levels of protective antibodies, as judged by the ability of their serum to prevent paralysis when injected into monkeys' brains together with extracts of poliomyelitic spinal cord.[22]

Brodie stayed ahead of Kolmer, and on 21 November 1934, he presented preliminary results from the first ever administration of a polio vaccine to humans, at the New York Academy of Science.[23] The reception was good, with nothing to hold up the next stage of large-scale clinical trials in healthy children. Brodie and Park must have worked very hard. Within months, they published a full paper in *JAMA*.[24] The conclusions were upbeat – the vaccine generated impressive levels of protective antibodies – and it had an eye-catching bottom line that was undoubtedly noted by those hoping for comeuppance:

> Formalin-inactivated virus is probably a perfectly safe vaccine inasmuch as no harmful effects have developed after more than 3,000 inoculations.

Kolmer had also pulled out all the stops, but Brodie and Park scraped in just ahead of him. In their haste, they had broken one of the golden rules of scientific research by calling a press conference to announce their findings in advance of publication. In July 1935, the *Literary Digest* featured the Brodie-Park vaccine as one of the 'weapons for the battle with this year's poliomyelitis epidemics' newly delivered by scientists. This claim was backed up with a photograph up with a photograph of the two pioneers and their reassurance that 'The experiment is perfectly safe' (Figure 8.4).[25] And in the absence of evidence to the contrary, it seemed believable.

The Brodie-Park vaccine quickly seized the public imagination around the world. This was what everyone had been waiting for; Flexner and the other prophets of doom were temporarily forgotten.

Figure 8.4 William Park and Maurice Brodie (1903–39), co-developers of an ill-fated polio vaccine based on formalin-inactivated poliovirus – a precursor of Jonas Salk's vaccine, some 20 years later. Image from *Literary Digest*, 20 July 1935, with grateful acknowledgements to UNZ.org

Figure 8.5 John A. Kolmer (1886–1962), microbiologist and Director of the Research Institute of Cutaneous Medicine in Philadelphia during the 1930s, when he developed an attenuated polio vaccine. This was abandoned in late 1935, at the same time as the Brodie-Park vaccine. Image courtesy of the History of Medicine, National Library of Medicine, National Institutes of Health

Photo finish

Meanwhile, 200 miles to the south in Philadelphia, John Kolmer (Figure 8.5) had almost caught up.

Midway in age between Brodie and Park, Kolmer was a highly published microbiologist with even more papers and textbooks under his belt than Park. He had stayed in Philadelphia since graduating from the University of Pennsylvania Medical School in 1908, and was now Director of the privately funded Research Institute of Cutaneous Medicine. The Institute's brief was wider than its name suggested, and had kept itself afloat financially during the 1920s by manufacturing the early antimicrobial compound, salvarsan, and selling it to the US government.[26]

Kolmer had made national headlines in 1924 when he was summoned to Washington, DC, to advise on Calvin Jnr, the 16-year-old son of President Calvin Coolidge.[27] The lad had blistered both feet badly by playing tennis in tight shoes without socks, and the wounds became infected with virulent staphylococci. Before long, he was fighting for his life against overwhelming septicaemia. There was little that Kolmer could do. Salvarsan was inactive against staphylococci, and effective antibiotics were still many years in the future. After several harrowing days in Kolmer's skilled but powerless hands, Calvin Coolidge Jnr died in agony, distressing the nation and almost unhingeing its leader.

Kolmer's association with polio dated back to 1917, when he had been one of the many in fruitless pursuit of 'poliomyelitic' streptococci. When polio hit Philadelphia hard in 1932, his interest was rekindled. Like Brodie, Kolmer became a believer in a polio vaccine, but one that was fundamentally different from Brodie's. His would use a live poliovirus, capable of limited replication in the human body, but strategically weakened so that it could not invade the central nervous system and cause paralysis. Kolmer borrowed Pasteur's adjective *attenué* to describe this state of carefully crafted impotence, and the name 'attenuated' has stuck to all modified 'live' virus vaccines ever since.

Kolmer's recipe for attenuating the poliovirus was perfected after many months of experimentation with his devoted assistant, Miss Anna M. Rule, who sometimes joined him as co-author on his papers. Kolmer used the 'MV' strain of poliovirus, isolated by Flexner's team and passed from monkey to monkey at the Rockefeller for over a decade; now that he was vehemently opposed to vaccination, Flexner might not have approved of the use to which his favourite virus was being put. MV-infected monkey spinal cord was steeped in glycerine for a week and then puréed with ricinoleate, a soapy derivative of ricin, the deadly toxin of the castor oil plant. The result was warmed to blood heat, then cooled to 12° C for two weeks, and given by subcutaneous injection.

Kolmer argued that his process introduced numerous layers of protection. MV, already weakened by repeated passages through monkeys' brains, would be crippled by the poisonous ricinoleate and be forced to survive in the hostile terrain under the skin. Kolmer's critics later ridiculed his 'kitchen chemistry', but his protocol was no odder than some of those for producing successful attenuated vaccines, such as the BCG, mentioned above.

Kolmer's testing schedule paralleled Brodie's and was not far behind. Initial trials in monkeys (twice as many as Brodie had used, but still only 42) led to the testing of doses on himself and Miss Rule ('without ill-effects and with good antibody responses'). His sons followed, and then 23 children of friends and colleagues.[28] Encouraged, Kolmer struck out into even more grandiose clinical trials than Brodie's. He sent 30,000 doses of his vaccine to over 700 physicians in 40 states across the United States and Canada, asking them to vaccinate children during the summer of 1935 and report back any problems. Just over 10,000 children were inoculated; meanwhile in Philadelphia, Kolmer and Miss Rule were frantically analysing the data,[29] now painfully aware that Brodie and Park were preparing to reveal their findings.

Just as Brodie and Park had done, Kolmer went public before his paper came out. Not to be outdone by Park a couple of days earlier, Kolmer's message was also one of undiluted positivity. This was again heavily embroidered by the press to make a good story great, and then transmitted around the world.

This was thrilling stuff: the excitement and immediacy of a major scientific breakthrough, long overdue and one which the experts said could never happen. Better still, this was a race between two groups from the East Coast of America, the home of the best polio research on the planet, and it had turned into a cliffhanger, and ended in a photo finish. One or both of the vaccines had to be the answer to polio, and the grateful public lapped it all up.

Brodie, Park and Kolmer had a more demanding hurdle to clear, that of peer review. Brodie had already reported his initial findings in New York the previous November,[23] and both parties were due to present their results to the American Public Health Association in Milwaukee in early October.[30] Even better, both had keynote addresses at a major meeting of the American Society of Public Health in mid-November.

In the early autumn of 1935, it looked as though Brodie and Kolmer had successfully broken Flexner's taboo and that their vaccines were destined to lift the curse of polio. But they had whipped up anger rather than admiration among some of their peers, and specifically the ones who pulled the strings in America. Trouble was brewing, and retribution was imminent; both vaccines were about to be blown out of the water and their inventors hung out to dry.

St. Louis Blues

The showdown took place on 19 November 1935 at the Annual Meeting of the Southern Branch of the American Society of Public Health in St Louis. Brodie and Kolmer apparently behaved as though they had no inkling that they were being handed the rope with which to hang themselves. Brodie presented results which showed his vaccine to be effective at raising antibodies, and entirely safe; so did Kolmer.

Kolmer was particularly expansive, cleverly throwing in quotations by Park (taken out of context) which seemed to support the supremacy of attenuated vaccines. He also had a sly dig at the deadness of the Brodie-Park vaccine:

> It is not unlikely that the completely inactivated vaccine cannot provide effective and durable immunity in this disease.[29]

Naturally, his own 'live' vaccine was immune to that criticism, and his extensive clinical trials in the United States and Canada allowed him to say, with some smugness:

> It is for these reasons that I have regarded the vaccine as safe for the immunization of human beings when prepared exactly as described.[29]

In fact, as Kolmer knew full well, this was not entirely true. His 'safe' vaccine had been clearly associated with complications, extremely rare but catastrophic. Of the 10,725 children vaccinated that summer, ten had developed classic paralytic polio a few days later. Five of them had died. Particularly striking was a cluster of three fatal cases in Plainfield, New Jersey. Among them was the two-year-old daughter of a doctor who, in deepening desperation, had injected her with large volumes of serum and even whole blood from convalescent polio victims.[29]

Kolmer did not believe that his vaccine could be to blame and speculated that these unfortunate children had already been incubating polio. On the other hand, some of the paralytic cases had partly regained their strength – evidence, he suggested, that his vaccine had rescued them from certain death. The strength of Kolmer's confidence in his vaccine may explain why he omitted any mention of these complications when he first submitted his definitive manuscript to the *Journal of the American Medical Association*. The ten cases appear in the paper as it was finally published, but only in an abstract which Kolmer added in the proofs.[29] Now, on his day of reckoning in St. Louis, Kolmer had to face the inquisition with nowhere to hide.

The level of risk with Kolmer's vaccine was very low – just one in a thousand. This was trivial when compared with other 'treatments' and 'cures' for polio that were in vogue at the time, but had somehow evaded clinical trials or proper scrutiny by the scientific community. However, it was enough to bring down the wrath of the medical establishment on Kolmer, and on Brodie.

A confrontation with the inner circle of polio experts had become inevitable. Brodie and Kolmer were interlopers who had dared to break Flexner's moratorium on developing polio vaccines. They had made things worse by demonstrating that the experts had been wrong, and by stirring up huge public support which made the experts look all the more foolish. Luckily for the establishment, Brodie and Kolmer had both rushed into human trials when further animal safety testing was still needed. The final nail in their shared coffin was the evidence that Kolmer's vaccine was dangerous, and that he had tried to cover this up.

So the establishment sent in two of its hardened hitmen: Tom Rivers, a renowned virologist and plain-talking bully from the Rockefeller, and James P. Leake of the US Public Health Service. Neither man had a track record in diplomacy; Rivers was renowned for his 'pyrotechnics' and, in his words, had been asked by unnamed people in authority to 'be the hatchetman'.[30] Both Rivers and Leake went to St. Louis fully briefed and spoiling for a fight.

It was evidently a lively session. After Brodie and Kolmer had presented their papers, the opposition waded in. Experts had checked out their data, they said, and the results – especially Brodie's – were nowhere as good as had been claimed. Brodie and Kolmer had both been dangerously irresponsible to inflict their vaccines on healthy children before they had been properly tested in animals; a few dozen monkeys was a pitifully weak base from which to make such a huge leap into the unknown. And above all, despite their reassurances, the vaccines were not safe.[31]

Brodie quickly saw that they were out for blood and realised that his would soon be on the carpet. He blurted out that it seemed that his vaccine was useless and that Kolmer's was dangerous. Kolmer, pinned down over the children who were paralysed and killed after receiving his vaccine, tried to fight back:

> I do not wish to be understood as merely seeking courses for excusing the vaccine as a possible cause of these ten cases of poliomyelitis.

– and then repeated that this was a one-in-a-thousand event and anyway the children had probably already caught polio, which even his vaccine had been unable to abort.

At this point, Leake lost his temper. The sanitised version of his interjection, later published in *JAMA*, suggests that this was pointed but measured:

> Paralytic poliomyelitis was not epidemic in any of the localities at this time ... The likelihood of a whole series of cases having occurred through natural causes is extremely small ... I beg you, Dr. Kolmer, to desist from the human use of this vaccine.[32]

What Leake actually said, according to those at the meeting, was that Kolmer was a murderer. At that point, Kolmer finally understood what was happening and tried to duck out as gracefully as he could:

> Gentlemen, this is one time I wish the floor would open up and swallow me.[31]

Much more than Kolmer was swallowed up. St. Louis marked the end of the trail for both vaccines and their inventors, and killed off any prospect of developing a polio vaccine for almost two decades.

The decisive deterrent was probably the brutality of the retribution meted out to the offenders, rather than the exposé of the many flaws in their science and ethical standards. All three retreated smartly from the limelight: Park into the overdue shade of retirement, Brodie to obscurity in the backwaters of Illinois, and Kolmer back to his Institute and his textbooks.[33] Their definitive vaccine papers could have been quietly retracted while still in press, but were published on schedule and are still there for all to see today. There is also a helpful commentary by Leake in *JAMA*, just in case a balanced view is needed.[32]

These papers by Brodie, Park and Kolmer were true landmarks, as they recorded the results of the first trials of human polio vaccines, conducted during the summer of 1935. However, like the heads of traitors stuck up around the walls of a medieval city, they were also a warning to others not to go down the same dangerous path.

Shame and fame

Brodie and Kolmer have been largely expunged from many histories of polio, except for fleeting mentions together with the word 'fiasco'. They were far from perfect, as pioneers usually are. Each was seduced beyond reason by the promise of his invention. But they were not the first to allow the scientist's respect for evidence to be crushed by a misplaced belief in their own infallibility, and they were not the last. A couple of decades later, both Salk and Sabin were protesting (wrongly) that their own polio vaccines could not possibly cause paralysis.

Did the Brodie-Park and Kolmer vaccines actually work? When given to healthy children, they certainly generated antibodies that prevented paralysis in monkeys that had infected spinal cord extract injected directly into their brains. However, we cannot tell whether or not this would have translated into the vaccinated children being protected against catching polio during an outbreak.

That possibility could only be tested in large-scale, placebo-controlled field trials of the type used in the mid-1950s to evaluate the Salk vaccine. The need is dictated by the patchy incidence of polio, even during a major national outbreak, and the variability of its clinical course. The 10,000 children given Kolmer's vaccine might seem a lot, but this is a small fraction of the numbers required for statistically robust proof that the vaccine actually cuts an individual's risk of being paralysed by or dying from polio. In the paper he presented at the fateful meeting in St. Louis,[34] Brodie suggested that 50,000 children would be needed. Even this number, which he apparently plucked out of the air, was not enough. The trials of Salk's vaccine eventually included 1.8 million children, and proof of its protective action hinged on just 90 vaccinated children who, statistically, should have caught polio during the summer of 1955 but did not.[35]

Some practical problems with the Brodie-Park and Kolmer vaccine trials also made it impossible to assess their true value. Kolmer's did not include any unvaccinated controls, viewed as an absolute necessity by most authorities – although not by Salk, who would later fight tooth and nail (and with some success) to have this tedious irrelevance removed from the trials of his vaccine.

Even if the vaccines produced serviceable antibodies, how long would protection have lasted? Monkeys were protected for up to three years, but this cannot be simply extrapolated to humans. It was later claimed that children who had received Kolmer's vaccine had remained free from polio during outbreaks 20 years later,[36] but these were second-hand observations with no hard supporting data. Even if true in those cases, full-scale clinical trials would still have been needed to prove that this was a genuine effect of vaccination rather than the generosity of random chance.

Both vaccines used just one strain of poliovirus. Brodie and Kolmer have been slated for their naiveté in assuming that their vaccines would provide useful coverage, as Macfarlane Burnet and Jean Macnamara had already reported (in 1931) that there were multiple types of poliovirus.[37] However, all that Burnet had shown was that there was more than one type. The full characterisation of poliovirus strains, and the demonstration that all were encompassed by just three antigenically different types, was a monumental undertaking which was not completed until 1954.[38] Brodie and Kolmer can be criticised on many counts, but not for the absence of clairvoyance.

Interestingly, the MV virus used in both vaccines was a mixture of two strains isolated at the Rockefeller in 1910. One was later shown to be a type 1 poliovirus, the type responsible for 80 per cent of outbreaks and 80 per cent of cases of paralysis and death. Any measure which could prevent such a high proportion of death and disability due to an untreatable infection would have been a major breakthrough.

How dangerous were the Brodie-Park and Kolmer vaccines? Both carried risks, which in the case of Kolmer's were clearly exposed in St. Louis together with his attempt to cover up the possibility that it had caused paralysis and death. The hazards of Brodie's vaccine were slower to emerge. John Paul made the enigmatic comment that it had been difficult to work out exactly what had happened during the trials of the Brodie-Park vaccine but that 'something went wrong'.[39]

In his *JAMA* paper of late 1935, Brodie mentions in a brief footnote that one child became paralysed with polio after receiving his vaccine.[24] He did not acknowledge the possibility that this was consequence rather than coincidence. Later, two more cases of paralysis (one fatal) were linked to Brodie's vaccine, presumably because his inactivation protocol had failed to kill all the wild-type virus. This foreshadowed a disastrous outbreak of paralytic polio when Salk's formalin-inactivated vaccine was first pushed into commercial production in 1956. (This so-called 'Cutter Incident' is described in the next chapter.)

It was also rumoured that Brodie's vaccine caused some cases of allergic brain damage. This is a potential hazard with any vaccine prepared in a monkey's brain, which contains antigens closely similar to those in the human brain. Antibodies provoked by monkey brain extracts in the vaccine can cross-react with components of the host's brain, causing a condition that resembles extensive multiple sclerosis and is often fatal. This 'postvaccinal encephalitis' can occur with any brain-derived vaccine. It was a widely feared complication of the early rabies vaccines, beginning with Pasteur's; this threat was lifted in 1946 with a novel rabies vaccine which was cultured in chick embryos, not mammalian brains.[40]

Levels of risk that are viewed as acceptable depend on contemporary expectations of life, death and health, as well as the statistics of individual diseases. The 1:50 chance of dying from smallpox after variolation was a gamble worth taking when smallpox infected one person in three and killed one in four of its victims. Similarly, most people bitten by a rabid dog chose to have Pasteur's vaccine, as there were no recorded survivors once rabies had taken hold.

The 1:1,000 risk of paralytic polio with Kolmer's vaccine was tiny, but made much starker because these were healthy children who might otherwise have been lucky and escaped polio altogether. At the time, though, the management of polio was fraught with risk, thanks to the many useless quack treatments which went in and out of fashion without any scrutiny at all. Some were undoubtedly at least as dangerous as polio itself. The efforts of Brodie and Kolmer need to be considered in this context. Ultimately, they were hauled over the coals for having tested their vaccines to a degree that many other inventors of new therapies managed to avoid.

In the calm before the furore, both Brodie and Parker presented their preliminary findings to the Section of Pediatrics at the annual meeting of the American Medical Association (AMA) in Atlantic City, on 11 May 1935.[41]

Speaking immediately before them in the 9 a.m. session in the Ballroom of the Convention Center was Dr George Retan, recounting 'the development of the therapeutic use of forced perivascular (spinal) drainage' to treat acute polio. As described earlier, Retan promoted his 'brain wash-out' therapy widely, while successfully avoiding any proper attempt to verify its effectiveness and safety – and even after it was proven to be both useless and dangerous. Moreover, Retan slipped away safely without having to face close questioning and criticism by his peers. Kolmer may have had ten crippled and dead children on his conscience; Retan should have had hundreds.

A momentous year

The year 1935 saw some notable entrances and exits. Polio vaccines might have been derailed but news was coming in from the Pasteur Institute in Paris of a new live vaccine against typhus, rolled out with apparent early success in Morocco.

In the hinterland of therapeutics, the AMA granted – just – a licence for a new laxative, while rebuking the manufacturer for having called it 'Dyn-A-Mite'. The AMA also licensed a diphtheria toxoid made by a small family-run company in Berkeley, California, called Cutter Laboratories – a name to watch for in the next chapter.[42] And in May, the AMA's Annual Conference in Atlantic City saw Bovril ('prescribed by European physicians for nearly half a century') being heavily promoted, alongside Hawaiian pineapple juice and fine cigars.[43]

The Conference promised a good time as well as professional refreshment. Attractions included the Boardwalk, the world's biggest pipe organ, the Betty Bacharach Home for Crippled Children, and the cream of American medical practice and science.

There were big audiences for two particularly grand old men, both aged 72, whose combined age was just two years less than the recorded history of polio.[44] Ludvig Hektoen talked about the 'specificness' (sic) of his beloved streptococci, while Simon Flexner reviewed advances in immunity. Flexner had just announced that he would retire in the following spring, but had by no means relaxed his grip on the polio research community.

On 1 November that year, Flexner used the forum of *Science*, one of the world's most influential scientific journals, to make it clear once again that the medical establishment had written off polio vaccines:

> No adequate evidence exists showing that physical or chemical agents can attenuate the poliomyelitis virus, to preserve its immunising power while depriving it of its potential paralysing power.[45]

The events at St. Louis a couple of weeks later graphically confirmed that these were not idle words.

The waters were poisoned for well over a decade. According to John Paul, vaccination against polio was still forbidden territory as late as 1947. When the notion was raised that summer of testing a new experimental polio vaccine in children, 'a veritable shudder went around the room' – and this was in the enlightened cloisters of the School of Hygiene at Johns Hopkins University in Baltimore, one of the world's foremost centres in public health.[46]

The vaccine, an inactivated polio vaccine based on formalin-treated extracts of infected monkey spinal cord, had been developed by Isabel Morgan, a 37-year-old scientist at Johns Hopkins. In 1948, she showed that her vaccine raised high antibody levels and protected against paralytic polio – but only in monkeys.[47] She was not medically qualified, and stopped short of taking her vaccine into humans. Shortly after, she married and gave up polio research altogether, abandoning her experimental vaccine at the stage which Maurice Brodie had reached over 15 years earlier. None of her colleagues at Johns Hopkins – including the brilliant and energetic David Bodian – had the interest (or perhaps the courage) to pick up where she had left off.

At that time, neither Salk nor Sabin knew that he would end up developing vaccines which would essentially rid the world of polio. But by then, Kolmer's notion of an attenuated polio vaccine had been resuscitated by a colourful and iconoclastic Polish refugee, who was working outside academia and had no fear of the curse of Flexner. Indeed, Hilary Koprowski had already given the first test doses of his vaccine to children – and by mouth, not by injection.

John Kolmer managed to rehabilitate himself after his public flaying in St. Louis, but Maurice Brodie was much younger and less resilient. In 1936, he moved to a mundane clinical post at Providence Hospital in Detroit. It was said that 'his grief and disappointment were great but his hope still greater'.[13] Unfortunately, his research never regained momentum and neither did he. By the time Isabel Morgan abandoned her own polio vaccine program in 1949, Brodie had been dead for ten years. His obituary records his sudden death from coronary thrombosis at the unusually early age of 36; John Paul mentions rumours that 'poor Brodie' killed himself.[48]

Naturally, neither Brodie nor Kolmer figure on the Wall of Fame at Warm Springs. But slightly to the right of centre of the line of seventeen bronze busts, suspended midway between Sabin and Salk, is Morgan.

Just a month after it published Flexner's article slating polio vaccines, *Science* carried a disturbing report of an internecine turf war. Male red squirrels had begun savaging their grey counterparts, leaping onto their backs from up to 15 feet away. The greys were not just seen off; examination showed that most had been castrated during the attack.[49]

It is best left to the reader to decide whether there are any parallels with what happened to Brodie and Kolmer.

9

Front Runner

Jonas Salk is one of the instantly recognisable figures of twentieth-century science: bespectacled, dark receding hair, earnest and usually unsmiling, and often posed in a white laboratory coat surrounded by test tubes (Figure 9.1).

Salk's name is forever stamped on the first polio vaccine that worked. This was rushed into clinical service in the spring of 1955 amid the euphoria whipped up by the success of the biggest clinical trial that the world had ever seen. Salk's vaccine swept the Western world, and within a few years had reduced by over 90 per cent the incidence of paralytic polio in the United States and Europe. Salk believed that his vaccine was perfect. It was not, and in the 1960s, it was largely superseded by the 'live' polio vaccine developed by Salk's rival and nemesis, Albert Sabin. But in a nice example of the cyclicity of medicine, Salk is now back in vogue, and his vaccine has turned the tables on Sabin in the United States and many other countries.

Salk could not have won the polio vaccine steeplechase without intelligence, doggedness and tunnel vision that often bordered on the pathological. A thick skin also helped to deflect criticism and the constant sniping, with lethal intent, by Sabin. Salk was a complex and paradoxical character. An obsessive experimenter, but so convinced that his vaccine would work that he fought off attempts to test it in a proper clinical trial. An odd man whom many found intensely allergenic, yet adored by the world at large and eventually by Françoise Gilot, Pablo Picasso's former mistress. As seen by his peers, a self-publicist and at best a methodical technician rather than an original thinker. And a research boss who reduced members of his team to tears because he did not even mention them in their greatest moment of glory.

Salk was born on 28 October 1914, the eldest son of a Russian Jewish family who had fled to America to escape the pogroms. Raised in East Harlem, New York City, Salk was bookish and ambitious from the start. In 1933, he won a scholarship to the Medical School at New York University. There he fell under the spell of the energetic and wise virologist Thomas Francis. Salk had his first taste of research during a summer vacation in Francis' laboratory.[1] Someone he did not encounter there was Maurice

Figure 9.1 Jonas Salk (1914–95). Reproduced by kind permission of the World Health Organisation

Brodie, as yet the untarnished golden protégé of William H. Park, who was pushing ahead with his formalin-treated polio vaccine.

Salk's medical career got off to a busy start. He qualified on 8 June 1939, and the next morning married Donna Lindsay, a student social worker and socialite whose father was prepared to let her to marry a doctor, but not a medical student. Salk's residency at Mount Sinai Hospital in New York was interrupted by what Franklin Delano Roosevelt called Japan's 'Day of Infamy' on 7 December 1941. Following Pearl Harbor, all American doctors of draft age and fitness were given the choice of serving with the US Armed Forces or remaining on home soil to further the war effort in other ways.

Salk applied to do research on influenza vaccines with Francis, who had been headhunted for a chair at the University of Michigan's new School of Public Health at Ann Arbor. Salk made a good choice. Protecting GIs against influenza was a high priority for the US military, who during the First World War had lost more soldiers to the Spanish flu than to enemy action. Salk's time in the Francis laboratory also gave him sound mentorship (which he often ignored), a solid introduction to the strange world of academic medicine, and hands-on experience in developing a successful flu vaccine based on inactivated virus.[2,3] Other important landmarks were the arrival of the Salks' first two sons, Peter and Darrell, both born in the run-down red brick Maternity Hospital near the Medical School. The building was soon to be decommissioned, but it reappears later in the story in a different guise.

By the time the War ended, fault lines had opened up in Salk's relationship with Francis. After six years and a dozen papers, it was time for Salk to move on, and in autumn 1947, he took his family to the dark satanic mills of Pittsburgh. There, the morgue in the basement of the University's Medical School was reincarnated as Salk's Viral Research Laboratory. In the beginning, there was little more than the grand name on the door, but Salk was good at coaxing funding out of benefactors and grant-giving bodies, and rapidly built up his team. Within months, his empire had begun to expand skywards.[4]

Way in

In 1949, Salk's research moved up a gear and acquired a clear focus on polio, although not yet on a vaccine. At this point, there was only one runner in the polio vaccine race, and nobody was paying much attention. Over in New York, Hilary Koprowski was quietly developing a 'live' vaccine to be given orally, but was largely invisible to the academic community because he worked on the dark side, in the pharmaceutical industry.

Salk's first major sortie into polio research was to join the flagship programme of the National Foundation for Infantile Paralysis (NFIP), under the inspired leadership of David Bodian from Johns Hopkins University in Baltimore.[5] This massive multicentre collaboration was sifting methodically through the hundreds of strains of poliovirus that had been isolated around the world, to determine how many types encompassed them. The work was much more than a worthy but dull taxonomic exercise; it was a complete intelligence offensive on the enemy and with a vital spin-off. To give comprehensive protection, a polio vaccine would have to cover all possible types of the virus. So far, there seemed to be just three types (which the project eventually confirmed). As well as being important, the project was grindingly tedious. Salk signed up to bring new money and people into his laboratory, and to make himself visible to the polio research community and the NFIP, which was awash with funding.

When the typing programme began, the methodology was laborious, expensive and terminally bad news for monkeys, which were the only assay system available. Samples of an unknown poliovirus strain (X) were mixed with aliquots of antiserum known to contain antibodies against either type 1, 2 or 3, and then injected into the brains of a series of healthy monkeys. A model result would be that X paralysed monkeys when mixed with type 1 or type 3 antiserum, but not when injected together with type 2 antiserum – indicating that X was a type 2 strain. Unfortunately, results were often uncertain, and experiments had to be repeated. The typing programme kept the NFIP's monkey import and distribution centre at Okatie Farms in South Carolina running at full stretch, and mopped up hundreds of thousands of animals. Salk's group alone went through 17,500 monkeys in their first two years with the project.[6]

This was where Salk's bright idea came in. It was inspired by the 1949 paper by John Enders, Thomas Weller and Frederick Robbins, which at the time was incubating the Nobel Prize for them.[7] They had succeeded in growing the poliovirus in cells cultured in the laboratory, rather than in live animals. Moreover, the viral attack could be followed, and the numbers of live viruses calculated, by counting the holes chewed in the carpet of cultured cells (the 'cytopathic effect' named by Enders). Salk neatly applied Enders' breakthrough to the conundrum of poliovirus typing. Samples of the unknown strain X, together with aliquots of antiserum against either type 1, 2 or 3, were now added to cell cultures. The result was interpreted as before, except that cells rather than monkeys were killed when the virus was not neutralised by the appropriate antibody. This was much better news for monkeys. One monkey's kidneys could seed 2,000 culture tubes, enough to spare 100 of its peers being tested to destruction with injections of poliovirus into their brains.[8]

Salk's right-hand man was Julius Youngner, a highly experienced virologist whose scientific pedigree included research into uranium toxicity during the Manhattan Project to develop America's atomic bomb. Youngner added an ingenious twist to the assay cocktail – a pH indicator dye, similar to litmus, which changed colour as the infected cells were killed and spilled their acidic contents into the culture medium.[9] This enabled Salk's team to crack through the drudgery of typing much faster and more cheaply than anyone else. The polio research community duly noted this industrious problem-solver, still a newcomer and not yet a threat to anyone.

The NFIP also marked Salk out as someone who delivered the goods and merited investment. This impression hit home where it mattered most, somewhere in the mid-Atlantic in late September 1951. The setting was the rarefied opulence of the *Queen Mary*, bringing back American delegates from the Second International Conference on Poliomyelitis which the NFIP had organised in Copenhagen under the watchful eye of the omnipresent Basil O'Connor. Enders had stolen the show, throwing in the tantalising hint that prolonged culture might weaken the viruses to the point where they could safely be used 'as specific immunising agents against poliomyelitis'. Salk's presentation of his progress with typing was seen as solid but not inspirational.[10]

On the voyage home, however, it was the earnest and quietly persuasive Salk who locked onto O'Connor's wavelength, initially by charming his polio-stricken daughter Bettyann. By the time they docked in New York, O'Connor was convinced that Salk was a man with whom the Foundation must do business.[11]

Salk had already started hatching plans to develop an inactivated polio vaccine. Fifteen months earlier, he had broached the notion with NFIP people a couple of tiers below O'Connor. Salk emphasised his success with flu vaccine while working with Francis, and pointed out that his golden touch had transformed the NFIP's typing programme. They sent him politely on his way. Now, having bent the ear of the man at the very top, it was a different story.

Shortly after returning to Pittsburgh, Salk sent the Foundation a massively ambitious proposal, and in early 1952 was awarded an equally massive $200,000 grant to make and test his experimental polio vaccine.

Within a couple of years, Salk had caught up with and overtaken Koprowski – a spectacular performance for a late starter in polio research. This was thanks to Salk's clever use of the usual enabling factors in research: money, good ideas, people of influence and more money.

Window of opportunity

In 1952, the time was right for a polio vaccine. It was feasible: the typing programme had determined that there were just three types of poliovirus, meaning that a 'trivalent' vaccine containing one representative of each type would protect against every single strain of the virus. Crucially, Brodie and John Kolmer were long off the scene, and the painful memories of their doomed vaccines had faded.

For the great American public, the polio vaccine was already overdue. Frightened by the NFIP's nightmarish rendition of polio, Americans had scraped together enough dimes to stretch from New York to Moscow. The NFIP was now in danger of being hoisted by its own petard. Of course it did great work – setting up rehabilitation centres, supporting families, airlifting iron lungs into epidemic hotspots – but this was no longer enough. Americans wanted an end to the nightmare, and they wanted it now.

Meanwhile, the NFIP's magical aura had dimmed distinctly. This was seven years after the death of Roosevelt, whose personal battle with The Crippler had been so inspirational and so lucrative. Now, a big hit was urgently needed to reward the generosity and patience of the American people, and to keep the dimes marching in. This big hit could only be a polio vaccine.

So the time was right, if not running out. Koprowski had been exploring prototype 'live' polio vaccines for some years, and Sabin was now dipping his toe into that same pond. Both insisted that full immunity could only be achieved with a vaccine virus that mimics the natural infection – multiplying in the host's intestine and even entering the bloodstream, but somehow stopping short of invading the central nervous system. Sabin was particularly vehement: any inactivated vaccine, incapable of replication, was dead in the water.

Salk took the opposite view. Low antibody levels in the bloodstream were known to protect against polio.[12] Surely these could be reached by giving further booster doses of an inactive vaccine after the initial shot, and the immune system would then be fully primed to mount an all-out assault should any poliovirus later try to break in. Salk's previous inactivated vaccines against flu were effective, although there was no evidence – just hope – that this approach would also work in polio.

Kitchen chemistry

Salk's eventual success hinged on a complicated recipe, every ingredient of which was later challenged by scientists, bureaucrats, lawyers and anti-vaccinationists.

Salk had to start by choosing one virus of each type for his fully comprehensive trivalent vaccine. Two were safe bets: MEF–1, isolated from an American soldier in the Middle Eastern Forces during the War, to represent type 2, and Saukett for type 3. No controversy here, except that 'Saukett' had been miscopied from a badly written label on the original sample from James Sarkett, a polio victim at the Municipal Hospital in Pittsburgh.[13]

However, the type 1 strain which he selected came back to haunt him. Mahoney, isolated from a family in Akron, Ohio, looked risky from the start. The eponymous family escaped with mild infections, but the virus killed three children next door. In monkeys, it caused paralysis and death more often than any type 1 strain, even when injected into a muscle – the route to be used for human vaccinations. Overall, Mahoney was by far the most virulent poliovirus known.[14]

To Salk, who believed that virulence tracked with the ability to stimulate immunity, this was a benefit and not a risk. As long as Mahoney's murderous tendencies could be tamed by inactivation, this was the ideal type 1 vaccine virus. An immune system toughened up by Mahoney would have no trouble in seeing off all other, lesser type 1 polioviruses – and type 1 was, after all, responsible for most epidemics, deaths and paralysis.

Inactivating the poliovirus was the next challenge – and a delicate balancing act that could topple Salk and his vaccine if, like Kolmer, he got it wrong. A new inactivation method, ultraviolet irradiation, had recently been reported, but Salk went for formalin, widely used and already familiar to him from his own flu vaccines. This was a brave choice, given the grim history of Brodie's formalin-based polio vaccine.

The marination in formalin had to be just right. Too much, and the vaccine would lack potency: the viral proteins would be so badly damaged that any antibodies they generated would not recognise and kill the wild-type virus. Incomplete activation, though, posed a much greater hazard: the potential catastrophe of enough 'live' virus remaining in the vaccine to cause paralytic polio or death.

Together with Julius Youngner, Salk conducted exhaustive experiments to work out how long a weak formalin solution (1:4,000) would take to kill off the poliovirus. Using Youngner's colorimetric method to count infective virus particles, they found that the number fell steadily during several days of incubation with formalin. Like many biological processes, the virus count seemed to decline in a straight line against time when their numbers were plotted on a logarithmic scale, in which each unit represents a factor of ten.

Salk could only count viral particle numbers accurately in the middle of the range, and so had to extrapolate his straight line down to the point where there would be so few viable viruses that they could not possibly pose a risk – for example, he calculated that six days in formalin would leave just one live virus particle in 100 million doses of vaccine.

However, these predictions depended critically on the slope of the straight line. Salk generally used only four data points, and drew his straight line through them by eye, using a ruler. This meant that a single rogue reading could skew the slope quite considerably. Also, his assumptions would break down if the decline in live virus count was not linear, but flattened off after a time – a crucial difference that Salk's four-point method could miss.

To cover these errors, Salk added a 'margin of safety' of two days to the 'safe' incubation time which he had calculated from his straight line. There was no clever science behind this; Salk plucked the number out of the air. Salk's final reckoning was that 12 days in 1:4,000 formalin would kill off all the polioviruses, while leaving their antigenic proteins recognisable to the immune system – even for Mahoney which, true to type, hung on for longer than the others.[13]

The highest primate

Salk's team, now 25 strong, made rapid progress with preclinical testing of the inactivated vaccine. Encouragingly, the vaccine stimulated antibody formation in monkeys and showed no sign of causing paralysis or other problems. Then, towards the end of 1952, it was time to venture into the hazardous territory of human experiments.

In his earlier abortive approach to the NFIP, Salk had suggested trying out a prototype vaccine on 'hydrocephalics and other unfortunates'. This seems artless and unethical to us now, but was not unusual for America at that time. The Nuremberg Medical War Crimes Trials, which unearthed many 'experimental' atrocities against the handicapped and helpless, had ended just five years earlier – but some of the lessons had not been learned in the nation which led the inquisition against Nazi 'medicine'.[15] In the 1950s, the inmates of long-term care institutions – and of prisons and reform homes for wayward women and their illegitimate children – were all perfectly acceptable guinea pigs for testing experimental drugs and vaccines.

Salk first injected his prototype vaccine into himself, his family and his team. He then moved on to two residential institutions outside Pittsburgh.[16] The D.T. Wilson Home for Crippled Children in Allegheny County, endowed by a rich banker to support 'destitute white children, crippled or deformed', had become a specialised rehabilitation centre for polio victims. The Polk School in Venango County housed mentally retarded boys and men. At both, Salk took great pains to explain the need to try out his vaccine, spending many hours with the residents and their families and drawing up detailed consent forms (including Spanish versions) with the aid of staff, parents and the NFIP's lawyers. He rewarded participation with his personal attention and a lollipop, the future currency of the full-scale field trials. The staff were evidently impressed, as 30 of them lined up alongside their patients to be given Salk's vaccine and to provide before-and-after blood samples.

The data from these pilot studies quickly built up a convincing picture. The vaccine induced brisk antibody responses to all three poliovirus types. As in the monkeys, there was no hint of serious side effects such as paralysis. Local reactions – soreness and occasional abscesses – were quite common at the injection site. Salk thought these were probably due to the mineral oil which he had included to boost antibody production, and he solved the problem by leaving it out.[17]

Two's a crowd

Salk wrote up his findings and submitted a bulky paper to the *Journal of the American Medical Association* (*JAMA*) in January 1954.[18] It was promptly accepted, with an accelerated publication date of 28 March. Excitement gripped the NFIP, and a meeting was hastily organised to consider Salk's findings. Also excited, Salk began to plan the massive field trial that would prove to the world that he had cracked the apparently insoluble problem of a polio vaccine. And at this point, things began to turn sour for him.

First, he now registered on Sabin's radar screen as a competitor, and therefore hostile. Sabin moved quickly to see off the interloper. The night before the NFIP meeting, he visited Salk on a fact-finding mission disguised as a friendly *dîner à deux*, during which Salk revealed nothing. The next morning and in front of everyone, Sabin was more blatant, launching a full-frontal assault on Salk that set the tone for their future relationship. Dead polio vaccines were a non-starter, insisted Sabin, and Salk's would be dangerous because his formalin inactivation protocol could not possibly work.[19]

Sabin's vitriol must have shocked Salk and those around the table, who had been expecting to celebrate a major milestone on the path to realising the NFIP's greatest ambition. However, Sabin failed to derail either Salk or the NFIP's support for him – probably because of his well-known vested interest. With his own oral polio vaccine still bogged down in monkey testing, and Koprowski's now beginning volunteer studies, Sabin was being left behind.

Publication of Salk's big *JAMA* paper was now only a few weeks away and excitement continue to grow. The NFIP's publicity machine was failing to contain leaks, mainly because it was not trying very hard. But it was Salk himself who revealed the secret just two days before the paper came out, on a prime-time radio show, *The Scientist Speaks for Himself*.[20] Salk wanted to temper expectations, and was careful to point out that his findings, although very encouraging, required wider confirmation and that a workable vaccine was still some time off. Predictably, that rider was lost in the general excitement sparked by his broadcast.

This was the start of Salk's media career, which made his face and voice familiar across the nation and landed him the covers of *Time* and *Life*, accolades reserved for those who really have arrived on the world scene. Sabin was disgusted, and for some other scientists too, this was the end of Salk's credibility.[17] Salk had broken one of the golden rules of the gentleman researcher by going public before the scientific community could read his big paper and reach their own conclusions about its strengths and weaknesses.

Salk's paper in *JAMA*, backed up by a strongly favourable editorial, was an instant and resounding success. Soon after, Salk tested an improved vaccine formulation in non-handicapped subjects; there was no shortage of volunteers, even though these trials were supposed to be secret.

However, there were still obstacles, all man made. Predictably, one was Sabin, who wasted no opportunity to try to torpedo Salk and his vaccine. At the annual convention of the American Medical Association (AMA) that June, which brought 38,500 delegates to New York City, Sabin scotched the 'rumours that a practicable vaccine against polio is imminent'.[21] Called as an expert witness before a Congressional Committee convened in October to review polio vaccination, he slated large-scale trials of any vaccine that had only been tested by its discoverer.[22] He mentioned no names, but he did not need to.

Also troublesome was Albert Milzer from Chicago, who told the November meeting of the American Public Health Association in New York City that he had followed Salk's recipe to the letter but could not inactivate the virus – as confirmed by the several monkeys which he had paralysed with his version of the Salk vaccine.[23] Salk reacted as though this was a personal attack and wrote a furious letter to the editor of the journal which had just accepted Milzer's paper, complaining about the 'irresponsible remarks' which had 'impugned experiments that were carefully conducted' and 'aroused fear'.[24]

Even worse, Henry Kempe, a paediatrician from the University of California Medical Center in San Francisco, wrote to the American Academy of Pediatrics, demanding that they withdraw all support for Salk's dangerous vaccine. His letter was sent on to the key players in the field, including Salk, who was apoplectic, and Sabin, who was cock-a-hoop. Sabin responded to say that he agreed entirely with Kempe's anxiety over this untested and potentially unsafe vaccine, and copied his letter to Salk. He included a covering note that was gratuitously bitchy, even for him:

> Dear Jonas,
>
> This is for your information so that you'll know what I am saying behind your back. This incidentally is also the opinion of many others whose judgment you respect.
>
> 'Love and Kisses' are being saved up.
>
> Albert[25]

All this was depressing and debilitating for the NFIP, but by now they had another overblown ego to deal with: Salk, whose stubbornness and bloody-mindedness were now evident to all. Salk had gone on the offensive. This was his vaccine, and he would run the trial his way. He did not need statisticians or people who designed clinical trials for a living. His vaccine was so good that it did not need placebo controls.

Salk was becoming a potential showstopper for his own vaccine. If he had his way, the big trial could fall at the first fence or yield results that would be uninterpretable.

Trial and tribulation

> If your experiment needs statistics, you ought to have done a better experiment.
>
> Ernest, Lord Rutherford (1871–1937)

The field trial of Salk's vaccine, which sneaked in just ahead of the polio season of 1954, belongs in the annals of superlatives. The biggest clinical trial ever conducted, backed up by the most intensive public-engagement campaign in medical history, and all made possible by one of the most lucrative fund-raising offensives of the twentieth century.

The headline figures read like an overblown Hollywood epic: a team of star scriptwriters, a central cast of 1.8 million with 325,000 supporting roles, and set in over 200 locations in 44 American States (not forgetting a few in Canada and Finland). This was a gargantuan, military-style operation that showed just how firmly the NFIP had wrapped its tentacles around American life.

Salk had to be quietly shunted away from the helm, as the NFIP had decided that he could not be trusted to do the job dispassionately and without alienating other key players. However, he still enjoyed some support at the top. His insistence that proper controls were superfluous had infected O'Connor, who despite reminding everyone that he was not a doctor or a scientist, managed to convince 30 state medical associations that Salk was right.[26] As a result, 300,000 vaccinated children ended up without placebo controls, and some of the country's best clinical trials experts walked away from the project. In the end, the protocol was written by Joseph Smadel, and then repeatedly rewritten after never-ending amendments by Salk. These tensions moulded the final design, which was a compromise and, in some respects, a mess.[27]

Drug trials are a breeding ground for wishful thinking, self-fulfilling prophecies and pliable statistics – which is why stringent safeguards must be written in. Subjects are randomly allocated to receive either the active drug or a look-alike dummy – the 'placebo' – to allow for the mysterious benefits that some people feel when taking part in a drug trial. Allocating the subjects to drug or placebo should be 'double-blind', meaning that neither the subject nor those monitoring the outcome know which is being taken. This is a defence against reporting bias – a natural but unhealthy tendency to round up a drug's benefits and minimise its hazards.

Purists such as Joseph Bell, clinical trials expert from the US Public Health Service, insisted that proper controls were needed for Salk's vaccine trial. Public and medical expectations were running particularly high after the 1952 polio epidemic, the worst in American history. Without proper blinding, reporting bias would be likely to give an overoptimistic impression of the vaccine. On the other hand, Salk argued that blinding was pointless, as this trial's major 'endpoint' – the numbers of children paralysed or killed – was impervious to any placebo effect. Reason did not prevail; O'Connor sided with Salk, and in October 1953 sacked Bell from his advisory role to the trial.[28]

Salk also tried to justify his opposition to placebo controls with the bizarre claim that he would feel personally responsible for every child who was given placebo and was then paralysed or killed by polio.[29] Luckily, school teachers and parents managed to persuade even six-year-old children that, for good scientific reasons, some of them would have to be injected with something that was not Dr Salk's life-saving vaccine, and that nobody would know how well it had worked until after the trial ended – and after polio had done its worst that year.

In the end, vaccine was given to 419,000 children and placebo to 330,000, under double-blind conditions. A further 232,000 were vaccinated and knew this, while another 800,000 received no injections and were simply followed up alongside the others to see what happened.[30] Choosing these populations was another numbers game. How many subjects were needed to prove that any benefits or side effects of the vaccine were not simply due to chance? Bidding had begun at 5,000, and risen quickly to 50,000. The final headcount was dictated more by feasibility than science, but luckily it was enough.

All the nation's children could not be covered, so where should vaccination be targeted? Polio epidemics were notoriously patchy, and there was no point in vaccinating thousands who would not have caught it that year. Fortunately, polio tended to recur for a couple of years in localities that had been badly hit. Gabriel Stickle, statistician and clairvoyant, scrutinised recent polio records through his crytsal ball and predicted with remarkable accuracy where the hot spots would be in 1954.[31]

The countdown to the trial was a tense race against time, with an unmissable deadline: the start of the polio season.

There were many cliff-hanging moments, mostly Salk-induced, with the six drug companies commissioned to produce the vaccine. The starting material consisted of the three wild-type viruses and a special culture medium (named '199', because the next attempt to get it right would have been the 200th), all shipped from the Connaught Laboratories at the University of Toronto. Each company had to follow a 50-page protocol written by Salk, with draconian safety checks to counter his paranoia that scaling up to industrial production lines would allow live virus into his vaccine. Each batch of vaccine was checked in-house, and by Salk's lab and a government agency, and only released if it and the previous 11 batches were completely devoid of live virus.[32,33] Most vaccine was made by three giants, Eli Lilly, Wyeth and Parke-Davis, with the smallest share contributed by Cutter Pharmaceuticals, a family business in Berkeley, near San Francisco.

There were also skirmishes with would-be saboteurs. For once, Sabin was outclassed, although he did try to persuade a psychologist to pressurise Salk into abandoning the trial because it would damage his mental health.[34] Real harm was done by radio broadcaster and gossip merchant Walter Winchell, in this instance acting as a mouthpiece for an unexpected snake in the grass. Winchell announced to his Sunday evening audience of over 50 million Americans that Salk's vaccine had paralysed monkeys in hushed-up safety tests. And the US government must think it was potentially lethal because they were secretly stockpiling thousands of 'little white coffins' to receive the victims of the trial. His source? Impeccable, but someone who had 'temporarily to remain anonymous'.[35]

There was no truth in the story. Some monkeys used to test the vaccine had become paralysed, but an exhaustive post-mortem study by David Bodian had concluded that, whatever the cause, it was not polio.[36] The 'little white coffins' rumour was simply a lie, deliberately fed to Winchell to discredit O'Connor. The perpetrator, it later turned out, was Paul de Kruif, writer and professional grudge bearer, who saw an ideal opportunity to get even with the boss who had sacked him from the NFIP. It is ironic that de Kruif, whose *Microbe Hunters* had idolised pioneers against infections, now had no qualms about sacrificing Salk and his vaccine in his own lust for revenge on O'Connor.[37]

Winchell broadcast his piece on 4 April 1954, just two weeks before the vaccine trial was due to start. It was clearly timed to inflict maximum damage, and had an immediate impact. The parents of 150,000 children withdrew them from the trial, and the Michigan State Medical Association came out against Salk's vaccine.[38]

Action

By now, however, the trial juggernaut was unstoppable. On Sunday, 25 April 1954, the NFIP steering committee decided to go ahead. Next morning at 9 a.m., at the Franklin Sherman Elementary School in McLean, Virginia, six-year-old Randy Kerr was heading the line of his classmates with one sleeve rolled up to expose the triceps muscle, and was given the first injection of the trial.[39]

Vaccination consisted of the initial dose followed by two others, over five weeks. Each shot was rewarded with a lollipop, and children who stayed the course received a 'Polio Pioneer' badge. This was a nice touch which helped to encourage compliance – as was the form which parents signed to request their child's participation, rather than a conventional consent form.[40]

Photographs of the trials show lines of hundreds of people snaking along streets across America. The lines converge on rows of tables, staffed by white-coated doctors and nurses, set up in classrooms, school halls and gyms, and grander settings such as the gothic magnificence of the Cathedral of Learning on the University of Pittsburgh campus. There are many fascinating vignettes from the time capsule of 1950s America. Rows of suspiciously neat children are facing a blackboard on which their smiling teacher has chalked 'We are really making history today. We are lucky'. Children, grinning bravely for the camera and with one arm bared in readiness, line up in Provo, Utah; all are white. In Jackson, Missouri, a similar line of children is all black. Elsewhere, black and white children are segregated into separate lines leading to separate rooms in a white school; the white children, but not the black ones, could be called by their first names and use the lavatories.

And at the centre of it all is the white-coated Salk, kneeling to vaccinate children (Figure 9.2). The publicity shots included him with Donna and their six-year-old son, Jonathan, perched on the edge of his chair wearing a smart bow tie and an apprehensive expression as the needle hovers over his arm.

There were a few glitches (such as the administrative error that nearly left out Denver), but thanks to around 20,000 doctors; 40,000 nurses; 60,000 teachers; and 200,000 volunteers, the trial ran with impressive smoothness.

The last injections were given in late June 1954, just before the annual upswing in polio cases. The real work then began, of analysing the 144 million pieces of information that the trial would yield – 160,000 of which were still missing by the end of August.[41] Salk was spared this task. Instead, the NFIP brought in a senior figure, universally admired for his integrity and even-handedness. He was also someone whom Salk respected, and crucially, who knew Salk and how to handle him: Tom Francis, Salk's former mentor, now recast in an even more demanding role. Francis set up his Vaccine Evaluation Center at Ann Arbor, coincidentally in a red brick building which had special resonance for the Salks, as this had been the maternity hospital where their sons had come into the world.[42]

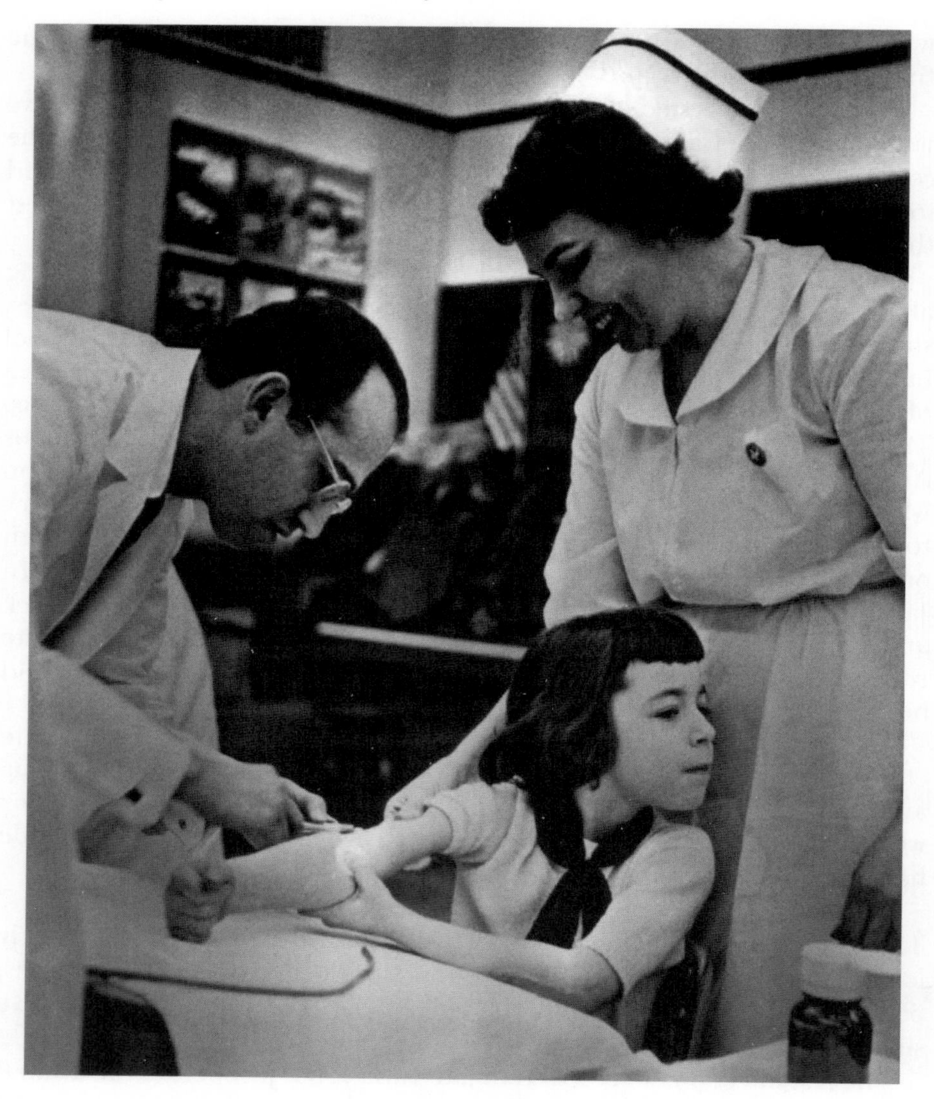

Figure 9.2 Jonas Salk, vaccinating a girl in Pittsburgh during the field trials of his vaccine, spring 1954. Reproduced by kind permission of the World Health Organisation

Francis worked tirelessly through the next several months, leading from the front even after his wife was badly injured in a car accident. Throughout, he wisely barred O'Connor, Salk and other potential troublemakers from his Centre. The final report of the trial bears his name, not Salk's, and it was Francis who eventually presented the findings to the world.

Despite its imperfect design, the trial's results were clear-cut. At their heart were 199 children whose lives had been terminated or wrecked by paralysis, whether or not they had been vaccinated. The outcome could be summarised by the bottom line of the press release which the NFIP later prepared: 'It works and is safe'.[43]

However, the vaccine did not provide complete protection. Fifty-seven vaccinated children had developed polio, compared with 142 cases among the unvaccinated controls. This equated to an 80–90 per cent reduction in risk after vaccination. Sixteen children had died of polio, none of them vaccinated. As expected, vaccination produced robust rises in antibodies against all three types of poliovirus. There were no significant side effects.

Laurels and tears

The Francis report was unveiled in the Rackham Auditorium on the campus of Ann Arbor, to scenes of high drama. This was on the morning of 12 April 1955, which happened to be the tenth anniversary of the death of Franklin Delano Roosevelt; the NFIP insisted that this was a coincidence, and nobody believed them.

Journalists and television crews (a rarity in those days) had staked out the auditorium from early morning and dived on the trolley carrying bound volumes of the report while it was being wheeled in by security guards. Copy was already filed, headlines were being set and newsreaders were on air by the time Francis was halfway through his 90-minute presentation. When Salk rose to speak, they gave him a standing ovation. The triumvirate was completed by O'Connor, who appeared tense and chain-smoked throughout the day (Figure 9.3). Outside the auditorium and across the nation, church bells were rung, grateful prayer meetings held, court hearings suspended, and radios turned up to broadcast the 'story of the year' to classrooms, factories and shopping malls.[44,45]

In some respects, the day was peculiar as well as momentous. Later that afternoon, the trial coordinators met government regulators and Oveta Culp Hobby, recently appointed as the first Secretary of the Department of Health, Education and Welfare. Mrs Hobby was a tough and capable politician who had led the Women's Army Corps during the War, but was arguably out of her depth in her new domain of science, medicine and industry. With less than three hours together and planes to catch, this ad-hoc group agreed that Salk's vaccine would be licensed for immediate use.[46] This decision came as a great relief for O'Connor, who months earlier had gambled another $9 million from the NFIP's

Figure 9.3 'It works and is safe'. Tom Francis, Basil O'Connor and Jonas Salk face the press in Rackham Hall, Ann Arbor, on 12 April 1955. Reproduced by kind permission of the March of Dimes

depleted coffers to buy 27 million doses of the still-unproven vaccine from the manufacturers.[47] These had already been produced, using a more flexible protocol and less rigorous safety testing, and were stockpiled in anticipation of a favourable ruling. Within hours, boxes prophetically marked 'RUSH' were being shipped out to the polio vaccination stations that were hastily set up across America.

As predicted, Salk ended the day a national hero, but others left Ann Arbor less than elated. Salk's own team from Pittsburgh had sat and listened to his long list of acknowledgments and thanks, only to realise that they had not been included. The chief's day of triumph ended literally in tears for some, crying uncontrollably on the train back home that evening.[48]

The day left a particularly bitter taste in the mouth of Julius Youngner, whose discoveries and inventions had lifted Salk's research to a higher plane and who was well known as a fair and level-headed man. Sixty years later, in 2005, Youngner still could not forgive Salk for having 'grandstanded' in Ann Arbor when so many others should have shared the limelight with him.[49]

Hero of a grateful nation

> Nothing short of the overthrow of the Communist regime in the Soviet Union could bring such rejoicing to the hearts and homes in America as the historic announcement last Tuesday that the 166-year war against polio is almost certainly at an end.
>
> Alastair Cooke, 18 April 1955

Sabin fervently hoped – and others had good reason to suspect – that Salk's miraculous vaccine would eventually come to grief. But few would have guessed how quickly the dream would turn into a nightmare.

For a few days, euphoria continued to reign. Back in the Viral Research Laboratory in Pittsburgh on the morning after Ann Arbor, Salk's team had dried their eyes and were putting on a brave face. In any event, they were too busy to brood. Spurred on by banner headlines – mostly variations on the theme of POLIO CONQUERED! – and unending radio coverage, correspondence addressed to Dr Salk flooded in from across the nation and further afield. The longest item, running to 70 yards, was a congratulatory telegram from all the inhabitants of Winnipeg, grateful that Salk had lifted the threat of the disease which had swept through their city three years earlier. And there was money, so much that three large bins were brought in for the triage of cheques, banknotes and coins.[50]

The aftershocks of that seismic day in Ann Arbor were already being felt around the world. The significance of this all-American breakthrough was appreciated just as keenly in Europe, Asia and Russia, and there were even hints from the White House that the vaccine could promote a useful thaw in the Cold War if shared appropriately with the Soviets (while reminding them of the superiority of American technology).[51]

The story of Salk, virtuous, selfless and dedicated scientist, captured more fanciful imaginations. Three Hollywood companies toyed with the notion of *Salk: the movie*, and someone in casting even suggested Marlon Brando for the leading role. That idea foundered, as did the more feasible one of the Salk dime, to complement the now-familiar one that carried FDR's profile.[52]

One plan that did move ahead was the recognition of Salk at the highest level on behalf of the American people. Designs for a Congressional Gold Medal, awarded exceptionally to the likes of Orville and Wilbur Wright, Thomas Edison and Irving Berlin, were hurriedly commissioned. In the meantime, a special citation was presented to Salk by President Dwight D. Eisenhower in the Rose Garden of the White House on the morning of 22 April – just ten days after the media circus had descended on Ann Arbor.

Photographs of the day show a surprisingly deferential president (whose voice reportedly trembled with emotion), shaking the hand of a smartly turned out Salk. Donna and their three sons look on, with Mrs Hobby and Basil O'Connor in attendance. All appear appropriately excited and happy. Behind the smiles, however, some of the players had other preoccupations.

Mrs Hobby had already attracted fierce criticism for having failed to seize ownership of vaccine distribution on behalf of the US Government. She had left this to free enterprise and the drug companies who, after all, were making the stuff and would have to cover their costs once the NFIP's $9 million advance purchase had run out.

Unfortunately, the companies were already talking about a post-NFIP price of over twice the 35 cents per shot which O'Connor had offered them, raising the spectre of a price-fixing cartel which could end up denying the poor and needy access to the vaccine. This would have been par for the course in America at that time – but was in marked contrast to the situation over the border, where the Canadian government had quickly swept in, taking responsibility for the vaccination programme and somehow managing to make it freely available to all their citizens. Mrs Hobby, supremely magisterial among her peers in the Women's Army Corps, suddenly found herself in the firing line, now billed as the token woman in the Eisenhower administration, and facing demands that she should resign to do something she could manage.[53]

And O'Connor was worried about holding together the untidy consortium of six competing drug companies, especially now that Salk was back in play and free to interfere. The day at Ann Arbor had exceeded the NFIP's expectations, but their prime asset was still a loose cannon.

Curse of the undead

Even if he was not concerned, Salk should have seen several omens that were beginning to coalesce into a threat. While the field trials were running, Salk had been working to improve his vaccine – which he believed would make it 100 per cent perfect. As yet, though, the new version had only been tested in-house. Some of the in-flight changes could eat into his margin of safety. He had decided to decrease the incubation time in formalin from 12 days to 9, backed up by careful filtration to remove any clumps of debris which could harbour live virus particles and protect them against formalin. In Salk's lab, vaccine was filtered through wads of asbestos, a laborious process that yielded a crystal-clear fluid which contained no detectable live virus.

Salk blended these changes into a new protocol, but decided to relax the technical constraints so that each company could adapt the basic process to suit their own production lines. We might assume that safety testing would be tightened up to cover the new laxity – and the involvement of four new drug companies which had never made the vaccine before. Strangely, though, testing for live viruses was also relaxed. The government regulators in the Laboratory for Biologics Control dropped the rule that the previous 11 batches would also be destroyed should one contaminated batch be found.[54]

Meanwhile, strange things had been happening around William Workman, the generally capable chief of the Laboratory for Biologics Control. Some weeks earlier, a trusted member of his team, Dr Bernice Eddy, had brought him shocking news: batches of vaccine prepared by the pharmaceutical giant Wyeth had failed routine safety tests. Indeed, the results would have vindicated the scurrilous claims broadcast by Walter Winchell. Seven of the ten batches that Eddy injected into the brains of healthy monkeys paralysed the animals. Live virus was clearly getting through, and this at one of the country's most respected leaders of Big Pharma.

Eddy pulled no punches, and her revelation should have set alarm bells clamouring with Workman. But, for whatever reason, he simply buried the bad news.[55]

In the end, it was Little Pharma that caused the catastrophe that had been waiting to happen.

Cutter Pharmaceuticals, which contributed the smallest share of the post-trial vaccine, was a family-run business set up in the early 1900s in Berkeley, California, to produce veterinary drugs.[56] Its first vaccine was against gas gangrene, an infection caused by clostridia bacteria which break down the muscles of livestock into a foul-smelling, bubbling liquid. Preparation of the vaccine, starting with infected muscles from newly killed calves, involved more butchery than pharmacology.

In the 1920s, Cutter had moved into human medicine, with antitoxins against the lethal toxins of tetanus and diphtheria. These were also raised in animals, and the company's cash cow (actually a horse) can be seen grazing in front of the headquarters in contemporary photographs. A notable first was achieved in 1947 with the production of a combined diphtheria-pertussis (whooping cough)-tetanus vaccine, DPT, which rapidly became a stock item in childhood vaccination programmes in the United States and many other countries.

Cutter was a small outfit, but its reputation was generally good. One black mark was the contamination of a batch of intravenous fluid, another of their big sellers, which killed several patients in California in 1948. Carelessness was evidently to blame, but the company got away with a reprimand and a fine that worked out at less than $100 per fatality.[57]

Cutter had bid successfully to produce polio vaccine (Figure 9.4), but doubts soon materialised over their ability to rise to the challenge.

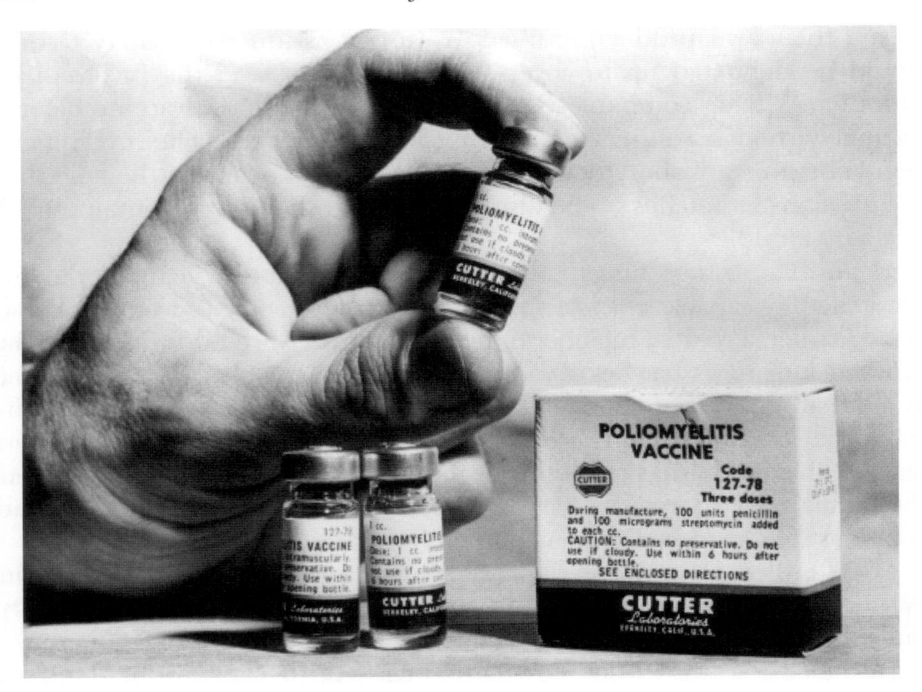

Figure 9.4 A catastrophe waiting to happen. Vials of Salk polio vaccine produced by Cutter Laboratories of Berkeley, California. Reproduced by kind permission of The Bettmann Archives/CORBIS

Uniquely, Cutter managed to produce several batches of 'bivalent' vaccine, a variant of their own invention in which they forgot to include the type 2 MEF-1 virus. More worrying was the phone call taken by Julius Youngner while on a trip to San Francisco. This was from the research director at Cutter, wanting to know how often live virus should get through into the vaccine. Alarmed, Youngner visited the Cutter Laboratories the next morning and was dismayed by what he saw: a shabby production line, with tubs of live virus close by, manned by people who seemed unaware of the safety precautions and why these were in place. Youngner thought it very likely that Cutter's vaccine could contain live virus.[58]

Youngner immediately reported all this back to Salk but, despite the obvious violations of his own protocol, Salk failed to react. Admittedly, Salk was heavily preoccupied: Youngner had flown to San Francisco on the evening of the frenzied day in Ann Arbor, and visited Cutter's production line in Berkeley the next morning. But Salk could never explain why he did nothing.

The full cost of ignoring these omens did not become apparent for a few weeks. The boulder began to roll on Sunday, 24 April, just 11 days after the first boxes of vaccine had been rushed out of the companies' stockpiles. Six-year-old Susan Pierce from Pocatello, Idaho, was diagnosed with paralytic polio. Cases were to be expected at the start of the polio season, but this was Idaho – one of the coolest States, and among the last to be struck each year. Worryingly, she had been given Salk's vaccine just ten days earlier, and it was her left arm – which had been injected – that she could not move. Paralysis quickly spread to her other limbs and then her respiratory muscles.[59]

In the three days that Susan Pierce took to die, many other cases sprang up. Some were in clusters, such as the five reported on 26 April from Oakland, California. There were enough common features to make a disturbing pattern. Most attacks began in the vaccinated arm and were brutal, with rapidly progressive paralysis. They stood out starkly against a blank background, as the polio season had not yet begun. Victims had been vaccinated during the preceding 7 to 12 days. Soon, similar cases were being recognised and within a week had reached 60, almost all concentrated in Idaho and California.

With this dawned the terrible realisation that these cases of polio had not occurred despite Salk's improved vaccine, but had been caused by it. Indeed, whichever live virus had sneaked into the vaccine was now behaving like a highly virulent wild-type strain and spreading to close contacts – such as the mothers of babies paralysed after vaccination, who developed polio themselves and died a few days later.[60–62]

The paralysing vials of vaccine were quickly traced back to their source. All came from Cutter Pharmaceuticals, which had evidently buried any concerns about its quality control and had churned out an impressive 380,000 doses of vaccine as its initial offering. Most had been shipped to California and Idaho, where an encouraging 98 per cent of eligible children had lined up to bare their arms. Further detective work using Salk's own typing tests quickly nailed the virus responsible. In every case, this was the type 1 strain whose virulence had persuaded Salk that it was ideally fit for purpose: Mahoney, reasserting itself as dangerous and especially liable to cause paralysis after intramuscular injection.

When the news broke, recriminations swiftly followed. Salk was devastated, both by the tragedies of paralysed and dead children and by the collapse of adulation for his vaccine. He was still protected by the armour of a national hero. After all, it was barely two weeks since Eisenhower had presented him with the expression of the nation's gratitude, and nothing like this had happened in all the hundreds of thousands of children given Salk's vaccine during the field trials.

Nonetheless, Salk must have been haunted by voices from the past. The first was his own, saying that he would feel personally responsible for any child who was paralysed or killed in the trial – after being given placebo. Now, the vaccine bearing his name had done exactly that, and he had broken the golden rule of therapeutics: First do no harm. Predictably, Albert Sabin, master of schadenfreude, was delighted to inform everyone that he had been right all along. Salk's formalin cookery was useless, and Mahoney was far too dangerous to use in any vaccine.

Another ominous voice was there too: Sven Gard, the 'father of Swedish virology' and world-famous polio expert, who had echoed Albert Milzer's dire warning that Salk's vaccine could not exclude live virus. Gard insisted that Salk's idealised straight-line graph was fiction; instead, he found that inactivation followed a shallow curve, which meant that nine *weeks* of pickling in formalin – not nine days – were needed to kill the virus. Gard had chosen his audience well to drive home his attack on Salk: the Third International Poliomyelitis Conference, in Rome.[63]

The Cutter crisis demanded calm and decisive action from the US Government. Unfortunately, this was lacking. All the evidence pointed to Cutter; backed up by O'Connor, the other five companies argued that their own products were absolutely safe and that vaccination with these must continue at all costs in order to keep a grip on public confidence. The Cutter vaccine was quickly withdrawn, with the soothing message that otherwise the vaccine was perfectly safe. Within a week, however, the entire vaccination campaign was put on hold until the fundamental questions about Salk's inactivated vaccine had been resolved.[64]

The combination of indecision, mixed messages and the U-turn killed off some promising political careers and shattered public confidence in what, three weeks earlier, had been the medical breakthrough of the century. The proportion of American parents wishing their children to be vaccinated plummeted from over 90 per cent to 37 per cent. The ripples soon spread across the Atlantic. The vaccination programmes planned around the Salk vaccine in Britain, France and Sweden were promptly abandoned.[65]

So what had gone wrong and who was to blame? The pieces of the puzzle were gradually teased out of the mess by the Infection Surveillance Unit, led by Neal Nathanson. The Unit had originally been formed within the Communicable Disease Center (CDC) by Alex Langmuir, the Director, to prepare for possible germ warfare attacks on the United States. Getting to the bottom of the 'Cutter Incident' – the Holmesian title coined by Langmuir – was their first real test. And they did an excellent job, as summarised in Nathanson's painstaking papers on the Incident.[60–62]

From the start, the finger had pointed unwaveringly to catastrophic safety lapses at Cutter. By the time the inspectors arrived, the company had smartened up the production line but could not hide their records (although they did try). These showed that Cutter's own tests had essentially confirmed Bernice Eddy's estimate that half of the batches of vaccine contained live virus. Incredibly, Cutter had done nothing to find out what was going wrong. Salk's 'margin of safety' did not apply to Cutter's own production line, because their filtration step (using a glass sieve instead of Salk's asbestos wad) failed to remove particles of debris that sheltered live virus from the formalin. Mahoney, being the toughest, was the one that crept through into the vaccine.[60–62]

Ultimately, the fault lay not with Salk, his inactivation process, the NFIP or even Mahoney, but with a company that was sloppily run, ill-equipped for the job and incapable of following instructions or reacting to its own obvious failures.

Fallout

The Cutter Incident lined up a beauty parade of scapegoats, some of whom were permanently damaged as a result. Heads that rolled included those belonging to Mrs Hobby, obliged to resign that July and return to her comfort zone in Texas, and the government regulators who had chosen to ignore the signs of danger. But nobody was fired, moved aside or even disciplined at Cutter.[66]

Salk, suddenly a fallen hero and a potential villain, had to endure a trial of scientific inquisition. Most gruelling was a Congressional hearing at which one expert after another laid into the straight-line plot of formalin inactivation which Salk had extrapolated into the grey zone of mathematical uncertainty. Wendell Stanley, virologist and chemist, told Congress that viruses simply did not work like that. Being a Nobel Laureate, his opinion had clout. Another Nobel Laureate, newly minted, agreed that there was no scientific foundation to Salk's 'quack medicine'. This really hurt Salk, as it was John Enders, one of Salk's great heroes and the man whose work had catalysed Salk's own.[67,68]

Those hit hardest by the tragedy were, of course, the victims of the Mahoney-contaminated Cutter vaccine. Mahoney showed its true colours as the virus which had jumped over the fence in Akron, Ohio, and killed the children next door. In all, there were 204 cases of paralysis, mostly irreversible, and ten deaths. Only 61 cases had actually been vaccinated. The others were close contacts, family members or friends.[60–62]

But the world moves on. Salk weathered the storm, as he had done so many times before, and the vaccination programme resumed. Initially, uptake was slow in many parts of the United States, but Salk's iconic status and the reassuring spectacle of other scapegoats being strung up undoubtedly helped to allay public anxiety. By the end of the year, some 5,394,000 American children had been given Salk's new vaccine, and with no new cases of post-vaccination paralysis.[69]

In late 1957, Cutter Laboratories was sued by victims' families, the test case being six-year-old Anne Gottsdanker from Santa Barbara,

California.[70] The hearing was classic High Court drama, boosted by the prosecution lawyer, Melvin Belli, widely known as the 'King of Torts' and famous for his 'avarice, immunity to logic and self-aggrandizement'. But even Belli's pyrotechnics and Salk's appearance to testify against Cutter failed to bring the clear condemnation that many had hoped for. After much agonising, Cutter were found to have caused Anne Gottsdanker's paralysis, but curiously were not held responsible for it. It was felt that they had taken reasonable steps to follow Salk's protocol, and that the accidental contamination of the vaccine was unfortunate, but just one of those things.

The Gottsdankers were awarded damages of $125,000. Another 60 cases had been waiting in the wings, and most decided to settle out of court when the Gottsdanker verdict was announced. In the end, Cutter paid out $3 million to the victims of its vaccine.

Some good came out of the Cutter debacle. Pharmaceutical companies were sharply reminded of their duty to produce drugs that contained only the ingredients on the label, and governmental regulations were tightened accordingly. The Infection Surveillance Unit had proved its worth and went on to take a leading role in tracking outbreaks of infectious diseases worldwide.

And it could have been much worse. The Cutter Incident prompted other vaccine manufacturers to fix the problems in their own production lines. To rephrase Oscar Wilde, to lose one manufacturer could be regarded as misfortune, to lose two would have looked like carelessness.

In fact, there were three, but their transgressions were dealt with discreetly and behind closed doors, to avoid undermining the Salk vaccination programme. A detailed report was prepared about live-virus contamination in the Wyeth vaccine; yielding gracefully to pressure from above in the CDC and the government, its authors decided not to pursue publication. Other evidence was also quietly buried. Three of America's biggest drug companies – Wyeth, Eli Lilly and Parke-Davis – all settled out of court and with well-crafted gagging clauses.[71]

Happy ending?

Salk's vaccine recovered almost as quickly as its inventor's reputation. Most other countries followed the lead of the United States when it was reintroduced in the summer of 1955, and it reigned alone and supreme for several years. In America, some 40 million doses were given between 1955 and 1962.

As the NFIP press release in Ann Arbor had said, it works. During this time, the annual spikes of polio in the United States were relentlessly squashed, and in many parts of the nation the disease ceased to exist. In

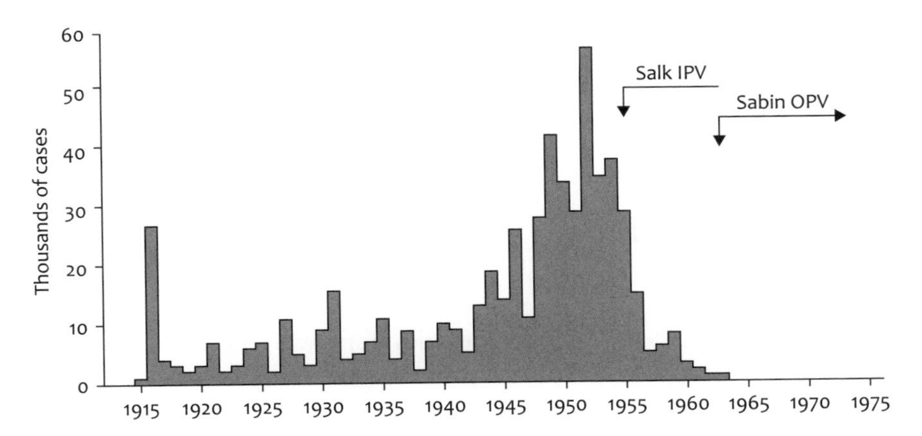

Figure 9.5 Annual incidence of paralytic polio in the United States between 1915 and 1975. Salk's inactivated polio vaccine (IPV) was introduced in mid-1955, and was replaced in 1963 by Sabin's oral vaccine (OPV). Data from the Centers for Disease Control

1952, the worst polio year in American history, there had been 58,000 cases of paralytic polio, with 9,000 deaths.

In 1957, the second full year of routine Salk vaccination, the numbers had fallen to 5,500 paralytic cases and 220 deaths. By 1962, there were just 910 cases and 60 deaths (Figure 9.5). Salk's vaccine had fulfilled its expectations, not just protecting individuals in their millions, but by depriving the virus of susceptible hosts, pushing out the disease itself. And since the spring of 1955, there has not been a single case of paralytic polio anywhere on the planet due to live virus contaminating Salk's inactivated vaccine.[72]

This came too late for the victims of the Cutter vaccine. Nonetheless, it is powerful confirmation that Dr Jonas Salk came up with a good recipe and that all you had to do to make it work was to follow the instructions.

10

Poles Apart

There is a long history of vaccine pioneers who believed that their invention was perfect. Edward Jenner, the 'father of vaccination', convinced himself and his disciples that a single inoculation with cowpox vaccine in infancy would protect against smallpox for life, and with no risk. Evidence rapidly proved him wrong: ten years after vaccination, people could catch smallpox and die from it, while blood poisoning and even syphilis were sometimes introduced together with the cowpox. Jenner and his supporters chose to ignore all this, blaming it instead on a conspiracy by his enemies.[1] And by turning his back on anything that defied his dogma, Jenner set the trend for many who followed in his footsteps, including both Jonas Salk and Albert Sabin.

By early 1956, when half the world was singing the praises of his polio vaccine, Salk had already fallen into the Jennerian trap of self-delusion. His improved vaccine, which he insisted was 100 per cent effective, was safely back in service. Since Cutter had been removed from the scene – which had quietly forced other, bigger players to clean up their own acts – the vaccine was absolutely safe. Salk also believed that, if intensively applied, it could rid the world of polio as well as protecting the vaccinated individual – just as Jenner had predicted that his vaccine would eventually exterminate smallpox.[2]

But other people, and not just Salk's many natural enemies, were already thinking beyond his vaccine. It reduced an individual's chance of catching polio during an outbreak by up to 90 per cent, and it had proved the concept that polio could, after all, be prevented by vaccination. Nonetheless, it was not perfect. Unlike naturally acquired polio, Salk's vaccine did not confer 100 per cent, lifelong protection. The initial course of four injections had to be topped up periodically and whenever polio threatened, such as during epidemics or visits to an endemic area. There was evidence that the shortfall was due to the lack of the 'intestinal immunity' induced during a natural attack of polio as the virus travelled down the gut. Intestinal immunity was provided by special antibodies secreted into the gut cavity by immune cells in the gut wall.[3] Salk's vaccine viruses, injected into the arm muscles, had no opportunity to cruise the intestines and generate this vital first line of defence.

There were practical drawbacks too. As a mass medication, Salk's vaccine was expensive. The on-costs of syringes and needles brought a course of four shots to about $1.50; when multiplied by millions, this was a serious headache for health economists in affluent Western countries and a cruelly impossible dream in the developing world. And of course, nobody likes injections.

It will already be apparent from the Punch and Judy exchanges between Jonas Salk and his chief tormentor, Albert Sabin, that Sabin believed that only an attenuated vaccine (and specifically the one he was working on) could really protect against polio. This would be given by mouth, delivering live but emasculated poliovirus which could infect the intestine, but never break into the central nervous system. This self-limiting infection would induce the all-important intestinal immunity as well as antibodies in the bloodstream. By faithfully mimicking a natural attack of polio, but with all three types of poliovirus participating, Sabin's vaccine would provide total and lifelong protection.[3]

Sabin did not have to wait long for events to back up his attacks on Salk's vaccine. The abstract mathematics of risk were given dramatic substance in 1959, just four years after the vaccine came into widespread use in North America. A polio epidemic swept through Boston and the East Coast of the United States and into Canada – areas that were supposed to be well vaccinated.[4] Paralytic polio cases in the United States jumped back up to 8,500 (with 450 deaths), while Canada saw 2,000 children paralysed. The balance was beginning to tip in Sabin's favour.

Like Salk, Sabin's name has gone down in history. However, Sabin was not the first with an oral, attenuated polio vaccine. That distinction belongs to Sabin's less well-known countryman, Hilary Koprowski, a flamboyant and multitalented showman-scientist. Koprowski also stole the march on Salk, having already given his experimental oral vaccine to human subjects many months before Salk had even lifted his eyes from the drudgery of poliovirus typing.

Sabin and Koprowski were kindred spirits: both refugees from Poland, both articulate and forceful evangelists for the supremacy of live over dead, and oral over injectable. And their actions spoke even louder. Each developed an oral vaccine that was given to tens of millions of European children during the first few years of the Salk vaccination programme in the United States.

Together, Sabin and Koprowski would have made a formidable double act that could have wiped the floor with Salk. But they were too alike and too competitive; as if obeying the laws of magnetism, these were two similar Poles that forever repelled each other. There could only be one winner in the race for an attenuated polio vaccine. We already know the result, but there was plenty of nail-biting and diversion along the way.

And in the final reckoning, it was Koprowski who came closer to dancing on his rival's grave.

Wunderkind

> Genius is one percent inspiration and 99% perspiration.
>
> Thomas A Edison (1847–1931). Quoted by Hilary Koprowski, 'History of Koprowski vaccine against poliomyelitis', in S. Plotkin, *History of Vaccination*, 2011.

Born in Warsaw on 6 December 1915, Hilary Koprowski enjoyed an embarrassment of intellectual riches from early childhood. Polyglot, polymath and a musical prodigy, he had to choose between training as a concert pianist at the Conservatoire or fulfilling his alternative ambitions at Poland's top medical school. He graduated as a doctor in 1939, just in time to carry his wife Irena overseas to safety; within months, many of his classmates found themselves walled inside the Warsaw ghetto with a cattle truck to Auschwitz as the only way out.[5]

Koprowski's ultimate target was America, but the couple moved first to Brazil and a research outpost of the Rockefeller Institute in Rio de Janeiro. There, he worked on yellow fever with John Fox and Max Theiler, one of the Rockefeller's all-time stars who had devised the first safe and effective vaccine against yellow fever.[6] This unpleasant infection, one of the great killers in human history, wrecks the liver; it is caused by a virus, not the corkscrew-shaped bacterium which tragically fooled Hideyo Noguchi. Liver failure, invariably lethal at that time, killed about one in four of those who presented with the ominous triad of fever, jaundice and dark, bile-laden urine. Noguchi's death from yellow fever in 1928 made headlines in microbiological circles, but was an insignificant statistic beside the hundreds of thousands killed each year in South America, Africa and Asia. Theiler's vaccine was a major breakthrough, recognised by the Nobel Prize committee in 1951.

The vaccine was based on live yellow fever virus, ruthlessly attenuated by being forced to survive inside the brains of mice and then in chick embryos – unsettling career moves for a virus that had evolved to thrive in liver cells. Theiler's vaccine was highly effective against yellow fever and showed no tendency to revert to attacking the liver. Koprowski's five years in Rio laid solid foundations for his future polio vaccine: experience at the bench with attenuating viruses, and growing them in rodents' brains and chick embryos, and lasting friendships with Fox and the great Theiler. Koprowski also made an early reputation for himself as an original thinker, a hard worker, and a forceful personality who ignored convention and regarded risk as an opportunity to be seized.

Figure 10.1 Hilary Koprowski (1916–2013, right) and Tom Norton at Lederle Laboratories, Pearl River, in 1949. Norton had worked with Max Theiler on the Nobel Prize-winning yellow fever vaccine, and became Koprowski's chief technician during the development of their oral polio vaccine. Photograph reproduced by kind permission of the late Hilary Koprowski

In 1944, Koprowski moved to New York (Figure 10.1). Unlike Noguchi, he was unable to persuade the Rockefeller Institute to take him on. Instead, he took a post in vaccine development at Lederle Laboratories, based in Pearl River to the north of New York City. Despite its laid-back name and rural setting, this was a powerhouse of drug development, fuelled by some highly successful antibiotics. Pearl River boasted a large, university-style campus and a research budget that rivalled the Rockefeller's own. But this was Big Pharma, not the Groves of Academe, and by accepting a job there, Koprowski instantly found himself relegated to the second division.

Koprowski's boss at Lederle was Herald Cox, a talented microbiologist who had defected to industry after an outstanding academic career (Figure 10.2). Together with Macfarlane Burnet he had nailed the cause of Q-fever, an infection first described from slaughterhouses in Australia. Q stood for 'Query', as the disease had baffled so many. The organism responsible, an odd bacterium that was notoriously difficult to culture, was named *Coxiella burnetii* to honour both men.[7] In his way, Cox was a colourful character

Figure 10.2 Herald Cox (1907–86), bacteriologist and polio vaccine pioneer, and Hilary Koprowski's chief at Lederle Laboratories in Pearl River, New York. Photograph reproduced by kind permission of the World Health Organisation

but in muted pastels; unlike Koprowski, he was measured in approach, preferring to look around corners rather than cut them.

Their collaboration began well and produced a new live rabies vaccine which Koprowski magicked up after 180 passages through chick embryos, accessed by drilling a hole through the shell of a fertilised hen's egg.[8]

This vaccine was safer and better than Pasteur's original, and proved to be a money-spinner for Lederle as well as a publication production line for Koprowski.

In early 1947, Koprowski's working relationship with Cox entered a new phase. While driving home from a meeting in New York City, Koprowski found himself thinking about a live polio vaccine, using polioviruses that he would attenuate in the laboratory just as Theiler had tamed the yellow fever virus.[9] Over the next few months, Koprowski quietly got on with his big new project – and concealed it from his chief.

Koprowski got off to a good start, thanks to a little help from his friends: advice from both John Fox and Max Theiler at the Rockefeller, samples of polioviruses from Fox, and cotton rats (at the time, the best non-primate host for the poliovirus) from Charles Armstrong in Washington, DC.[10] Best of all was the loan of Tom Norton, Theiler's chief technician at the Rockefeller, who had been instrumental in bringing on the yellow fever vaccine (Figure 10.1).

Attenuation was a laborious grind that took many months, and which made Salk's 12-day formalin inactivation protocol look like a stroll in the park. However, by the end of 1948, their first attempts to attenuate a poliovirus had succeeded. This began with the type 2 virus called Brockman, grown up as in Landsteiner's era in the spinal cord of a monkey and then passaged eight times through the brains of mice. For good measure, this was then pushed through another three passages in the brains of cotton rats.[11]

The end product, named 'TN' for Tom Norton, looked promising. It was safely attenuated, with no sign of causing paralysis when injected directly into monkeys' brains. It raised circulating antibodies when given orally to chimpanzees, and protected them against infection with type 2 polioviruses. Moreover, the chimpanzees excreted the TN virus in their stools. This was an additional benefit, as the vaccine virus could theoretically spread to infect and protect other non-immune people who had not themselves been vaccinated. It also set ringing a faint alarm bell that, within a few years, would nearly sound the death knell for Koprowski's vaccine.

These experiments brought Koprowski to the same point of no return that Brodie and Kolmer had faced a quarter of a century earlier, and which Salk would not confront for another two years. The vaccine appeared effective and safe in animals, and it was time for Koprowski to test the strength of his conviction by giving it to man. Koprowski and Norton began with themselves, knocking back a cocktail of TN-infected cotton rat brain and spinal cord which they whizzed up in a kitchen blender on the laboratory bench one afternoon in January 1948. Apart from tasting like cod liver oil, it had no undesirable side effects and both men showed encouraging rises in antibodies against type 2 polioviruses.[9,11]

For his first clinical trials, Koprowski selected a population which would bear some similarities to Salk's pilot subjects. This was Letchworth Village in Rockland County, not far from Pearl River. Its full title was 'Letchworth Village for the Epileptic and Feeble-Minded'. The children of Letchworth had various diagnoses that caused severe behavioural problems, including a fondness for throwing things at each other. Staff therefore removed all projectile items, which left just one option – faeces – and created an environment which, in the view of Letchworth's Research Director, George Jervis, was conducive to the spread of gut-borne infections. To quote Koprowski:

> [Jervis] proposed that we try to protect his patients against the possible introduction of polio virus into the wards of the institution. Permission was obtained from the institutional authorities as there were no human subject committees in those days![11]

So in February 1950, the TN-infected cotton rat brain cocktail was fed to 20 children in Letchworth. There were no obvious complications, and the results reinforced the earlier positive findings. All the children (17 in all) who had initially lacked antibodies against type 2 poliovirus now had them, and also excreted the TN virus in their stools. Furthermore, attempts to reinfect them with TN failed in all but a couple of cases. As proof of concept for an attenuated polio vaccine – and the first vaccine ever to be given orally – it was compelling and exciting stuff.[11]

But the experimental setting soon brought Koprowski rebukes and hate mail. The attacks started at a conference on 'Immunization in Poliomyelitis' which the NFIP organised in mid-March 1951 at Hershey, Pennsylvania. For most of the polio experts around the table, the idea of a polio vaccine was still speculation. The chairman summed up the situation by asking, 'Are we justified, on the basis of present-day knowledge, in undertaking limited and well-defined experiments in human beings?'[12]

At that point, nobody knew that Koprowski had already been there and done that. He had been invited to talk about his rabies vaccine, but after sitting through a long lunch, Koprowski decided to tantalise the gathering with a summary of his experiments at Letchworth. The post-prandial torpor was such that it took some time for this to sink in. Tom Rivers assumed that Koprowski had been talking about monkeys; it was reportedly Salk who put Rivers right.[13] Rivers was bitterly opposed to doing experiments in 'mentally defective children' and later protested that 'many of these children did not have mommas or poppas, or if they did, their mommas didn't give a damn about them'.[14] Sabin attacked Koprowski on the spot, demanding to know why he had done such a thing.[13] He may have been taken by surprise at the boldness of what had been done, but this was the first of Sabin's many personal assaults on Koprowski.

Later, the publication of Koprowski's findings in the *American Journal of Hygiene*[15] provoked further outrage from across the Atlantic. In a hostile editorial in the *Lancet*, an anonymous commentator laid into Koprowski for describing the children of Letchworth as 'volunteers' when they clearly lacked any capacity to understand what was being done to them, let alone confirm that they were happy to cooperate.[16] In his paper, Koprowski had not been expansive about the children's condition, but had let slip that a couple of them had to be tube fed. If these children were 'volunteers', thundered the *Lancet*, then so were laboratory mice.

Koprowski ploughed on regardless. His proof that that there was life after Brodie and Kolmer stimulated much more than anger. Within a few months, both Salk and Sabin had set out to develop their own vaccines.

The typing studies had already shown that type 1 was the prime mover and the most profitable target for a polio vaccine. As Salk would later do, Koprowski selected Mahoney, but chose to mix it with Sickles, another type 1 strain. This 'SM' blend was pushed through some challenging hoops, with dozens of passages through mouse brains, monkey kidney cells, mouse brains again, chick embryos, and finally a trip through the human bowel. The latter belonged to a child in another residential home for retarded children, in Sonoma, California.[11]

The virus collected from the child's stools was further attenuated by another three transits through the intestines of other children in the home. The end result was named CHAT, after 'Charlton', the surname of the little girl who was the fourth and final attenuator.[17] CHAT proved to be exceptionally good at inducing antibodies, and quickly became Koprowski's choice for the type 1 virus in his vaccine. Koprowski was inordinately proud of CHAT – even when, years later, it dumped him into the centre of a furore that nearly destroyed his reputation.

By now, viruses were being routinely grown in cultures of monkey kidney cells, and TN, with its cod liver oil tang of mashed cotton rat brain, was soon to be passé. It was superseded by a better type 2 strain, named after John Fox, who sent it to Koprowski. The value of CHAT and Fox in combination was demonstrated during 1955 with the help of the lady inmates of Clinton Farms, an open prison which was so open that a steady stream of babies was produced.[11] For the first time, Koprowski gave his vaccine to infants and showed that they generated good antibody responses to both CHAT and Fox.[18]

Meanwhile, the clock had been ticking and Salk, from a standing start just four years earlier, had caught up with Koprowski and comfortably overtaken him. Koprowski's experiments at Clinton Farms during the summer of 1955 were barely noticed in the frenzy around Salk's vaccine, shown to be safe and effective by the massive trials in 650,000 American children during the previous year.

Salk's success made life even more difficult for Koprowski, who now decided to cast his net further afield and set about forging new alliances beyond the United States. Out there, across the Atlantic, he could sidestep the regulation, competition and approbation which bedevilled his work in the United States. Salk's vaccine had a stranglehold on America, leaving little possibility of finding a large enough virgin population who would be prepared to undergo experiments with an alternative vaccine. Koprowski was also desperate to return to academia and increasingly aware that, for him, the sun was going down over Pearl River. He had concealed his experiments at Letchworth from Herald Cox and had not asked his permission to report the findings to the National Foundation for Infantile Paralysis – which Cox only found out about secondhand and some time after the event.

All this set the stage for Koprowski to go global.

Into the mouths of babes and sucklings

Britain seemed a promising setting for further tests of Koprowski's vaccine. Northern Ireland in particular caught his eye: self-contained, good infrastructure and in those pre-Troubled days, a feeling of provincial solidarity that enabled imaginative things to be done quickly and without getting tangled up in the red tape of Whitehall. In late 1955, Koprowski contacted George Dick, Professor of Microbiology at Queen's University in Belfast, and proposed a clinical trial of his combined TN/SM oral polio vaccine, fully funded by Lederle.

Dick leaped at the opportunity. Together with his senior lecturer, David Dane (later famous for describing the 'Dane particle' structure of the hepatitis B virus), Dick set up a trial to begin in early 1956 (Figure 10.3). Dick's vision of the protocol differed in an important respect from Koprowski's own. To Dick, the vaccine was still experimental, and the trial presented an opportunity to confirm that it was safe as well as effective. Koprowski simply took those things for granted.

Koprowski later dismissed Dick as 'a microbiologist'. In fact, George Dick was much more: a well-published researcher who had trained at the Rockefeller and Johns Hopkins University (which later made him one of their Heroes of Public Health).[19] Dick already had an international reputation in virology, including insect-borne infections and polio, which he had seen at first hand during a particularly vicious outbreak on Mauritius. Dick was a tall, confident Scot with a big personality and a reputation for decisive action; once, when he missed a mouse and injected

Figure 10.3 George Dick (1914–97), Professor of Microbiology at Queen's University, Belfast (left), with David Dane and Moya Briggs. Dick stopped the trial of Koprowski's oral polio vaccine in 1956 when he discovered that the vaccine virus had regained its power to paralyse monkeys. Image by Northen Ireland Government Information Service

rabies virus into his thumb, he calmly picked up a scalpel and sliced off the end of the digit.[20]

Dick also had well-developed powers of persuasion. Strategic application of those skills, notably over lunch in the Staff Club at Queen's, quickly drummed up the necessary recruits. They included Dick's own four-year-old daughter, and the son and daughter of Professor of Geology Alwyn Williams. 'We didn't sign anything', explained Williams's wife Joan many years later. 'We didn't need to. George said that this vaccine deserved to be tested, and we all believed him'. I would not dare to contradict her; after all, she was my mother.[21]

Being one of Professor Dick's guineapigs was not that bad: a few blood tests, and daily stool samples deposited into waxed card containers that were lined up in our coal shed for collection each week. And the whole thing was rewarded by a visit to Dick's laboratory, an Aladdin's cave full of wonderful machines, busy people in white coats and awe-inspiring monkeys.

Together with those monkeys, the children in Dick's trial came up with an unexpected result that wounded Koprowski and left a stain of suspicion on his vaccine.

Ever cautious, Dick injected samples of the vaccine into monkeys' brains in the time-honoured test to confirm that 'neurovirulence' – the ability to paralyse – had been completely stripped away by attenuation. Fresh from the vaccine vial, both TN and SM behaved exactly as Koprowski claimed and caused no paralysis. However, some of the viruses isolated from the stool samples in those waxed cartons showed a sinister change. Somehow, passage through a child's intestine had made the vaccine viruses regain some of their original neurovirulence, because they now caused paralysis when injected into healthy monkeys.[22] Compared with wild-type, unattenuated polioviruses, the paralysis was relatively mild and the histological damage in the monkeys' spinal cords was less severe. However, this was enough for Dick. He stopped the trial immediately and informed Koprowski that he would have to tell the world.

Koprowski later wrote that Dick saw this as 'the signal to campaign in the newspaper against live attenuated virus vaccination'. In fact, Dick simply did what any conscientious scientist should do. He reported exactly what he found, in three measured papers and a thoughtful overview in the *British Medical Journal* (*BMJ*), and drew conclusions that most would have regarded as inescapable.[22–24] Koprowski's vaccine undoubtedly worked, in that it induced healthy levels of protective antibodies. Unfortunately, it had failed what Dick called 'the necessary theoretical requirements of safety'.[22] Children given the vaccine were probably not at risk, as they had already produced antibodies by the time the vaccine virus had undergone its ominous transformation in their intestine. But the re-energised virus could possibly spread to infect others, with the risk of causing an outbreak of paralytic polio.

Dick's findings killed off the already obsolete TN, together with any immediate prospect of an oral polio vaccine supplanting Salk's vaccine in Britain. Koprowski took this 'diatribe' as a personal affront, but this was not Dick's intention. Six years later, in July 1961, he published another paper in the *BMJ*, again showing that an oral polio vaccine virus could not be trusted to remain stable and safe while passing through the bowels of children in Belfast. This time, he took on an even more formidable adversary, as the virus in question was the attenuated type 2 strain in Sabin's vaccine.[25] For once, the two mutually repelling Poles came together. Sabin, like Koprowski, refused to believe Dick's results and denied any possibility that his vaccine carried the slightest risk of causing paralysis.

The Belfast trial signalled more than the demise of TN. In 1957, the 42-year-old Koprowski left Lederle to become Director of the Wistar Institute in Philadelphia, leaving behind an irate Herald Cox, a large hole where the polio vaccine project had been, and accusations that the polio strains which Koprowski took with him were not his to plunder.

The Wistar Institute posed challenges of a different sort. Long famous as America's oldest independent medical research outfit, it had degenerated into an unhappy blend of museum and mausoleum. It now housed the world's largest collection of human and animal skeletons, together with countless formalin-pickled brains and some brass urns containing the ashes of Casper Wistar (founder) and various other incinerated luminaries. In 1906, the Institute had presented the world with the Wistar rat, which rapidly became the laboratory standard; arguably, little of substance had happened since then.

Koprowski was exactly what they needed, even though some soon realised that this was not the same as wanting him. Under his 22-year reign as Director, the Wistar regained momentum, pulled in international stars and grew up several of its own, and became a world-class centre for virology, cancer research and much else.[5]

From the start, Koprowski led from the front and pushed ahead with his own research into oral polio vaccines. By now, Sabin was in the race and closing the gap. Koprowski applied his considerable charm and networking skills in the Old World, and between 1957 and 1961 set up several highly productive collaborations.[11] A private paediatrician in Bern provided an entrée to Eastern Switzerland, where 34,000 children were entered into a trial of CHAT combined with Fox. Croatia soon followed, with 1.3 million recruits. His native Poland was an obvious target, and he proved adept at sidestepping Cold War froideur, communist bureaucracy and the attentions of the secret police. The result: an astonishing 14 million children vaccinated with CHAT. In all these trials, the vaccine behaved impeccably: antibodies were raised, and there were no cases of paralysis or other side effects. As further proof that it worked, the number of cases of paralytic polio across Poland fell from 1,122 in 1959 to just 28 four years later.[26]

Koprowski also moved into Africa, and specifically the Belgian Congo, the region that now embraces Zaire, Rwanda and the Democratic Republic of Congo. In 1957, polio was endemic out in the bush and was moving in to threaten the European expatriate populations in the cities of Stanleyville and Leopoldville (now Kisangani and Kinshasa, respectively). Koprowski was well known there: he had visited several times and tested vaccines on chimpanzees in a special laboratory at Lindi, near Stanleyville. Not surprisingly, he was asked to bring in his oral vaccine.[11]

So he did, and in spring 1958, 250,000 Congolese children, mostly living along the Ruzizi River, were vaccinated with CHAT. This was achieved in just six weeks – a tribute to the energy of the vaccination teams and the combined pulling power of Koprowski, his vaccine and the traditional drums that thumped out the message throughout the length of the Ruzizi Valley.[27]

At this point, we have to return to the United States to catch up with the rapid progress being made by Sabin. In the event, even though millions of doses had been given successfully, Koprowski's vaccine would be condemned to death within a year. But its story does not end there.

Koprowski's hard-earned CHAT strain, given to the 250,000 Congolese children, had set ticking a time bomb with a 20-year fuse. When the explosion came, it would pull the world's attention to an even greater plague than polio and would come close to destroying Koprowski.

All will be revealed in due course. In the meantime, remember that the vaccination campaign in the Ruzizi Valley took place in spring 1958, and that CHAT had been named in 1953 after a handicapped girl in a residential home in Sonoma, California.

He who shouts loudest

> An able, tireless and articulate medical scientist, who did not stand on ceremony and seldom dealt delicately with situations which from his viewpoint required direct, immediate, and strong action... Yet in my estimation no man has contributed so much information to so many aspects of poliomyelitis, as Sabin.
>
> John Paul, *A History of Poliomyelitis,* 1971

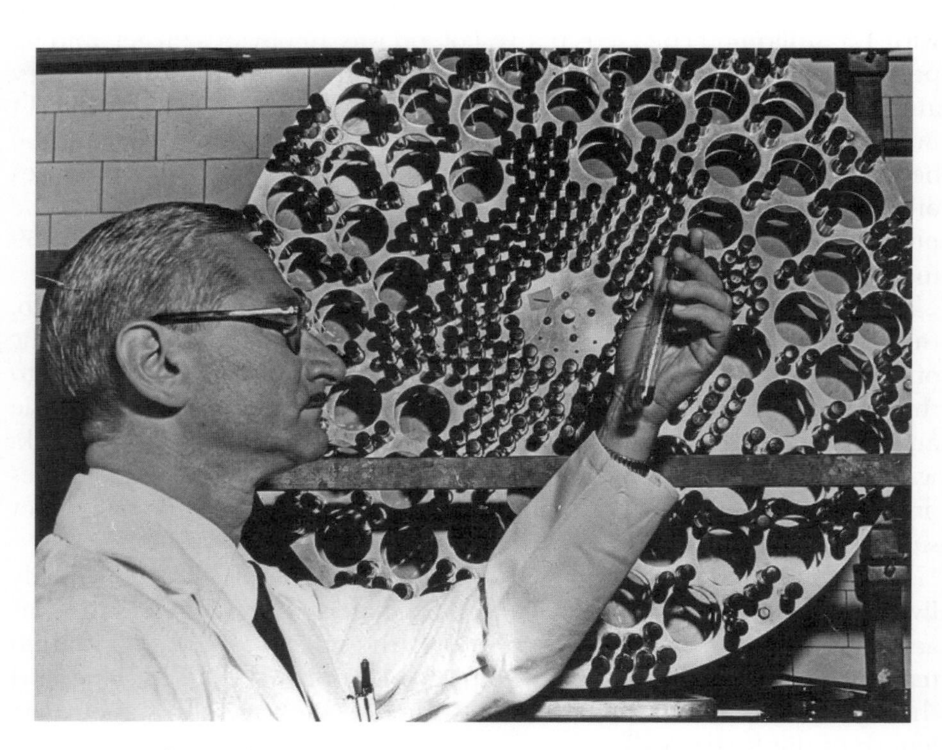

Figure 10.4 Albert Sabin (1906–93), who developed the oral polio vaccine which superseded Salk's inactivated vaccine during the 1960s and has all but delivered the *coup de grâce* to polio across the world. Reproduced by kind permission of the World Health Organisation

Of the polio vaccine triumvirate that comprised Salk, Koprowski and Sabin, it was Sabin who had the strongest track record in polio research (Figure 10.4). He was also the one who channelled the greatest energy into undermining his competitors and displayed the most obvious pleasure from seeing them run into misfortune.

There are few gentlemen in the history of polio, and Sabin was not one of them. His peers' admiration for his achievements was often given grudgingly, whereas they were much more liberal with their dislike of him. Audiences at scientific meetings broke into applause when someone had the guts to stand up to him.[28] Eminent vaccinologists who knew him well have described him variously as 'a mean bastard' and someone whose obituary everyone would have enjoyed writing,[29] and 'a total a***h***'[30] (readers wishing to clarify the expletive will find a Latin translation in the title awarded by the Skeptics Forum to the anti-vaccination propagandist Viera Scheibner).[31] Even the Sabins' family dog went for him, during a shouting match between him and his two daughters.[32]

Young man in a hurry

Sabin's origins bore similarities to those of both Koprowski and Salk. Ten years older than Koprowski, he was born Abram Sapersztejn on 26 August 1906. He grew up in the city of Bialystok, which is now proudly Polish, but at the time lay within the Russian Empire.[32] Like the Salks, his family emigrated to America to avoid the risk of being massacred in the anti-Jewish pogroms. Arriving in the hard winter of 1920–21, they settled in Paterson, New Jersey, where his father, a weaver with particular skills in silk, quickly built up a good business in women's clothing. His son, already bilingual in Yiddish and German, added English to his repertoire in record time and bulldozed his way through high school with his sights set on medicine.

A doting uncle offered to support him through dental school, which was generous, but not close enough to young Sapersztejn's ambitions. He was fired up by the dental course in microbiology, a passion later reinforced by reading Paul de Kruif's newlypublished *The Microbe Hunters*, a chatty and often melodramatic account of the great bacteriologists in history.[33] By contrast, the drudgery of hand-crafting dentures did not appeal, and he left the dental course in 1926, aged 20. Deserting the oral cavity also meant the end of his uncle's support, and Sapersztejn had to survive on casual work and charity. He soon homed in on a strategic target, who was both a big name in bacteriology and the Dean of the Medical School at New York University. This was William H. Park, expert in diphtheria and polio, who had not yet heard of a bright young Canadian called Maurice Brodie. Exploiting both chutzpah and pathos, Sapersztejn persuaded Park that his only aim in life was a career in microbiological research. Park duly found him a place in the medical school and a string of casual jobs and scholarships to help him pay his way.[32]

Park had made an astute investment. His protégé worked hard and showed initiative and originality in the laboratory. In 1928, he published a full paper about the physical properties of the gels used to grow bacterial cultures – an outstanding achievement for a second-year medical student. By now, Abram Sapersztejn had mutated into Albert Sabin, in preparation for taking up American citizenship.

While still a student, Sabin followed up his first publication with a series of papers on an improved method that he had invented for typing the pneumococcus bacteria which commonly cause pneumonia.[34] At the time, before penicillin, the only treatment for pneumonia was a series of antisera, each of which only killed the specific type of pneumococcus against which it had been raised. The standard procedure for typing a patient's pneumococcus was to inject a sample of sputum into the abdominal cavity of a mouse, leave the infection to brew for a couple of days, and then test samples of pure pneumococcus from the mouse's abdomen against a battery of antisera to see which one reacted. Sabin worked out how to stain pneumococci directly with antisera, in sputum samples smeared onto a microscope slide. This saved time, effort, dollars and lives, and was good news for patients, doctors and mice. Sabin's method rapidly became the diagnostic standard across New York, and then throughout the United States. Park was greatly impressed, and others began to take notice of the tall, confident young man.

From the start, Sabin was more comfortable in the lab than with patients. After qualifying in 1931, he delayed starting his residency in the Bellevue Hospital to do his first project on polio. This was timely, as polio had broken out in New York that summer. Up to 110 new cases of paralysis presented in a single day, keeping Park busy with a trial of immune serum treatment.[32]

Then Park saw (but did not believe) a report from Claus Jungenblut, the highly respected professor at Columbia University who later insisted that vitamin C could cure polio. Jungenblut claimed to have invented a simple skin test that could identify children susceptible to polio. There was a precedent, in the 'Dick test', in which a tiny amount of diphtheria toxin injected into the skin produces a characteristic reaction if the subject has antibodies to the toxin, and is therefore protected against the infection. Park assigned Sabin to repeat Jungenblut's experiment – which he did, and quickly showed that the so-called diagnostic test was worthless. Park brokered a diplomatic solution, with a joint paper from Sabin, himself and Jungenblut, which gently popped the myth.[35]

The following year, Sabin seized another opportunity and milked it thoroughly, eventually producing six papers about a previously undiscovered neurovirulent virus. This began uncomfortably close to home, when William Brebner, a dynamic young medical researcher and personal friend of Sabin, was bitten on the hand by a monkey during an experiment. The wound seemed trivial at first, but Brebner went rapidly downhill with flaccid paralysis and loss of feeling that began in his feet and soon spread to both legs. The clinical diagnosis was ascending myelitis, inflammation that cuts across the entire width of the spinal cord and marches inexorably upwards, like slicing a salami.

This was harrowing to observe and must have been terrifying for Brebner, who must quickly have worked out how things would end. Eighteen days after being bitten, the paralysis reached his diaphragm, and he was transferred to the still-experimental iron lung. Twelve hours later, the diagnosis of myelitis was confirmed down the microscope, and pieces of Brebner's brain and spinal cord were preserved in glycerine, so that attempts could be made to see if a virus had been responsible.

Sabin took on the project and successfully isolated the virus that killed his friend: a novel herpesvirus, which he called 'B' to commemorate Brebner.[36] The B virus turned out to be common in monkeys' body fluids and harmless in those species, but almost always lethal if inoculated into man. The rigour of Sabin's investigation brought him more attention, as did the forcefulness with which he attacked anyone (including senior professors) who challenged his conclusions.

This episode helped to launch Sabin off on a one-year scholarship from the National Research Council to the prestigious Lister Institute in London. This went well, and lined him up for a junior research post at the Rockefeller Institute on his return to America. Tom Rivers, the 'father of American virology' and author of the near-biblical *Filterable Viruses*, had spotted the young man's potential. Unfortunately, their personal chemistries were incompatible, and the interaction exposed Rivers' well-known bigotry. Sabin brought back more from London than a sound understanding of neurotropic viruses and fond memories of his bohemian digs in Chelsea. Rivers told him bluntly that fancy English clothes and a phoney transatlantic accent just did not work for 'a cheap East-side Jew'. Sabin took this in his stride; at least Rivers referred to him as 'the smart Jew', while Salk would later be stuck with Rivers' label of 'the young Jew'.[37]

Work-life balance

In January 1935, Sabin joined the Rockefeller to work under Peter Olitsky, a patricianly (and Jewish) expert in neurotropic viruses. Olitsky's natural curiosity had been stifled by Simon Flexner's *idées fixes* about polio. Sabin initially fell into line, but was a rebellious spirit and eventually found himself challenging the intellectual dictatorship of the man at the top. During his four years at the Rockefeller, Sabin published over 40 papers.

Most of the first dozen supported the Flexnerian party line that polio entered the brain through the olfactory nerves at the top of the nose – at least when massive doses of Flexner's favourite poliovirus, the MV strain, were repeatedly squirted high up a monkey's nostrils.[38]

Sabin worked hard during his first year, but had other things on his mind that summer. On 12 September, Olitsky received a brief telegram from Chicago. Two of its five words were not about work, which was unusually relaxed for Sabin. But this was a significant day for him and for Sylvia Tregullis, whom he had met just a month earlier:

MARRIED HAPPY BACK NEXT WEEK.[39]

The following year, Sabin got into his stride with a prophetic paper reporting that poliovirus could be grown in tissue culture – but with a crucial limitation.

The concept of culturing poliovirus in vitro rather than in living animals was not new, and had been tried repeatedly since Levaditi's first failed attempts in Paris during the 1920s. A few successes had been reported, but with results so fabulous that they could only have been fabricated. Sabin's paper of 1936 pre-empted by 12 years the genuinely transformative discovery by Enders, Weller and Robbins that the poliovirus could replicate in a range of cell types maintained in artificial culture media. Enders' team marched away with the Nobel Prize, while Sabin, despite getting there first, had to content himself with a single paper in the *Journal of Experimental Medicine*.[40]

The main problem was his unlucky choice of poliovirus. He used the Rockefeller's in-house strain, the MV ('mixed virus') originally derived in 1916 by Flexner himself from two different isolates injected simultaneously into a monkey's brain. MV had been lovingly nurtured and passed repeatedly from one monkey's brain to another's over 20 years, because Flexner believed that this was the natural habitat of the supposedly neurotrophic poliovirus. This forced adaptation had become a self-fulfilling prophecy. MV was now so comfortably ensconced in the central nervous system that it had lost the ability to grow in any other tissue. Sabin tried a range of other, non-neural tissues from three- and four-month fetuses that had been aborted at Bellevue Hospital, but MV grew only in fetal brain tissue. As a result, Sabin's paper was pigeonholed as elegant confirmation of Flexner's dogma, rather than the portal of entry into the brave new world of virus research which Enders would open up in 1948.

During the summer of 1936, Sabin enjoyed a busman's holiday at the Marine Biological Laboratory in Woods Hole, Massachusetts. He took Sylvia, together with hundreds of microscope slides of olfactory bulbs from experimental animals which had died of paralytic polio.

Sabin had dared to think laterally, and was looking to see whether the poliovirus had rampaged through the olfactory bulbs (which receive the olfactory nerves from the nose) when the virus was introduced into the body by routes other than up the nostrils. These included injection into the abdominal cavity, muscles, nerves and (jump to the next line if you are squeamish) the eyeball.

Sabin searched hard, but found only pristine olfactory bulbs in animals which had been infected by non-nasal routes. He wrote a paper concluding that olfactory bulb lesions were merely a consequence of intranasal injection of the virus.[41] Flexner, perhaps not yet sensing the danger to his doctrine, warmly congratulated Sabin on his 'beautiful, beautiful' histological studies.[42]

Throwing caution to the winds, Sabin then extended his animal work by pulling together a collection of olfactory bulbs from human subjects who had died of polio. He claimed that 'practically no attention has hitherto been paid to the olfactory bulbs in human poliomyelitis'. In fact, several others had already looked, and had all found either no damage or minimal non-specific changes. Sabin now searched systematically for evidence of olfactory damage, to test Flexner's conviction that humans caught polio through the nose. This was timely, as even some Americans were now expressing interest in the 20-year-old notion from Sweden that polio was primarily an infection of the gut rather than of the central nervous system.

Like the earlier investigators, Sabin found no significant damage in the human olfactory bulbs.[43] The conclusion was obvious. As Sabin remarked, 'the olfactory pathway concept is not a good deal'. The wider implications were even more obvious. Sabin had killed off the boss's pet theory which had wasted years of Olitsky's and others' time, but had also helped to build the Rockefeller's reputation as a centre of excellence for polio research and Flexner's own status as a leader in the field. Sabin described Flexner's reaction to his findings as 'not very happy'.[44] This may have explained why Sabin, usually so fearless and forthright, sat quietly on the fence when discussing the ominous tranquillity within the olfactory bulbs of human polio victims:

> At present, no direct evidence exists for or against the belief that any particular site is a portal of entry of the virus in man.[43]

One research topic that did not attract Sabin's interest during his time at the Rockefeller was anything to do with a polio vaccine. His return to New York coincided with the rising excitement over the Brodie-Park and Kolmer vaccines, but this was quickly followed by their spectacular and painful falls from grace, orchestrated in part by Rivers. Besides, Flexner still maintained that polio vaccines were a non-starter, and had published his dismissive article to this effect in *Science* in November 1935.[45]

By 1939, Sabin had grown tired of Flexner's dirigiste regime. He had become hot property, and overtures were made to attract him to the Karolinska Institute in Stockholm and the Medical School in Cincinnati. Inducements included the chance to work with Sven Gard and other world-class virologists in Stockholm, versus a salary hike, 'open access to clinical material' and a packet of tobacco from the Dean in Cincinnati. The latter proved more persuasive. In April 1939, Sabin became an associate professor of paediatrics in Cincinnati – an intriguing job title for a man who actively disliked children and whose main clinical experience (4,000 post-mortems) had provided rather limited opportunities for honing his communication skills.[46]

Dealing directly with 'clinical material' and their anxious families soon confirmed Sabin's conviction that he did not want to be a paediatrician. Instead, he dug himself into polio research, and especially the pivotal role of the intestine, which both nurtured the virus and generated immunity against it. The importance of the gut in transmitting polio was soon graphically highlighted when his technician, Barbara Johnson, used her bare hands to clear up the mess after an infected monkey's intestines disintegrated while being washed out. She was struck down a few days later by paralytic polio, ending her scientific career.[47]

Overall, though, Sabin's star was rising. He celebrated New Year 1940 with a $7,000 grant from the NFIP. This marked the start of a steady stream of funding that soon swelled to an average of $200,000 per year – undermining his perpetual gripe that the NFIP gave all its money to Salk.[48]

America's entry into the War propelled Sabin into the military. He began on the home front as a civilian adviser on viral infections of the central nervous system – encephalitis and polio – which were cutting down British and American troops in Egypt while leaving the local population strangely untouched. Then, despite protests of neglect from the still-childless Sylvia, he was commissioned as a major in the US Army and sent to North Africa. There, Sabin picked up useful skills while helping to develop an experimental vaccine against insect-borne viral encephalitis.

Killer instinct

After demobilisation in early 1946, Sabin returned to Cincinnati and to Sylvia. His research now had a new focus, mentioned for the first time in a letter to the NFIP in November 1946: an oral polio vaccine, which would mimic the natural intestinal infection and produce complete, lifelong immunity. But it was not until mid-1952 that funding from the NFIP got him started.[49] This was over a year after the meeting in Hershey where Hilary Koprowski had revealed his experiments with his own oral vaccine and had faced Sabin's vitriolic tirade.

Sabin caught up quickly, thanks to an unremitting two-pronged attack. One prong was an intensive laboratory programme to find and attenuate suitable strains of poliovirus; the other was a vicious campaign to sabotage his competitors. His main target was Salk, at the time steadily pulling ahead and basking in the approval of the NFIP. Sabin's killer arguments were that dead vaccines were useless and that Salk's 'kitchen chemistry' was particularly dangerous because it could not guarantee inactivation, especially of the treacherous type 1 strain, Mahoney.[50] To torpedo the rival oral vaccine, Sabin played up the nightmare spectre of Koprowski's strains reverting to virulence and causing outbreaks of paralytic polio.[51]

Sabin's own strategy for attenuating polioviruses would not have been tolerated at the Rockefeller. It cut across all the anti-vaccination diktats that Flexner had restated in *Science* in 1935 while laying the ground for Rivers and others to annihilate the Brodie-Park and Kolmer vaccines. Flexner had insisted that attenuation by repeated passage through monkeys' brains could never work. The result would either be too feeble to stimulate a useful immune response, or so un-attenuated that it would break out and cause full-blown polio. The only evidence to support Flexner's prejudice was that he himself had tried to pull off this trick but had failed. Sabin now set out to prove him wrong. He eventually did so in 1956, after four years, over 20,000 monkeys, 500 chimpanzees and nearly 200 human volunteers – all of which were personally inoculated, bled and documented by Sabin.[52]

Given the ferocity of his attacks on Salk and Koprowski, there was some irony in Sabin's plan of attack. Three strains of attenuated poliovirus eventually emerged from the monkey production line and went into his vaccine. These were given complicated in-house names: Sc-Li KP33 for the type 1 vaccine strain, Y-SK-KP51 for type 2, and Leon-KP34 for type 3. It is not immediately obvious that Sc-Li KP33 was derived from Mahoney, the villain of the Cutter Incident and (according to Sabin) a catastrophically foolish choice by Salk.[52]

Sabin's first ideas for human trials also smelled of hypocrisy. Having watched Sabin savaging Koprowski at Hershey, Pennsylvania, in March 1951 for using institutionalised, mentally incompetent youngsters, the NFIP committee were bemused to receive a proposal from Sabin just a few months later. This was to test his prototype vaccine in 'mentally defective children ... under constant observation in an institution over long periods of time'.[53]

The institution Sabin had in mind was Willowbrook School on Staten Island in New York, which housed children with Down's syndrome and other conditions causing severe learning difficulties. Willowbrook later became notorious for its repeated violations of the Nuremberg Code that governs human experimentation. Particularly unsettling were studies which proved in the most obvious way that different forms of viral hepatitis are spread by the faecal-oral route or the injection of infected blood.[54]

Sabin was dismayed when the NFIP turned his application down, but he quickly found an alternative captive population. This was in the Federal Reformatory at Chillicothe, Ohio, about 90 minutes by road to the northeast of Cincinnati. These studies also fell in a grey ethical zone, although less disturbing than those at Willowbrook would have been. The prisoners, all male and mostly aged under 30, could give informed consent, but were nudged towards participation by the promise of $25 and a few days knocked off their sentences.[55]

The studies at Chillicothe showed that Sabin's vaccine, given orally, raised highly respectable antibody levels against all three types of poliovirus. Also, the vaccine virus was excreted in the subjects' stools and spread easily to infect others who had not been vaccinated – and who themselves went on to show similar rises in blood antibody levels. Better still, those who had been infected by a vaccine strain, either directly or caught from vaccinated subject, were then protected against subsequent re-infection by that strain. This was shown by the fact that they did not excrete the poliovirus in their stools after being re-challenged with the vaccine virus.[56]

Here was living proof of the 'intestinal immunity' which Sabin had been chasing in his monkeys. It was also a major selling point for his vaccine, especially as Salk's vaccine, injected into a muscle rather than being ingested in the natural way, was incapable of weaving this crucial coccoon of protection within the gut.

Breakpoint

Sabin's fiftieth birthday on 26 August 1956 would have been an interesting day for him to survey the landscape. His vaccine was ready to be rolled out, but for one apparently insurmountable obstacle: Salk had got there first.

The Cutter Incident was last year's story (except for its victims), and the benefits of the first full year of Salk vaccination were becoming apparent in the tumbling numbers of American children paralysed or killed by polio. Thanks to the combination of public adulation for Salk and relentless pressure from the NFIP, Salk's vaccine seemed well on course to saturate America and was already infiltrating other countries.

This was a daunting monopoly to break, even to find a large enough Salk-free population in which Sabin could test his own vaccine, let alone push it into routine clinical use. But Sabin was convinced that his vaccine was perfect and that he had a duty to rescue mankind from Salk's inferior product. Sabin's other competitors were much less of a threat. Hilary Koprowski had just embarked on his ill-fated incursion into Ulster with George Dick, while Herald Cox was trying desperately to resurrect his own oral polio vaccine from the wreckage left behind when Koprowski had abandoned Lederle for the Wistar Institute.

Sabin's reputation was not as powerful as he would have liked, but he may not have had enough insight to realise this. To the public, he was still a minor player, eclipsed by the recently beatified Salk. Sabin also came across as unpolished, brusque and occasionally odd. As well as Salk-style images showing him holding pieces of laboratory glassware up to the light, Sabin released a publicity shot of himself and one of his favourite monkeys (personally trained by him), both smoking cigars.[57]

Sabin cast himself as a commanding Janus-like figure who could look both ways across science and medicine and see how basic knowledge could be exploited to solve clinical problems. Unfortunately, he somehow failed to impress his peers on both sides of the divide. Many doctors were suspicious of the man who made no secret of preferring monkeys and the laboratory bench to sick children and the clinic.

Many scientists were also dubious, despite Sabin's massive list of publications. Major misgivings had surfaced earlier that year during a meeting which Sabin organised in New York. He set out to bring together brilliant minds to focus on the cutting edge of polio vaccinology – as represented by his own research. Participants included Macfarlane Burnet and America's top biologists and molecular geneticists, notably the Nobel Laureates-in-waiting, Renato Dulbecco and Joshua Lederberg; Salk and Koprowski were not invited. Putting himself in the limelight was a risky strategy for Sabin. The meeting yielded some helpful collaborations, but also exposed large gaps in Sabin's grasp of molecular biology and immunology, leaving many of the big names with lasting doubts about his credibility as a serious scientist.[58]

At home, Sabin's reputation was also far from perfect. The Sabins now had two girls aged four and six; they were pressed into service to test the oral vaccine, but generally saw little of their father. Sylvia's feelings of neglect increased in step with the rising trajectory of her husband's career. On the day in question, 26 August 1956, their marriage had exactly ten years left to run.

Despite all these problems, the planets were lining up for Sabin. His vaccine undeniably had promise, and the notion that an oral vaccine could outperform Salk's was gaining momentum. Dorothy Horstmann at Yale University, applying the same approach that Sabin had used in Chillicothe, had shown that a natural attack of polio provided lasting protection against a poliovirus of the same strain. When the virus was given experimentally to subjects who had recovered from the disease, it was unable to gain a foothold in the gut and was not excreted in the stools.[59]

Salk's vaccine failed to provide this crucial aspect of immunity. It could pick off polioviruses that enter the bloodstream and so reduce the individual's risk of paralysis or death, but could not break the chain of faecal-oral transmission during an outbreak. Sabin's vaccine did all this, and more. By infecting and immunising contacts, its benefits would rapidly spread far beyond those vaccinated. It would be much more effective at stopping outbreaks and, in theory, might eventually even wipe out the wild-type virus. And as well as providing better immunity, it was much easier to give.

As far as Sabin was concerned, his vaccine won hands down.[3]

From Russia, with love

Further afield, world-class scientists from overseas were also taking Sabin seriously.

A small delegation from the Russian Academy of Sciences had visited the United States early that year to find out about vaccines against polio, which was erupting in ever-expanding epidemics across the Soviet Union, just as it had done in America 40 years earlier. The high point of the Russians' mission was to be a pilgrimage to the shrine of the saintly Salk in Pittsburgh, but Sabin managed to hijack them to Cincinnati for two days. Salk's welcome was lukewarm; Sabin made a huge fuss of them, punctuated by Russian, dredged up from his childhood.[60]

The visitors were led by Anatoli Smorodintsev from the Virology Department at the Institute of Experimental Medicine in Leningrad, together with Mikhail Chumakov, director of the Institute for Poliovirus Research in Moscow, and his former research assistant and now-wife, Marina Voroshilova. All were noted virologists in their own right. Chumakov and Smorodintsev had studied insect-borne encephalitis viruses which ravaged parts of Siberia and the Far Eastern USSR; each had published several hundred papers and been awarded the Lenin Prize for his work. Chumakov had a more lasting memento of the tick-borne encephalitis which he had discovered in Khabarovsk, namely complete deafness and paralysis of his right arm.[61] Significantly, Chumakov was also the man in charge of the newly established campaign to eradicate polio completely from the Soviet Union (Figure 10.5).

Sabin and the Russians all got on famously, and one thing led to another. Sabin was invited as a guest of honour to the 13th All-Union Congress of Epidemiologists, Microbiologists and Hygienists in Leningrad in late

Figure 10.5 Mikhail Chumakov (1909–93), Russian virologist who took Sabin's oral polio vaccine behind the Iron Curtain to immunise millions of children in the USSR and Czechoslovakia in 1958–59. Chumakov discovered many viral illnesses in Russia and developed several vaccines of his own. His name is commemorated in the viral research institute of the Russian Academy of Medical Sciences, and in Asteroid 5465

June 1956. There, he met the man at the very top: Viktor Zhdanov, academician and virologist, and the politically slick Deputy Minister of Health who represented Russia at the World Health Organisation (WHO). Unusually for the Cold War era, Zhdanov was highly regarded both at home and abroad. His Russian connections established, Sabin returned to Cincinnati and the shadow of Salk.

On 6 October 1956, Sabin was invited to speak at the Ohio Valley Section of the Society for Experimental Biology. His title, 'Vaccination against poliomyelitis – present and future', pulled in the press, and Sabin's confident prediction that his oral vaccine would soon blow Salk's out of the water made headlines that were picked up across the United States.[62] But the American establishment remained in the thrall of Salk, even though interest in Sabin's vaccine was now gaining momentum in Holland, Italy, Mexico and, of course, the USSR.

All this set the scene for Sabin's grand European tour the following summer: Bucharest, Rome, Leningrad, Moscow and Geneva. En route, he dished out samples of his vaccine, blithely contravening the export limits set by US Customs. In Geneva, Sabin reported the results from his first 110 human volunteers to the Fourth International Poliomyelitis Conference, where a large audience was duly impressed and excited. Back in Cincinnati, Sabin sat back to await developments.

They took his bait on both sides of the Atlantic, although more hesitantly in America. The Pan-American Sanitary Bureau decided to back Sabin, and ordered 1 million doses of his vaccine. This was counter-balanced by a typically caustic put-down from Rivers, who effectively told Sabin to cut out the middleman and dump his vaccine straight into the sewers, where it belonged.[63]

But Russia came up trumps. Chumakov requested more vaccine 'for further studies', together with permission to set up a manufacturing centre in Russia. Sabin was delighted to approve this, and by the end of 1958, Chumakov's production line had churned out 10 million doses which were already being pressed into service with the draconian efficiency at which the Soviets excelled.[64] By spring 1959, Sabin's vaccine had been given to 2.5 million children in selected Russian states and to 350,000 children in Czechoslovakia. These were not formal clinical trials, as there were no placebo controls – in fact, exactly the design that Salk had fought for unsuccessfully in the field studies of his own vaccine a few years earlier.

Updated results were reported in triumph to the First International Conference on Live Poliomyelitis Vaccines, held in Washington, DC, in late June 1959.[65] With this conference, the NFIP began to lose its grip on the polio research community. The meeting was funded by the unlikely coalition of the Pan-American Sanitary Bureau and the Sister Kenny Foundation.

Smorodintsev, Voroshilova and their Czech colleagues presented the data from a total of 6.25 million children inoculated with Sabin's vaccine. The results were so clear-cut that, arguably, no statistics were required to interpret them. Numbers of polio cases had plummeted where vaccination had been done, but not in the 'control' unvaccinated states. Amazingly, not one of the several million doses of vaccine had caused any complications whatsoever. Sabin, who scripted, orchestrated and stole the show, also put the knife into the opposition. He claimed (accurately, as it later turned out) that Koprowski's vaccine carried another hidden hazard. A virus, previously unrecognised and presumably picked up from monkeys somewhere in production, was potentially dangerous. Coming from the discoverer of the B virus, this was taken seriously, and helped to kill enthusiasm for Koprowski's vaccine, which was about to be rolled out in Poland and Croatia.[66]

To some in America, the results of the pseudo-trial from beyond the Iron Curtain appeared too good to be true. Pressure was applied to Sabin and his Russian colleagues for an independent expert to be flown in to observe the continuing vaccination campaign. This had to be a top-rank scientist, respected by everyone including Sabin, and of unimpeachable integrity.

The eventual choice could not be faulted. Dorothy Horstmann of Yale was sent to visit vaccination centres in the USSR and Czechoslovakia during the summer of 1959. She declared herself impressed with the record keeping and quality of the data, and confirmed that both uptake and effectiveness of the vaccine appeared to be extremely high, that side effects were vanishingly rare, and especially that no cases of paralytic polio had followed vaccination.[67]

Horstmann also had the chance to appraise Sabin's vaccine in action and in real time, when a polio outbreak in Uzbekistan that summer was quickly stamped out by a mass vaccination campaign. Her report, filed in the United States in early 1960, painted a powerfully positive picture of Sabin's vaccine. It was close to 100 per cent effective, absolutely safe, and so easy and innocuous to take that, even without Soviet-style incentives to comply, entire populations could be comprehensively vaccinated.[68]

Horstmann's report did more than bolster Sabin's credibility. It also highlighted the stark and unsettling contrast with Salk's vaccine which, now that the hype had died away, was seen to be struggling. Indeed, several of Sabin's Cassandra-like prophecies were being fulfilled. At best, Salk's vaccine provided only 80 per cent protection against paralysis or death, compared with Sabin's 100 per cent, at least in American and European populations. And because it did not multiply in the gut and stimulate intestinal immunity, Salk's vaccine could not immunise the unvaccinated or break the chain of spread so effectively during outbreak.

Much more important than theory and statistics, Americans were voting against Salk's vaccine with their feet, and with their shirt sleeves obstinately rolled down. Even in 1957, when mass vaccination had only been running for a couple of years, the magic of Salk's triumph had evidently shrivelled away for much of the American population. Of those who needed vaccination, only 40 per cent had received the full course of four injections – and another 40 per cent had not been vaccinated at all.[69]

Salk's vaccine still appealed to middle-class, White America, but uptake was generally low in the Southern states and rock bottom in many deprived, predominantly Black populations. And to confirm that even the best medicine will not work if you do not take it, polio was making a comeback. Cases were still far fewer than the 20,000–30,000 annual tally in the pre-Salk era, but the initial awe-inspiring fall had not been sustained. In 1958 there had been 6,000 cases, of whom 3,000 were paralysed; the following year, to everyone's consternation and dismay, the incidence had bounced back up to 8,500 cases with over 5,600 paralysed, thanks to the outbreak that hit Boston, the East Coast and Canada.[4]

Homecoming

Horstmann's endorsement of the Russian campaign was helpful, but Sabin desperately needed home-grown credibility. This came from a trial which he conducted in 1959 in his home city of Cincinnati and the adjacent Hamilton County.[70]

With just over 179,000 children and 3,000 adults, this was minuscule compared with the Soviet bloc experience, but the data were all-American, and therefore more compelling for the medical establishment and the regulatory authorities in the United States. Uptake was enthusiastic and rises in protective antibodies impressive. It soon dawned on Americans, as it had already had on tens of millions of people behind the Iron Curtain, that polio could be better fended off with a spoonful of pink, cherry-flavoured syrup than with several injections deep into the triceps muscle.

The Cincinnati-Hamilton trial also threw up some results that proved to be the shape of things to come. These should have sounded the alarm, but Sabin, firmly in command of the data from his trial, simply made the omens disappear as smoothly as a Soviet airbrush might have removed disgraced dignitaries from photographs of the May Day parade.

Three children became paralysed within ten days of receiving Sabin's vaccine. Any concerns that this was a rerun of the Cutter Incident were promptly quashed by Sabin, who looked into all three cases in detail. His investigations included post-mortems, as all three children had died. Sabin pronounced that the cause in each case was Guillain-Barré syndrome, not polio – and certainly nothing to do with his vaccine, which as everyone already knew from millions of Russians, was 100 per cent safe. And even though Sabin had so studiously avoided dirtying his hands with 'clinical material' and could never have been described as an experienced diagnostician, they all believed him.[71]

Over the next couple of years, the balance tipped steadily in Sabin's favour. The Russians stepped up their vaccination programme to include everyone under the age of 20 – all 77 million of them. The outcome, reported to great excitement at the Second International Conference on Live Poliomyelitis Vaccines in 1960, was even better than Sabin had hoped for. Paralytic polio had virtually been swept away by vaccination, vindicating Chumakov's outrageous ambition to eradicate the infection from the whole of the USSR.[72]

By contrast, America remained bogged down in the impasse created by the inability of Salk's vaccine to penetrate the hard-to-reach populations of the deprived, the uninformed and the reluctant. In March 1961, the gloomy comment was made that:

> Polio seems far from being eradicated. The dreamed-for goal has not been achieved.[73]

The speaker was Alexander Langmuir, Chief of the Communicable Disease Center, testifying before Congress. Coming from the man who drove the entire American public health machine, this was a devastating admission of defeat.

As if to make the point, polio broke out around Syracuse in upstate New York, in August 1961. The region had ostensibly been Salk-vaccinated, but uptake had been so patchy that the numbers protected remained far below the threshold to achieve herd immunity. The outbreak – 64 paralytic cases in all – was promptly snuffed out by Sabin's oral vaccine, which was accepted gratefully by many who had rejected Salk's.[74]

After that, things moved quickly, pushed ahead by Sabin's hectoring and clever footwork. In June 1961, the American Medical Association (AMA) approved the recommendations of its Vaccine Committee that Sabin's oral vaccine should replace Salk. Salk and O'Connor were outraged by Sabin's 'corrupt' lobbying. Sabin merely said that he had not even known that the AMA had a Vaccine Committee (perhaps forgetting that he had sent some of his vaccine to the Committee's Chair a few months earlier).[75]

The certainty that Sabin's day was dawning was reinforced in spring 1961, when the newly elected President John F. Kennedy requested $1 million from Congress to buy stocks of Sabin's vaccine to deal with future epidemics. The media now dropped Salk and turned to Sabin, laying the ground for the new weekend tradition of 'SOS', or 'Sabin Oral Sundays', which were heavily promoted by the Communicable Disease Center (CDC). This neighbourhood-centred vaccination campaign began in early 1962 and was an instant hit; the child-friendly syrup

Figure 10.6 Queues for Sabin's oral polio vaccine, outside the Municipal Auditorium in San Antonio, Texas, 1962. Photograph by Stafford Smith, reproduced by kind permission of the Centers for Disease Control, Atlanta, Georgia

made a palatable and homely contrast to the tears or brave smiles for those waiting in line for Salk's needle (Figure 10.6). And it was cheap: just 25 cents, with government support to make it free to those who really could not pay.[76]

Checkmate

The tables finally turned in August 1962, with a decision by the Food and Drug Administration to withdraw Salk's vaccine and resume mass vaccination using Sabin's. Predictably, Salk was furious, and continued his bitter tirade about Sabin's duplicity and conspiracy. Salk's accusations were grounded on evidence, but unfortunately so was the conclusion that his vaccine had failed to deliver and now had to be replaced by something better.

Also furious were Sabin's rivals, whose oral attenuated polio vaccines had already been consigned to the dustbin of drugs that had failed to make the grade. Entirely reasonably, Koprowski accused Sabin of underhand dealing and reputational sabotage. But Sabin had insinuated himself into pole position and had pointed to George Dick and others who, like Sabin, had felt it their duty to spell out the hazards of Koprowski's vaccine – and especially the risk that it would switch back to a virulent neurotrophic form which could paralyse.

Herald Cox was also out of the running. The coup de grâce had been delivered by a rushed trial of his vaccine on 40,000 children in Dade County, Florida, conducted in 1959 at the same time that Sabin began his own trial in Cincinnati and Hamilton County. Cox's vaccine was dead from the moment that the first of six children developed paralysis within a few days of vaccination.[77]

This time, as Sabin and others were quick to point out, the resemblance to the Cutter Incident was too blatant to be ignored. Any fool could see that this was paralysis caused by Cox's vaccine virus reverting to its paralysing ancestor. The authorities agreed, and rejected Cox's application for further clinical trials. As the final insult to Cox, one of the first companies to sign up to produce Sabin's oral vaccine was Lederle Laboratories, of Pearl River, New York.[78]

So Albert Sabin, Tom Rivers' 'smart Jew', came out on top after a protracted battle with Hilary Koprowski and especially Jonas Salk. In her biography of Sabin, Angela Matysiak wrote, 'This is a classic American story of the power of science, the virtues of the competitive free-market economy, and good old Yankee ingenuity'.[79] An alternative interpretation is that it was a no-holds-barred brawl between the big egos of refugees from Eastern Europe, in which the sanctity of science and the rules of fair play did not figure prominently. In this instance, though, the end would appear to justify the means.

In late 1962, things appeared to have ended happily for all except for Salk, Koprowski and Cox. However, like all good stories, this one still had a couple of further twists.

11

In the Opposite Corner

Towards the end of his life in 1823, Edward Jenner could have looked back on his achievements with some satisfaction. In barely 25 years, vaccination had made the world a better place. Millions of people had been spared death or mutilation by smallpox, and there were already signs that his prophesy (made in 1801, and based on aspiration rather than evidence) might eventually come true:

> It now becomes too manifest to admit of controversy that the annihilation of the Small Pox, the most dreadful scourge of the human species, must be the final result of this practice.[1]

By the end of the nineteenth century, the area of the planet's surface in the grip of smallpox was clearly shrinking, and entire countries – beginning with Sweden and Puerto Rico – had been swept free of 'the most dreadful scourge'. But not everyone was happy.[2]

From its earliest years, vaccination had mustered enemies as well as supporters. To the growing army of those opposed to vaccination, Jenner was no saint; instead, he was a sloppy experimenter, a cynical exploiter of human misery, and the murderer whose 'disastrous illusion' killed tens of thousands of babies each year. As the twentieth century dawned, a coordinated anti-vaccination movement was gaining momentum, especially in Britain, Europe and North America, and had claimed its first palpable victories. Whole cities, ranging from Jenner's own county town of Gloucester to Montreal, had said no to vaccination and had become shining beacons for all those determined to flout the tyranny of Jenner's invention.

Vaccines have remained a potent source of unhappiness and a unifying force for all those who, for reasons of their own, refuse to submit themselves and their children to the medical ritual of immunisation. Polio vaccines are no exception; hostility and suspicion were stirred up within weeks of the introduction of Jonas Salk's vaccine in 1955. Opposition to polio vaccines has had real consequences: thousands of children paralysed by polio outbreaks that should never have happened, the murder of vaccination workers in Pakistan and Nigeria, and the derailing of the last push to exterminate polio for all time.

Why would people take against something with such obvious benefits? Their motives are powerful, personal and as diverse as humanity itself, from an unshakable trust in the Scriptures to the belief that vaccination is a pointless and dangerous charade whose hazards are covered up by the medical establishment.

Strength of belief

> Whether I have my children vaccinated or not does not matter to me, because I don't believe in it. I believe that if God wants to spare my children from an accident, then He will spare them from it.
>
> A Dutch Orthodox Protestant parent, interviewed in 2009.[3]

Religious objections to vaccination run back to the turn of the nineteenth century, when some men of the church slated Jenner's vaccine as a sacrilegious attempt to override the Divine will. God alone had the right to determine how and when each of His subjects would die. If He had lined you up to meet smallpox, the 'Angel of Death', then so be it.

During the intervening two centuries, religions have generally come to accept the benefits of vaccination. Some devout Catholics refuse vaccines prepared in cells from aborted fetuses, but most Christians have no objections. The same is true of mainstream Islam; indeed, pilgrims can only join the Hajj to Mecca if they have been vaccinated against meningitis and polio. There remain a few exceptions, Christian and Muslim – and these have proved conclusively that vaccination works, but only if you have it.

Christians who refuse vaccination are highly concentrated in the Dutch *Bijbelgordel* (Bible Belt), a broad seam of mostly rural districts that runs north-east to south-west through the heart of Holland (Figure 11.1). The Belt is home to about 250,000 Orthodox Protestants, roughly 1.5 per cent of the Dutch population. Most Orthodox Protestants adhere to the Heidelberg Catechism of 1563 and the principle that the works of man must not interfere with Divine providence. They resisted smallpox vaccination from the start, and many applied the same veto when vaccines became available against polio, measles, mumps and rubella. Vaccination rates generally exceed 97 per cent in the general Dutch population, but average only 60 per cent among the children of the Bible Belt, 40 per cent of whom are not vaccinated at all.[4]

In public health terms, this was a disaster waiting to happen. In 1978, a polio epidemic swept through the Bible Belt, paralysing 110 children in seven months. Most were under the age of two, and none had ever been vaccinated. The value of herd immunity was illustrated by the complete absence of cases in the general Dutch population.[5]

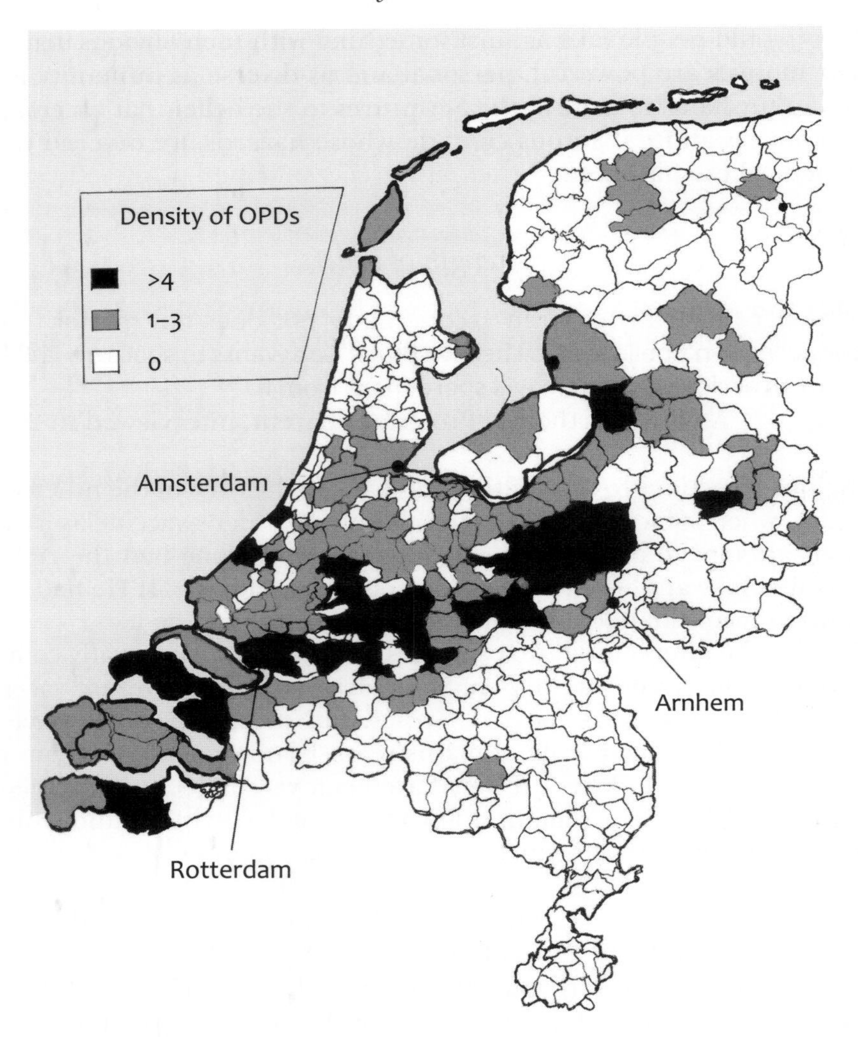

Figure 11.1 The Dutch 'Bible Belt', where opposition to vaccination is common among Orthodox Protestant Demoninations (OPDs). Here, the number of OPD congregations is shown for each of the districts in Holland. [5] Reproduced by kind permission of Dr Helma Ruijs, Nijmegen

Members of the Bible Belt community unwittingly repeated the experiment a few months later when they visited relatives tucked away in unvaccinated pockets of Old Dutch Protestantism in North America. They left a trail of paralysed children in their wake: 11 cases in Canada, followed by 10 more scattered across four American states. Again, there were no cases in the well-vaccinated general population. These transatlantic outbreaks were no coincidence: molecular fingerprinting showed that the same wild type 1 poliovirus was responsible, and was still circulating back home in the Bible Belt.[6]

During the 1978 epidemic, the Orthodox Protestant Churches relaxed their opposition to vaccination, leaving the choice to parents and their consciences. However, the message was mixed, as many preachers still openly condemned vaccination. These were clearly testing times for believers. Some parents had their children vaccinated, while others kept to their faith. Across the Bible Belt, the vaccination rate remained low and was almost zero in some communities.[3,4]

The predictable happened again in 1992–93, with a further epidemic of polio (71 children paralysed), followed by measles (1999–2000), rubella (2004–05) and mumps (2007–08).[7] As before, the non-Belt population was spared. Polio (this time, a Type 3 strain) was again taken across the Atlantic to meet the brethren in Canada. Several children in Alberta picked it up, but luckily, none was paralysed.[8]

Today, there are signs of change in the Bible Belt. Increasing numbers of parents are breaking with tradition and viewing vaccination in a different light, as a gift of God that can be used with gratitude and without guilt. With this comes the hope that history will not repeat itself.

Another battleground between Christianity and vaccination is in Southern Africa. Here, a much harsher interpretation of the Scriptures is being played out against the backdrop of immunisation campaigns to stamp out measles, a seasoned killer of African children.

In September 2010, six parents from Chibila in Binga District, Zimbabwe, were arrested for hiding their children from vaccinators. All belonged to the Johanne Marange Apostolic Church. Measles struck their small village with a vengeance, with 35 cases and five deaths.[9] It could have been worse. A few weeks earlier, another member of the same church, Elvis Tapiwa Magusa from Magunje, was jailed for 18 months after police broke into his home and freed three children so that they could be vaccinated. Unfortunately, the police arrived too late to save Magusa's other four children, who had died of measles in the house while their father watched and dosed them with holy water.[10]

The Apostolic Church was founded in 1932 by Johanne Marange after God spoke to him and endowed him and his followers with supernatural powers. We can only guess how Marange would have divided his blessings between two of his disciples, Jeremiah and Beauty Makumbe from Soti. On 24 September 2010, Beauty gave in to her maternal instincts and took their children to be vaccinated against measles. When she returned home, Jeremiah beat her to death with an iron rod.[11]

Needles and poisons

Many of today's arguments against polio vaccines can be traced back to the hostility that greeted Jenner's original cowpox inoculation, over two centuries ago. Many saw vaccination as an evil whose dangers were covered up by Jenner and his supporters. Jenner was a godsend to conspiracy theorists. He rejected solid evidence that vaccinated people could catch smallpox (and die from it), claiming that this was a lie invented by his enemies. His disciples were no better. In 1870, *The Times* printed a letter from 500 highly respected doctors, denying that vaccination could transmit syphilis – even though many of them must have seen the well-publicised reports which proved that this occurred.[12]

Mistrust grew during the nineteenth century, as vaccination became big business and an obligatory passport for entry to foreign countries, the armed forces and primary school. The English government became a co-conspirator when the Vaccination Acts of 1840–71 made vaccination compulsory. The Acts were deeply unpopular and socially divisive and, by the time they were repealed in 1907, had spawned a militant anti-vaccination movement.[12]

The previous year – 1906 – saw the publication in America of *Crimes of the Cowpox Ring*, an impassioned exposé of a cartel that supposedly comprised vaccine producers, doctors and government.[13] The author was Lora C. Little, an energetic natural health campaigner and anti-vaccinationist from Portland, Oregon. *Crimes* (subtitled *Some moving pictures thrown on the dead wall of official silence*) was a compendium of 300 medical catastrophes following vaccination, made all the worse by the brutality and deviousness of the authorities. Little had a personal interest to declare; Case Number 30 was her only son, Kenneth, who died at the age of seven after being hauled out of class and held down to be vaccinated (Figure 11.2).

The anti-vaccinationists' main weapon was to exploit public fears that vaccination was dangerous, and that the hazards were deliberately concealed. Risks were inflated or invented, sometimes backed up by bent statistics. Just a few of the catastrophes in Little's *Crimes* could reasonably be blamed on vaccination (for example, blood poisoning from contaminated vaccine). The rest, including her son's death – from diphtheria, seven months after being vaccinated – were tragic and shocking, but coincidental.

Another tactic was to deny that smallpox was caused by a virus, which at a stroke exploded the whole rationale for vaccination. It was therefore claimed (and sometimes endorsed by tame doctors) that smallpox was caused by twisted vertebrae in the neck, emotional shocks or sunspots. It therefore followed that vaccination had nothing to do with the disappearance of smallpox. Like cholera and typhoid, it had been swept away by sanitation and the other processes of civilisation.[14]

The enemies of polio vaccines were already lying in waiting for the results of Salk's field trial in 1955. Their objections plugged neatly into many of the themes that had been so well rehearsed with smallpox

KENNETH MARION LITTLE.

Only child of Mrs. Lora C. Little. Vaccinated Sept., 1895. Died from the effects, April 10, 1896. Age 7 yrs. 3 mos. (No. 30.)

Figure 11.2 Kenneth Little, only son of Lora C. Little, who died aged 7 in April 1896 'from the effects' of vaccination. From *Crimes of the Cowpox Ring*. Reproduced by kind permission of the Gladstone Library, Hawarden, North Wales

vaccine. The natural descendant of Little's *Crimes of the Cowpox Ring* was *The poisoned needle*, a lengthy anti-vaccination tract published in 1957 by Eleanor McBean.[15] Like Little, McBean was an activist on many fronts – anti-sugar, anti-smoking, anti-doctor and above all, anti-vaccination. She was also a gifted purveyor of conspiracies, as suggested by some of the chapter titles from her book: 'A planned medical conspiracy', 'Vaccination deaths not reported' and 'Gangsters in the Government'.

McBean devotes an entire chapter of *The poisoned needle* to the new polio vaccine.[16] She mocks Albert Sabin for 'toying with the idea of a virus just between the dead and the live variety... [which] would be laughable if it were not so pathetic'. But she concentrates her vitriol on Salk, 'a confused little man with a batch of putrid serum who thinks *he* can improve on nature'.

McBean's attack is multi-pronged. The vaccine is pointless, because viruses have 'nothing to do with the cause of polio'; Landsteiner's experiments are only believed by 'backward medical doctors' who are still 'floundering around in the delusion of virus causation'. Instead, polio is caused by sugar, cola, insecticides, cyanide and internal poisons generated by negative emotions – and the polio vaccine itself.

To hammer the point home, McBean wheels in medical 'experts', one of whom is already familiar to us: Dr Ralph R. Scobey, the poliovirus agnostic who believed that pesticides and other poisons caused polio. Scobey was convinced that the medical establishment was out to suppress his theory – hence his paper entitled, *The poison cause of poliomyelitis and obstruction to its investigation*. According to Scobey, the 'so-called poliovirus' is just a 'biochemical substance produced by the disease'; it might paralyse monkeys, but is irrelevant to human polio.

To McBean, the scientific concept of 'poliomyelitis' is also nonsense. Grey matter, McBean insists, is found not just in nerve cells but in every cell of the body. This explains why polio is identical to the 170 other disorders (including pellagra and cholera) listed by the good Dr Scobey in that prestigious scientific journal, the *Archives of Pediatrics*.

Next in the firing line are ordinary doctors – ignorant, arrogant and dangerous. She quotes Dr Shelton, hygienist, who explains that physicians 'kill and cripple their patients with their oppressive measures and then blame the results on an unknown virus'. By contrast, natural therapies, based on a proper understanding of the real causes of polio, will cure virtually all cases. McBean mentions a survey in 1946, which showed complete recovery rates in polio patients of 91–100 per cent with 'hygienic treatment' or chiropractic and 72 per cent with Sister Kenny's therapy – as compared with 35 per cent when nature takes its course unassisted and a mere 17 per cent when doctors interfere. Unfortunately, McBean does not provide the source of this information.[16]

McBean was one of a loose alliance of 'alternative' practitioners, united in their mission to destroy vaccination against polio and other infections. Their tactics were broadly similar, and echoed those of Lora C. Little: playing up fears of a doctors' conspiracy in order to undermine vaccination and conventional medicine, while offering better, safer 'natural' alternatives.

One such is chiropractic, based around a 'scientific' dogma formulated in the 1890s which blamed all human afflictions on 'luxation' (misalignment) of vertebrae in the neck. Luxation supposedly interfered with the flow of energy through the nerves; all diseases, even the horrors of smallpox, could be cured by tweaking the troublesome vertebrae back into line. By contrast, orthodox medicine and vaccination were frauds to exploit the gullible. From the start, the founders of chiropractic were determined to 'give the death blow to the vaccine poison swindle'.[17] The chiropractors who touted for trade during the San Angelo outbreak in 1949 were therefore simply following doctrine.[18]

Homoeopaths were in there too. Their version of the truth about polio is revealed in *Homoeopathy and epidemic diseases* (1967), which was ghosted for Dr Dorothy Sheppard (deceased) by Mrs G.E. Robinson.[19] Regarding polio, 'homoeopathy provides an absolutely safe method without any serious after-effects, both for protection and cure of this dreaded disease'. Dr Shepherd feels it is important to 'pass on our knowledge to that great majority of our medical brethren who have not heard of it', because of the 'useless effects and sometimes fatal results ... [of] drugs and the other recognised modern means of treatment such as the serum and vaccines'.

Shepherd does not discuss the cause of polio; the taboo word 'virus' is mentioned only twice in passing, in quotations from others. Instead, she reminds readers of Samuel Hahnemann's Law of Similars (1796), which states that a substance which causes particular symptoms will cure someone suffering from those symptoms. The key to understanding polio lay with the Indian grass pea, *Lathyrus sativus*, which causes paralysis (this is true: the legume contains a neurotoxin which destroys motoneurones).

According to Dr Grimmer of Chicago, *Lathyrus* given in doses of '30th or 200th potency once every three weeks' was 'the most perfect antidote for prevention and cure' of polio. The '200th potency' means a 10-fold dilution in water, repeated 200 times, with or without a mystical tap ('succussion') on the test-tube at each step. Cynics might feel that the resulting solution is too weak to do much good: all the water on the planet would not be enough to produce a one in 10^{200} dilution from a single grass pea. However, *Lathyrus* had enjoyed 'one hundred percent success during the last 30 years in many epidemics'. Shepherd adds, 'Does this sound too good to be true? Try it my friends and see'.

If they did, they did not get around to writing it up.

Risks and benefits

'Primum non nocere' – first do no harm – is the guiding principle of therapeutics. Unfortunately, the medicine that works but has no side effects is yet to be discovered.

The complications of polio vaccines are vanishingly rare, but they must be acknowledged and confronted. The failure to do this has been, and still is, exploited by those opposed to vaccination.

Ideal for propaganda purposes is the 'own goal' tragedy of previously healthy children paralysed by polio vaccines. The Cutter Incident was a gift on a plate, as was the dawning realisation that Sabin's vaccine could also cause paralysis. The risk of being paralysed by Sabin's vaccine is extremely low (just one in 750,000), but in America during the 1970s more cases of paralysis were caused by the vaccine than by naturally acquired polio.[20] This was, of course, because vaccination had wiped out wild-type polio across the United States – a fact that is carefully ignored by anti-vaccinationists. However, vaccine-induced paralysis does occur, and is devastating for victims and their families. Denying the possibility – as Sabin did until the end of his life – does no service to them or to the reputation of science.[21]

Another pervasive fear is that vaccines contain toxins or other viruses, and that the authorities are concealing this knowledge. Here, too, the establishment has helped to give credibility to conspiracy theorists. In 1960, it was discovered that batches of Salk and Sabin vaccines were contaminated with a virus that had lain undetected in cultures of monkey kidney cells. The virus was named Simian (monkey) Virus 40 (SV40), as it was the fortieth virus found hidden away inside apparently healthy Rhesus monkeys.[22] SV40 survives formalin treatment better than the poliovirus, and inevitably entered some vaccinated subjects. It is a DNA virus with worrying properties, killing cells in culture and causing cancers when injected into hamsters.

Long-term follow-up studies of subjects who received SV40-contaminated polio vaccines have not found any excess cancer risk – but during the 1960s and '70s, this lay in the future and was no comfort for the many parents who were terrified that their children had been given a cancer-causing virus. It was also easy to accuse the authorities of covering up the risk, because they did. In 1959, the US government laboratory responsible for monitoring vaccine safety had prevented Dr Bernice Eddy from publishing her research on a cancer-inducing virus which she found in the kidneys of healthy Rhesus monkeys. This virus later turned out to be SV40.[23]

The big one

This brings us to one of the great medical controversies of the last half-century. It had everything: vivid characters and a supporting cast of millions, exotic locations, enough credibility to shake up serious scientists, no end of reasons to suspect a cover-up – and a jaw-dropping storyline. The plot in a nutshell: Hilary Koprowski's oral polio vaccine infected African children with a monkey virus that mutated into HIV, and so unleashed the acquired immune deficiency syndrome (AIDS) pandemic.

The discovery of SV40 had proved that polio vaccines prepared in cultures of monkey kidney cells (the norm until human cell cultures replaced them in 1963) could not be guaranteed free of monkey viruses. In 1967, the 'Marburg' virus which lived invisibly in African green monkeys proved that this loophole could be deadly. This previously unknown virus broke into man from monkeys imported from Uganda into the Behringwerke laboratory in Marburg, Germany, with catastrophic results: 31 cases of fulminating haemorrhagic fever and seven deaths.[24]

Five years before that, Koprowski's vaccine had been abandoned in favour of Sabin's. It would have slipped quietly into oblivion had it not been for another area of virus research which expanded exponentially during the late 1980s. The viruses in the ascendancy were HIV-1 and HIV-2, the two strains of human immunodeficiency virus which cause AIDS.

AIDS appeared to be a new disease, with an origin in the recent past. The first cluster of cases, in homosexual men around Los Angeles, was reported in 1981. Once the clinical pattern was recognised, other cases were identified retrospectively. From there, the trail seemed to run back to Africa – and apes. The human HIV-1 and HIV-2 viruses turned out to have counterparts in monkeys, the simian immunodeficiency viruses (SIVs). Like other monkey viruses, SIVs lived in harmony with their host but caused a fatal AIDS-like disease if they entered other monkeys.[25]

Molecular genetic analyses showed that the human HIVs had descended from two distinct SIVs in African monkeys: HIV-1 from SIVcpz (in chimpanzees) and HIV-2 from SIVsm in the sooty mangabey. From the average mutation rate of these viruses, an educated (but intrinsically imprecise) guess was made that SIVs had turned into HIVs some time around the mid-twentieth century.[25]

Apart from accidents in research laboratories, how could a monkey virus get into man? Possibilities included monkey bites and handling or eating infected monkey meat – a suggestion backed up by gruesome photos of butchered chimps and other monkeys sold as 'bush meat'.[26]

Figure 11.3 Mass vaccination in the Belgian Congo in early 1957, using Hilary Koprowski's oral polio vaccine which contained the attenuated 'CHAT' virus. Children in the Ruzizi Valley are being vaccinated by Koprowski's colleague Dr Agnes Flack. Reproduced by kind permission of Hilary Koprowski

And as God's gift to conspiracy theorists, there was also Koprowski's oral polio vaccine.[27] Produced at the Wistar Institute in Philadelphia in cultures of monkey kidney cells, this had been given to over one million children in the Belgian Congo between 1957 and 1960 (Figure 11.3). The stage was set for a drama that badly damaged public confidence in doctors and scientists, and came close to wrecking Koprowski and his career.

In autumn 1991, two scientists from San Francisco, Baine Elswood and Raphael Stricker, let slip that they were working on a hypothesis paper with the ominous title, *Polio vaccines and the origins of AIDS*. Word soon filtered out, but not from them – partly because the paper which they sent to *Research in Virology* had been held up for closer scrutiny by the editor.[28]

The story was broken by a freelance journalist, Tom Curtis, who had spotted the potential for a scoop. On 9 March 1992, he published a piece in *Rolling Stone*. Back in the 1960s, *Rolling Stone* had begun life as a free-wheeling, hippie-friendly magazine and had not quite made the transition into mainstream respectability. Curtis' title hints at the kind of reader he hoped to snare: *The origin of AIDS. A startling new theory attempts to answer the question 'Was it an Act of God or an Act of Man?'*[29]

Curtis came down firmly in favour of an Act of Man, and specifically an Act of Koprowski. He theorised that HIV-1 had arisen in humans who had accidentally been given an SIV from infected monkey tissues. Curtis had a time, a place and a suspect: 1957–60, the Belgian Congo, and 'CHAT', the modified Type 1 poliovirus in Koprowski's vaccine. The timing of Koprowski's mass vaccination campaign fitted nicely, and Curtis' suspicions deepened when he interviewed Koprowski – who laughed off the notion but then could not remember where the monkey kidneys used to grow CHAT had come from. Curtis ended his article: 'If the Congo vaccine turns out not to be the way that AIDS got started in people, it will be because medicine was lucky, not because it was infallible'.

Curtis got his scoop and created a furore, which soon dragged in scientists and provoked a string of articles in *Science*. The opening shot was a caustic piece which poked fun at *Rolling Stone* and Curtis' hypothesis, which contained 'not one picogramme of evidence' (a picogramme is one-millionth of one-millionth of a gram). This was followed by a spirited riposte from Curtis, challenging scientists to prove him wrong, and then a letter from Koprowski, which was erudite but unfocused and, to some, smacked of a cover-up.[30]

The Wistar Institute – of which Koprowski was the director – was eventually goaded into investigating Curtis' allegations. Its verdict, delivered in October 1992, was guaranteed to fan the flames of suspicion. The probability that Koprowski's vaccine had introduced an AIDS precursor virus into man was 'extremely low'. However, the origins of the monkey kidneys used to prepare the vaccine were 'unlikely ever to be determined with certainty', while 'the possible presence of SIV ... in the vaccine preparation cannot be discounted'.[31]

In December 1992, Koprowski sued *Rolling Stone* for libel. The magazine later printed a statement that absolved Koprowski of being the father of AIDS, but left dangling the possibility that Curtis had been right after all.

Enter Edward Hooper, one-time BBC correspondent and author of best-selling books about AIDS, which he had researched intensively in Equatorial West Africa. Hooper was 'riveted' by Curtis' article, because some areas covered by the CHAT vaccination campaign coincided with foci of AIDS that he knew at first hand. Also, he believed that Curtis had missed a vital lead. Hooper speculated that SIV-infected kidneys had not come from any old monkey, but from chimpanzees, the natural bearers of the HIV-1 precursor, SIVcpz. Crucially, chimpanzees had been in a special laboratory compound at Lindi, eight miles north-west of Stanleyville, where Koprowski's team had tested polio and other vaccines.

In 1999, after several years' research, Hooper published a massive 1,070-page book entitled, *The River: a journey back to the source of HIV and AIDS*.[31] Hooper believed that he had sorted 'the wheat from the chaff', but some obvious chaff remained, including the bizarre claim that both William H. Park and Maurice Brodie had 'very possibly died of poliomyelitis, contracted from their own inadequately refined vaccine'.[32]

Hooper maintained that chimpanzees' kidneys, contaminated with SIVcpz, had found their way into the CHAT vaccine production line at the Wistar. Naturally, Koprowski strenuously denied that chimpanzees' kidneys, rather than those routinely obtained from rhesus monkeys or vervets, had ever been used to make vaccines. So did Koprowski's colleague, Stanley Plotkin, a highly published virologist and senior author of *Vaccines*, the leading textbook in the field which weighed even more than *The River*. So did everyone else directly associated with vaccine production at the Wistar or the experimental chimp facility at Lindi.

However, Hooper would not give up. Gaps in the 78-year-old Koprowski's recollection of events 35 years earlier seemed to be a cover-up.[33] Even the name 'CHAT' fell under Hooper's suspicion. Koprowski and Plotkin insisted that this was short for 'Charlton', the surname of the little girl in the residential home for disabled children in Sonoma, from whose stools the vaccine virus had been isolated in 1956. Hooper located the girl's father and found that he was indeed Mr Charlton, but remained convinced that 'CHAT' held a secret code which, when cracked, would prove that Koprowski was lying. Hooper suggested that CHAT stood for '**CH**impanzee-**A**dapted and **T**ested.[33]

Sales of *The River* prospered, and the controversy refused to go away. Finally, the scientific establishment decided to bring the matter to a head, by giving both sides a fair hearing. A two-day meeting, 'Origins of HIV and the AIDS epidemic', was convened on neutral ground at the Royal Society in London on 11–12 September 2000.

For most of the participants, this was a chance to see in action the combatants whose names had become familiar in the lay media and the scientific press. For Koprowski, now aged 84 and painfully aware that being blameless did not guarantee a happy outcome, it might well have felt like being in the dock at the Nuremberg Medical War Crimes Trial.

The programme was well planned, beginning with expert overviews of the history of AIDS, updated family trees of HIVs and SIVs, and accounts of previous catastrophes when animal viruses had crossed into man. Session II, 'Oral polio vaccines', pitted Hooper against Plotkin; Koprowski had the penultimate word before the summing-up on the second afternoon. Hooper's supporters could argue that it was impossible for him to condense a 1,000-page book into a 25-minute talk and 10 minutes for questions[34] – but that was all that every speaker was allowed.

When Hooper presented his case, he put forward a completely new accusation: the Wistar had no record of using chimpanzee kidneys, because the vaccine had actually been made locally in African veterinary laboratories, newly tooled up for that purpose. Koprowski and his supporters had not prepared for this wild card, but they had everything else covered. Meticulous combing of records at Wistar and Lindi confirmed that chimpanzee kidneys had never come anywhere near vaccine production lines. Samples of Koprowski's oral polio were found to contain DNA from monkeys, but no trace of chimpanzee DNA.[35] For good measure, it had been shown that SIV-1 could not survive in the CHAT vaccine medium. Finally, accurate molecular dating now pushed the transition from SIV to HIV back to 1930 or earlier – a quarter of a century before Koprowski brought his vaccine to the Congo.[36]

Later, Plotkin wrote a blow-by-blow account of the evidence that demolished each of Hooper's claims which, for many, drew a line firmly under the episode.[37]

Game, set and match to Koprowski? Not if you believe that the establishment will do anything to cover-up mistakes and malefaction. When *The River* was published in 1999, Hooper claimed that he was '95 per cent persuaded of the merits of the vaccine theory' by 2006, his index of persuasion had risen to 99.9 per cent.[38]

The virus that never was (reprise)

> Today the World Health Organisation still credits Landsteiner and Popper as having found the polio virus ... Why it does so is inexplicable.
>
> Janine Roberts, *The Ecologist*, May 2004[39]

The world moves on, but not necessarily at the same speed for everyone on board. Even in the twenty-first century, it is still easy to find people who have not noticed that, sometime during the last 200 years, germ theory somehow became germ fact.

Today, a few clicks of the mouse will transport you into a strange parallel world where the flow of history has been frozen at carefully selected moments: the poliovirus remains a figment of scientists' imagination; Benjamin Sandler's claim that low blood sugar precipitates polio is uncontested; and Hilary Koprowski is still guilty of bringing AIDS to mankind.

Here, Eleanor McBean and her like live on as heroes, without the threat of challenge or rebuttal. Thanks to the global penetration of the Internet, their audiences are now hundreds or thousands of times larger than during their lifetimes. Their works are propagated from website to website and occasionally embroidered by their present-day disciples, like restorers patching over moth damage in a priceless but time-worn tapestry.

The poliovirus denialists also celebrate Dr Morton Biskind's courageous pursuit of the 'unfunded, ostracised theory of poison causality' and reaffirm his conviction that polio is caused by pesticides. Polio disappeared from America only because DDT – 'the greatest mass poisoning of all time' – was withdrawn; polio vaccines are as irrelevant now as they were in the late 1950s.

Others concede that the poliovirus may exist after all, but still has nothing to do with polio. Biskind's 'innocent bystander' theory of 1952 has been updated by Jim West, engineer, musician and anti-vaccinationist.[40] According to West, the poliovirus is created as a byproduct of pesticide poisoning by 'accelerated genetic recombination', an enigmatic process not known to anyone else, which somehow spins viral RNA out of human DNA.

To West and others like him, the words of poliovirus agnostic doctors such as Biskind and Scobey are gospel – especially when blessed by scientific peer review and printed in 'established medical journals'. West slavishly reproduces Scobey's paper in the *Archives of Pediatrics*, listing the 170 diseases which Scobey believes to be the same as polio. However, West's rendition is not entirely faithful: a transcription error transforms the all-powerful 'cure for polio', British Anti-Lewisite (BAL), into PAL, the well-known British dogfood. Intriguingly, 'PAL' is a watermark that can be traced into many anti-vaccination websites, confirming that it is much easier to cut and paste than to try to understand what is on the page.

What about the biggest polio conspiracy of all? After the Grand Inquisition at the Royal Society in September 2000, mainstream science lost interest in the myth that Koprowski's vaccine caused the AIDS pandemic. Now that the ticking of the mutation-rate clock has been calibrated more precisely, the emergence of human AIDS from its SIV precursors has been put back to between 1902 and 1921.[41] This interval happens to include 6 December 1916, the day on which Koprowski was born. Moreover, the variant of SIVcpz circulating in Congolese chimpanzees has been typed in detail and cannot have given rise to HIV-1.[42] Nonetheless, it is easy to find websites that still vilify Koprowski as the father of AIDS – although some acknowledge that Koprowski might not have been entirely responsible, as 1957 was also the year of peak atmospheric fallout from nuclear weapons testing.

As in Lora C. Little's day over a century ago, it is a short step from anti-vaccination to 'natural' and 'alternative' therapies. Prowling the Internet today are natural healers, homoeopaths, chiropractors and nutritional and lifestyle advisers. Forget orthodox medicine, the conspiracy which exploits the gullible and does all harm and no good. Instead, trust in natural solutions: an authentic 'paleodiet', as eaten back in the Stone Age; an electrical 'zapper' which kills 'all types of worms, bacteria, viruses and fungi'; books such as *The cure of all diseases*; or internal cleansing therapy, shown to be effective by colour photographs of satisfied clients' bowel movements.

And polio, even though it had long since disappeared from the Western world, is still out there. We can thank Benjamin Sandler, killer of ice cream sales in North Carolina, for drawing attention to yet another hazard of sugar, alongside epilepsy, hyperactivity in children, haemorrhoids and cancer.[43] Depending on which website you visit and on which day, polio may be ranked 69th out of 78 reasons to avoid sugar, or 113th out of 124, or 124th out of 157.

Front line

Polio vaccines have been used successfully in over 50 Muslim countries and have wiped out polio as effectively there as elsewhere in the world. At the time of writing, there were just three exceptions: Nigeria (especially its northern States of Kano, Zamfara and Kaduna), Pakistan and Afghanistan.

Polio has long had a firm foothold in northern Nigeria, largely because of the tropical heat and difficult terrain which make it difficult to keep oral vaccine viable during the long treks to reach remote communities. There is also a general reluctance to have children vaccinated, stemming from a deeply ingrained mistrust of Westerners and their medicines. This was not helped in 1996 when the American drug giant Pfizer experimented in Kano with a new antibiotic against meningitis. Several children died, and the trial was later judged both illegal and unethical.

In 2003, Nigeria contained 45 per cent of all polio cases worldwide, and polio was exported across its porous frontiers to eight other African countries in that year.[44] Something had to be done. The 'Kick Polio out of Africa' campaign led the way in October 2003, aiming to immunise 15 million children in Nigeria and neighbouring countries. This 'final onslaught' soon ran into fierce resistance in northern Nigeria, and especially the Muslim state of Kano. Vaccination was boycotted by religious and political leaders, including the Kano State government and the hard-line Islamist Supreme Council for Sharia in Nigeria.[45]

The confrontation had little to do with religion; instead, conspiracy was a key strand of the propaganda. The polio vaccine was not a gift from the West; it was a weapon designed to weaken and kill Muslims. It contained hormones that make men impotent and both sexes infertile, together with HIV and other viruses that cause cancer. These supposed contaminants played directly into fears that were widely held in Nigeria: death from AIDS, impotence and the fall of Islam. The State government put things into perspective: it was much better to lose some children to paralytic polio than to have 'perhaps millions of girl-children' rendered sterile. This stark message was given medical credibility because it was endorsed by Dr Datti Ahmed, a senior physician in Kano. On 11 February 2004, he said:

> Polio vaccines are corrupted and tainted by evildoers from America and their Western allies...We believe that modern-day Hitlers have deliberately adulterated the oral polio vaccines with anti-fertility drugs and viruses which are known to cause HIV and AIDS.[45]

Ahmed had another interest to declare: he was head of the Supreme Council for Sharia in Nigeria.[45]

The boycott was successful, vaccination foundered and polio flourished. A world-famous Islamist scholar, Sheikh Yusuf Al-Qaradawi, protested that it was 'as clear as sunlight' that polio vaccine was lawful under Islam. He was ignored. The impasse was only broken the following summer, after the World Health Organisation (WHO) and the United Nations Children's Fund (UNICEF) arranged for the vaccine to be tested independently, watched by observers from Kano, and was proved to be free of hormones and HIV. The deciding factor was a clean report from a laboratory in Indonesia, also a Muslim country, which then took over manufacture of the polio vaccine destined for Nigeria.

The boycott was lifted in July 2004 after 11 months, but has cast a long shadow. In 2005, the Type 1 poliovirus prevalent in Nigeria caused an outbreak in Yemen and Saudi Arabia, probably spread by migrant workers or pilgrims on the Hajj; 1,500 children were paralysed. Closer to home, the same virus broke out in Botswana and Sudan. And in 2006, Nigeria accounted for 80 per cent of all cases worldwide, an even higher proportion than in 2003.[46]

The other hot spot of confrontation embraces Afghanistan and Pakistan and particularly the Tribal Area, the ragged strip of high mountains and deep valleys along the border between the two countries – a region that is dominated by the Taliban.

The offensive against polio vaccination began with the now-familiar message that the vaccine was an American plot to sterilise young

Muslims and subvert the will of Allah, broadcast by clerics by radio and loudspeakers on mosques. When mere words proved too weak, the Taliban turned to intimidation, threatening violence to vaccinators and all those who allowed their children to be ensnared by the conspiracy. And when kidnappings and beatings failed to kill off the vaccination campaign, the Taliban brought their people to heel in the usual way.

In February 2009, Abdul Ghani Marwat, head of the polio vaccination campaign in Bajaur in the Tribal Area, was killed by a bomb on his way home from negotiations with a cleric near the Afghan border.[47] At last, the message got through. Vaccinators and health workers wore black armbands and went on strike for three days; some were too frightened to return to work afterwards.[48]

And the parents of an estimated 24,000 children refused to have them vaccinated.

Flip-side

To round off this chapter of beliefs and intrigue, we return to America – and a man who inspired even more conspiracy theories than the pioneers of polio vaccines.

This was Franklin D. Roosevelt, a towering figure in the history of polio and in the much wider arena of twentieth-century American and world politics. Roosevelt provided the suspicious with plenty of promising material: his clumsy attempt to cover-up his involvement in the homosexuality scandal at the Newport naval base in 1921; the intricate deceptions to conceal his illness and its aftermath, all the way from Campobello to the White House and Yalta; and his bellicosity and pro-British leanings which dragged America into the Second World War.

The conspiracy aficionados rose to the challenge. Grandest by far was the 'Mother of Conspiracies', in which Roosevelt goaded the Japanese into bombing Pearl Harbor to legitimise America's entry into the War – and deliberately did nothing to stop the Japanese attack, which he knew about in advance.[49] More modest are the claims that FDR's paralysis in 1921 was caused by toxic chemicals released into the pristine waters of the Bay of Fundy from factories on the coast of Maine. Failing that, he was paralysed by flu vaccine (this would involve time travel as well as conspiracy, as flu vaccine did not become available for another 15 years).[50]

Even small change is drawn into the web of conspiracy surrounding FDR. The Roosevelt dime has been in circulation continuously since the day in 1946 that would have been his sixty-fourth birthday (Figure 11.4).

Figure 11.4 The reverse of the Roosevelt dime. This was first issued in silver on 30 January 1946, which would have been Roosevelt's sixty-fourth birthday

The reverse of the coin carries Roosevelt's profile. Below it, barely visible without a hand-lens, are the tiny initials 'JS'.

John Sinnott, the US Mint's master engraver who designed all American coins at the time? Wrong. Thanks to the Soviet agent who had infiltrated the Mint, this could only be the fellow conspirator who sat at Roosevelt's left hand at Yalta. JS: Josef Stalin.[51]

12

Loose Ends and a Gordian Knot

When I began writing this book in spring 2011, it seemed that the story of polio would soon reach a satisfying conclusion, with all the strands neatly tied up and the disease itself safely consigned to the past. Instead, it has turned into a cliffhanger, and one that may leave us dangling in a frustrating limbo for years or even decades to come.

Apart from that final page, waiting to receive whichever ending destiny comes up with, the other chapters in the history of polio have essentially been closed. A vast body of knowledge has been built up about polio, and many landmarks in scientific and medical understanding deserve to be celebrated. A good place to begin is with the original villain of the piece – the virus which, unlike the species it targets, does not contain an atom of malice.

State of the art

We may have known a lot about the poliovirus in 1960, but that was before the explosion of knowledge about molecular biology, which has allowed us to rewrite the genetic code as well as to read its messages. Now, the last secrets of the poliovirus have been laid bare, and it has even been built from scratch using chemicals off the shelves of biochemistry laboratories.

The poliovirus is an RNA virus, like those that cause influenza and hepatitis A, and the 'winter vomiting bug', the norovirus. The poliovirus was the first animal virus to have its genome sequenced, in 1981.[1] The single RNA molecule coiled up in its heart consists of 7,440 nucleotides and contains just 11 genes (Figure 12.1). Compared with the 200,000 nucleotides and 200 genes of the variola virus which caused smallpox, this is short and primitive. However, the poliovirus is a triumph of simplicity. Locked away in those 11 genes is all the information the virus needs to remain in suspended inanimation until prompted to lock onto a target cell, force its way inside and coerce the cell into making thousands of replicas of itself.

Four of the poliovirus's eleven proteins make up the subunits of its outer coat (Figure 12.1). The other seven 'control' proteins are stealth weapons that reprogramme the cell's machinery to churn out copies of the virus's RNA and proteins. Hot off the production line, the viral RNA and protein

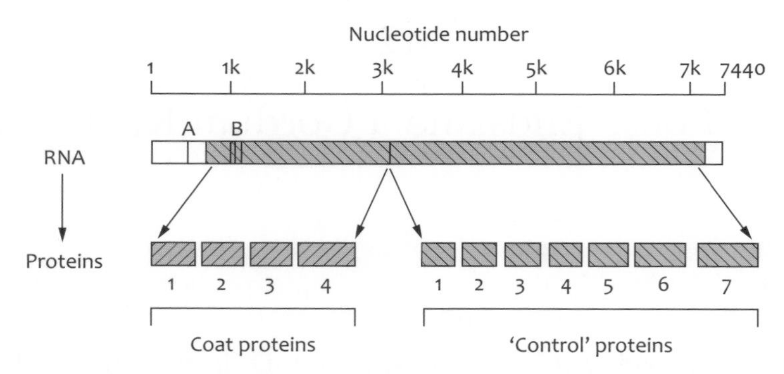

Figure 12.1 The RNA genome of the poliovirus comprises 7,440 nucleotides and contains just 11 genes. Four genes encode the proteins that make up the outer coat of the virus. The other 7 'control' proteins force the cell to synthesise thousands of copies of the virus. A and B indicate two key sites where mutations can cripple the virus's ability to paralyse – the basis of the attenuated viruses used in oral polio vaccine

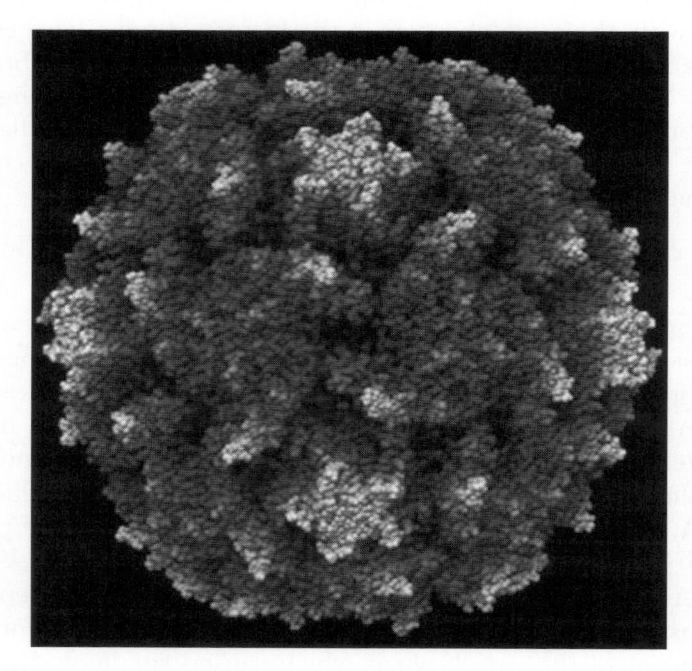

Figure 12.2 Computer-generated image of the poliovirus, with the coat proteins highlighted in different shades. The particle has icosahedral (20-sided) symmetry

molecules rapidly assemble themselves into complete new viral particles. The whole process, from a virus breaking into a cell to that same cell exploding and liberating up to 10,000 new viruses, takes just six hours.[2]

We now know the fine details of poliovirus's portrait, thanks to computer-generated imagery which can paint in the three-dimensional structure of its proteins. The coat proteins are not simple spheres, like the 60 ping-pong balls which Crick and Watson glued together in 1956 to make their 20-sided model virus. They have complex shapes that intertwine with each other and are stitched together into a thick coat that fits snugly around the RNA core. The result is an intricately sculpted sphere, and an object of beauty (Figure 12.2).

The fine fretwork on its surface holds the key to how the poliovirus invades its target cells.[2] Crucially important are clefts between the protein subunits called 'canyons' – a grand name for structures that span barely five nanometres (5 millionths of a millimetre). Each poliovirus has 60 identical canyons, spaced regularly across its surface. The canyon is exactly the right size and shape for a molecule called the 'poliovirus receptor' (PVR) to plug into it. The PVR is found on the outside of only a few cell types, notably the motoneurones, the cells lining the gut and the tonsils.[3]

The act of invasion is a collusion between the virus and its host (Figure 12.3). Imagine the poliovirus as a ball, indented by its 60 peculiarly shaped canyons, rolling across its target cell. The cell's surface is studded with hundreds of different molecules, but only the PVR is the right shape to lock into a canyon. Once captured by the PVR, the whole poliovirus is somehow flipped through the cell membrane and inside the cell. There, its coat proteins fall away, and the RNA molecule snakes out and sets about producing the 'control' proteins which hijack the cell's production lines to produce new viruses.[2]

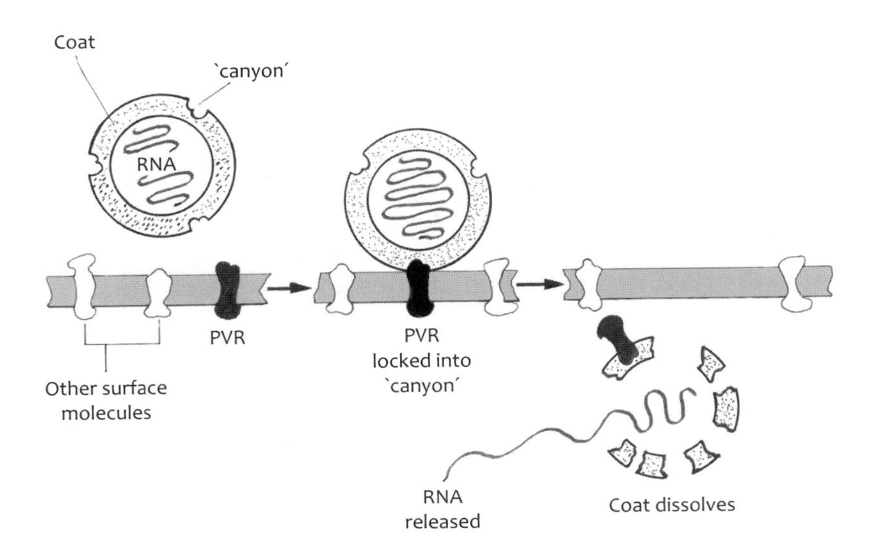

Figure 12.3 How the poliovirus invades its target cell. The poliovirus receptor (PVR) on the outside of the cell locks into a 'canyon' on the surface of the virus, and the virus is pulled inside. The coat proteins break down, releasing the RNA to start the process of replicating new viruses

The PVR is crucial; without it, the poliovirus cannot enter cells. The PVR is found only in primates, with the closest matches to the human PVR in the chimpanzee and other great apes. This explains why lower species cannot catch polio – except for cotton rats and mice, which possess a PVR-like molecule and are susceptible to some modified Type 2 poliovirus strains.

The importance of the PVR has been confirmed by creating 'transgenic' mice, genetically engineered to express the human PVR.[4] Unlike normal mice, these are dramatically susceptible to polio, being paralysed and killed within 24 hours of exposure to the virus. PVR-bearing transgenic mice are now used to check batches of polio vaccine for 'neurovirulence' – a critical test which previously involved injecting vaccine samples into the brain of a monkey and waiting to see if it became paralysed.

All this begs an intriguing question: why do humans produce a molecule which invites disaster by pulling polioviruses inside cells? The PVR is one of a family of 'adhesion molecules' which anchor cells to the matrix on which they sit. The PVR is probably the result of an unlucky mutation from a 'useful' adhesion molecule, and may have first locked onto an ancestor of the poliovirus – a happy union for the virus, as this provided an environment in which it could thrive and evolve.

Out of sequence

The RNA sequences of many types of poliovirus have been determined and have solved some old mysteries. There are clear differences between Types 1, 2 and 3, with more subtle variations among the various strains in each type.[5]

The molecular basis of 'neurovirulence' and 'attenuation' has been revealed by comparing RNA sequences between the non-paralysing strains used in vaccines and the original, 'wild-type' polioviruses from which they are descended. Viruses mutate spontaneously, and the alien environments used in attenuation are designed to select mutants which drift progressively further from the wild type and eventually lose the capacity to paralyse.

Albert Sabin proudly proclaimed that it took four years, 20,000 monkeys and 500 chimpanzees to produce his three attenuated strains. The genetic baggage accumulated during that process has now been pinpointed. The transformation of the virulent Mahoney into Sabin's Type 1 vaccine virus involves changes in just 57 of the 7,440 nucleotides.[6] Many of these mutations are concentrated in one of the coat-protein genes (B in Figure 12.1), and alter that protein's structure. Another key mutation is a simple substitution of the 480th nucleotide (A in Figure 12.1).

This site lies 'upstream' of the genes, and so does not affect any protein's structure. Instead, it may influence the amounts of protein produced, or the copying of the RNA itself.

In 2002, the poliovirus again broke new ground when it became the first virus to be completely synthesised from scratch – a technological tour de force by Jeronimo Cello's team in New York.[7] The task was laborious, but devoid of mystery. Cello used the published sequence of poliovirus RNA, commercially available reagents and the services of biotech companies as required. The result was chemically perfect, to the last atom of $C_{332662}H_{492388}N_{98245}O_{131196}P_{7500}S_{2340}$. As final proof that the RNA sequence is all that the poliovirus needs to work from, the artificial virus could infect cells in the usual way.

Not everyone sees this as a triumph of molecular biology. For some, it is a grave error of scientific judgement; the deliberate synthesis of a virus that can paralyse and kill is a Pandora's box that should never have been touched other than to nail its lid down.

The RNA genomes of other enteroviruses have also been mapped, and we can now see where the poliovirus may have come from.[8] The enteroviruses can be plotted as a family tree, whose branches represent the extent to which their RNA sequences differ (Figure 12.4).

As expected, the three types of poliovirus are crowded together at the end of one branch. The nearest twigs to them are two type A Coxsackie viruses. It seems likely that the polioviruses originated from a Coxsackie

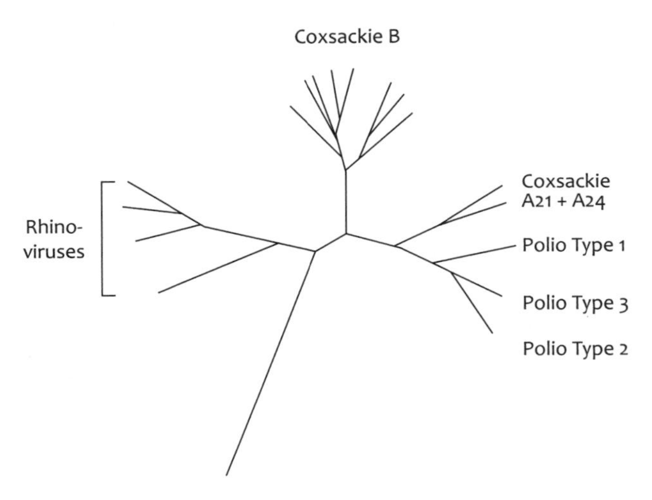

Figure 12.4 'Family tree' showing relationships between the RNA sequences of various enteroviruses. The three types of poliovirus most closely resemble two type A Coxsackie viruses, A21 and A24, and may have arisen by mutation from an ancestral type A Coxsackie virus

A virus precursor, probably through a series of mutations affecting the coat proteins around the edges of the canyon, enabling the mutant to lock onto the PVR and so enter human cells.

Survival of the fittest

A killed virus must be safe 100 times out of 100.

Albert Sabin, letter to Basil O'Connor, 1 August 1955[9]

We left the Salk-Sabin battlefield in 1963, when Jonas Salk's inactivated polio vaccine (IPV) had become a time-expired miracle and was being replaced in the United States by Sabin's oral vaccine (OPV). However, this was far from the end of the trail for either vaccine.

In 1963, the OPV certainly seemed to have the edge. Its 'live' viruses infect the human gut, just like the wild-type polioviruses from which they were derived, and so induce the all-important 'intestinal immunity'. This first line of defence should mop up wild polioviruses before they can enter the bloodstream and home in on the central nervous system. Salk's vaccine does not produce intestinal immunity; Sabin's therefore provides better protection. Also, being 'live', the OPV viruses can infect close contacts of those vaccinated, inducing immunity in them and spreading the benefits of vaccination even further.[10] Finally, the 'O' in 'OPV' was a major selling point for needle phobics, hard-to-reach populations and economists. Many of those who shrank away from Salk's intramuscular injections saw no problem with two drops of pink liquid, while there were huge savings from writing off the costs of syringes and needles.

The Soviet Union had led the way with Sabin's OPV, giving it to millions of people while its inventor was still frozen out of Salk-obsessed America. Various countries followed the United States' example in switching from IPV to OPV.

In Britain, IPV was introduced in 1955, but with a variation on Salk's theme: the highly virulent Type 1 strain Mahoney was replaced by the more benign 'Brunenders' (derived by John Enders from the strain named after Brunhilde, the well-endowed female chimpanzee). Initially, there was little public interest in Britain – perhaps because of the absence of NFIP-inspired terror – but uptake increased markedly after the much-publicised death from polio of Jeff Hall, 29-year old member of the England football team.[11] The British Ministry of Health mistrusted oral vaccines after the unfortunate episode with Koprowski's vaccine in Belfast. Eventually, however, Sabin's OPV won them over, and it replaced IPV in early 1962.

Other countries, such as Japan and Cuba, had bided their time and began their vaccination programmes directly with Sabin's OPV. It was highly effective. In Cuba, polio was wiped out in record time after just two mass vaccination sweeps; public health authorities in America, the Land of the Free, may well have been jealous of the nearly 100 per cent compliance with vaccination that was achieved under Cuba's Communist regime.[12]

By contrast, Sweden and Holland were unmoved by arguments favouring OPV and stayed with their own versions of IPV. Sweden had never signed up to Salk's vaccine, because its chief opinion leader, virologist Sven Gard, constantly attacked Salk's formalin inactivation protocol as unsafe. Having done his best to discredit Salk, Gard developed his own IPV, which featured longer incubation in formalin and Brunenders instead of Mahoney. Gard's IPV worked as effectively as Salk's, and chased polio out of Sweden in 1963. This was helped by his country's very high vaccination rate, nearly matching that in Cuba, which would have shown any vaccine in a good light.

The excellent outcomes in Cuba (with OPV alone) and Sweden (IPV alone) indicate that either type of vaccine is effective if given to a high proportion of the population. Conversely, the lacklustre results with Salk's vaccine after a few years in America prove that no vaccine will work if people are not prepared to take it.

Back in the United States, Sabin's vaccine enjoyed a monopoly during the final years of the American reign of polio. The last cases were reported in 1979, when OPV had been used exclusively for 16 years. However, Salk's vaccine had already done most of the work between 1955 and 1963, and the switch to Sabin's did not obviously accelerate the decline in numbers of paralytic polio cases (Figure 12.5).[13]

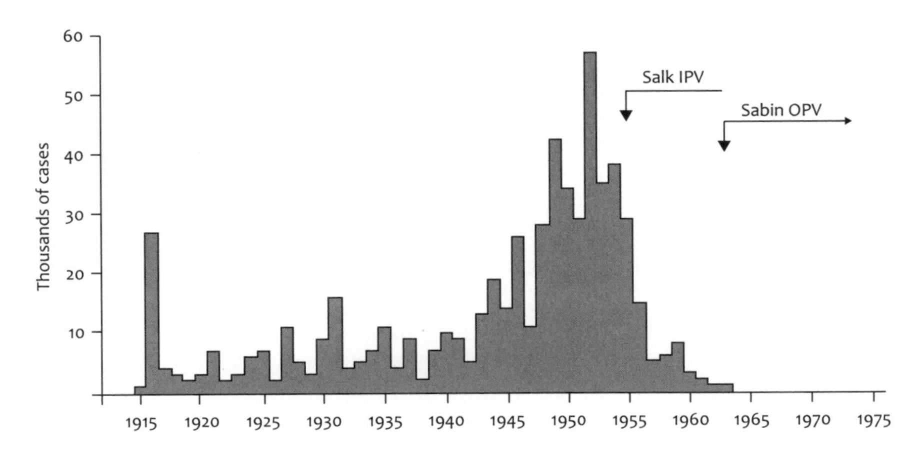

Figure 12.5 The disappearance of polio from the United States after the introduction of Salk's IPV in 1955, followed in 1963 by the switch to Sabin's OPV. Data from the Centers for Disease Control

A non-problem

Soon after its introduction in the United States, hints emerged of a rare but serious complication of Sabin's OPV. Ironically, this was the same curse that had nearly killed Salk's vaccine: paralysis.

In 1963, D.A. Henderson of the Poliomyelitis Surveillance Unit of the Communicable Diseases Center in Atlanta, Georgia, was sent to investigate. Henderson looked into 60 cases where paralytic polio had struck within three weeks of the patient taking OPV; wild-type polio was not around at the time, and the poliovirus isolated from each case looked suspiciously like one of the attenuated strains in Sabin's vaccine (Type 3 was the most common). Henderson published his findings in the *Journal of the American Medical Association* (*JAMA*) in August 1964, with the non-accusatory title, 'Paralytic disease associated with oral polio vaccines'. However, he did not shy away from the conclusion that 'at least some of these cases were caused by the vaccine.[14]

Sabin, not mentioned by name anywhere in Henderson's article, saw things differently. Even if his vaccine viruses were found at the scene of the crime, they were blameless and had only been found 'guilty by statistical probability'. *JAMA* printed Sabin's counterattack straight after Henderson's paper. Sabin's opening shot, 'The items in this report with which I disagree', is followed by three pages of quibbles: the methods used to identify the paralysing viruses were dubious, some cases must have been incubating polio when they were vaccinated, and others did not even have polio.[15]

Sabin was a lone voice. Everyone else accepted that 'vaccine-associated paralytic polio' (VAPP) was caused by the attenuated viruses in Sabin's OPV, which somehow had regained the wild type's capacity to paralyse. VAPP was exceedingly rare. The risk was about 1 in 750,000 with the first dose of OPV, falling to 1 in 13 million with subsequent doses; the overall incidence was less than 1 in 2 million doses.[16]

But VAPP did occur. In the United States between 1965 and 1985, there were between eight and eighteen cases per year. After wild-type polio was pushed into extinction in the United States in 1979, VAPP was left high and dry as the only cause of paralytic polio. Even then, as Roland Sutter put it (possibly with tongue in cheek), 'Some have challenged the existence of VAPP'.[17]

For 'some', read 'Sabin', the only person who still denied that his OPV could ever cause paralysis. In 1985, Sabin wrote, 'To some, the extraordinarily rare "vaccine-associated" paralytic case ... has come to be synonymous with "vaccine-caused" – in my judgement, uncritically'.[18]

Today, there are 250–500 cases of VAPP worldwide per year.[16] Thanks to advanced RNA sequencing methods, the viruses responsible can be quickly identified and their derivation from the OPV strains worked out. These 'vaccine-derived polioviruses' (VDPVs) reacquire the power to paralyse in various ways. Some arise when an OPV strain is hit by a mutation in one of the sensitive areas that encode some aspect of neurovirulence (such as A or B in Figure 12.1). This effectively undoes all the benefits of attenuation at a stroke. The Type 3 virus in Sabin's vaccine is closest to its wild-type ancestor, and so reverts relatively easily to neurovirulence – which explains why this is the most common type associated with VAPP.

Most of the VDPVs cause isolated cases of paralysis, mainly in recently vaccinated children and occasionally spreading to close contacts. They can infect people with rare immunodeficiency conditions (luckily, not acquired immune deficiency syndrome [AIDS]), and may persist for ten years or more, creating a potential hazard to others, as the virus continues to be excreted in the stools.[19]

More sinister is the creation of entirely new hybrid viruses from 'recombination' between an OPV strain and a wild enterovirus (especially Coxsackie) in the gut of a vaccinated child.[20] These hybrids are neurovirulent and can take on a life of their own, spreading rapidly in poorly immunised communities – hence their name 'circulating VDPV' (cVDPV). cVDPV caused outbreaks of paralysis in Egypt (which smouldered on for five years, between 1988 and 1993) and the Caribbean island of Hispaniola (which comprises Haiti and the Dominican Republic), in 2000–01.[21] Fortunately, these outbreaks affected only 20–30 cases and were eventually controlled with intensive vaccination using OPV – which, paradoxically, protects against cVDPV.

VDPVs may be exceedingly rare, but their existence has fundamentally changed the approach to vaccination. In countries that are free from polio, the tiny, one in 2 million risk of being paralysed by Sabin's OPV is now unacceptably high, compared with the zero risk of being paralysed by wild-type polio. For this reason, many countries including the United States, Canada, the United Kingdom and France, have now replaced OPV with IPV for routine vaccination. Salk might have played second fiddle to Sabin since the early 1960s, but his vaccine has not stood still. Thanks to improvements in cell culture and virus harvesting, the Salk vaccine is now potent and consistent – and devoid of any risk of causing paralysis.[22]

In the United States, the tables were finally turned in Salk's favour in 1999, after 37 years of Sabin monopoly. As a final insult to Sabin, his OPV was withdrawn so that no physicians can be tempted to use it. Since then, there have been no cases of paralytic polio among American children.

Sabin was spared knowledge of this final humiliation, as he had died in 1993.

Exeunt

The history of polio was built around a cast of colourful characters, with a few true heroes, a few real villains and many who were a bit of both.

Some stayed with polio for the rest of their lives, while others moved on. The story is incomplete without knowing what happened to some of key players.

Robert W. Lovett (1859–1924)

As Professor of Orthopaedics at Harvard University and Chief Surgeon at Boston's Hospital for Crippled and Deformed Children, Robert Lovett became a world authority on polio and curvature of the spine. He also brought science to orthopaedics, and was decades ahead of his peers by making 'kinesiotherapists' (the forerunners of today's physiotherapists) full members of his multidisciplinary team.[23]

Lovett was always busy and in demand. In July 1921, he answered the summons to Campobello Island to rescue Franklin D. Roosevelt from the clutches of local doctors. At the time, he was working on a massive two-volume textbook, *Orthopaedic Surgery*, which he wrote with his like-minded friend and colleague, Sir Robert Jones of Liverpool. Lovett and Jones had known each other for decades, but when their book was published in late 1923, their friendship had less than a year to run.

In July 1924, while crossing the Atlantic to attend a conference in London, Lovett fell seriously ill with septic pericarditis. He was carried off the liner and rushed with his wife to the nearest port – which was Liverpool. With no hope of treatment in that pre-antibiotic age, Lovett was taken to Jones's house to be kept comfortable. Nursed by his wife and Jones, Lovett lasted four days, giving him enough time to discuss with Jones his latest thoughts about curvature of the spine.[23]

A week later, Lovett's widow was again on board ship, accompanying her husband's embalmed body on the grim return voyage to Boston. And back at the Presbyterian Church in Liverpool, Jones read the lesson at the memorial service for his co-author and friend, in front of an altar draped in the Stars and Stripes.

Simon Flexner (1863–1946)

Flexner was brilliant, phenomenally hard working and productive – and arguably as two-faced as Janus. His name is immortalised in the bacterium *Shigella flexneri*, one of the main causes of bacillary dysentery, which he discovered in the Philippines in 1899. He was most famous for his meningitis antiserum, which slashed the death rate by two-thirds and remained the standard therapy for over 30 years until sulphonamide antibiotics became available during the 1940s. Flexner was also outstanding as the founding director of the Rockefeller Institute for Medical Research, transforming architects' plans into a world-class institution in less than a decade.[24]

But Flexner had faults, and did more harm than good in the field of polio. The comment that Flexner's 'logic was severe'[25] was a diplomatic understatement. Flexner's personality and reputation were both heavyweight. and few were brave enough to challenge him, even when he was obviously wrong. He was strongly prejudiced against non-American scientists, even when they were obviously right. Overall, Flexner's grip on American medicine, exerted through his heading the Rockefeller and editing the *Journal of Experimental Medicine*, stifled progress in polio research and especially vaccine development until well after his retirement in 1925.

Flexner's last public pronouncement on 'epidemic poliomyelitis', in a lecture delivered in February 1938 at the University of Cambridge, gives some fascinating insights. Here, Flexner looked back on his prime, when he had controlled research into the disease that targeted America 'on a scale exceeding any other country'.[26] It was he who had found the poliovirus (Landsteiner is not even mentioned). The virus entered the human body via the olfactory nerves; nasal sprays failed to protect against polio only because their use had been 'roughly improvised'. There was only one type of poliovirus (Macfarlane Burnet was wrong in believing in multiple strains). And polio vaccines were going nowhere, as recently confirmed by the fiasco that brought down Kolmer, Park and Brodie.

Just as revealing is Flexner's obituary, written by Peyton Rous, his long-standing friend and co-editor of the *Journal of Experimental Medicine*.[24] Expansive in praise and devoid of criticism, Rous shows us a humane Flexner, tough but fair and imbued with 'a self-mastery that was supreme ... serenity that verged upon fortitude' (whatever that means). Rous admits that Flexner could be ruthless 'in that he let no worker become ensconsed who did not really belong', but insists that he acted only in the Institute's best interests.[27] He glosses over Flexner's misadventures in polio, while pointing out that Flexner and Lewis found the filterable virus 'independently of Landsteiner, who was making similar but less comprehensive discoveries in Vienna'.

Photographs of Flexner show a man with a piercing gaze who seems unlikely to tolerate fools or contradiction (Figure 4.4). John Paul, who fell under Flexner's spell while still a medical student, acknowledged that he had faults, but added, in one of the more cryptic remarks in the *History of Poliomyelitis*: 'He can be forgiven for making mistakes about poliomyelitis'.

Simon Flexner died on 2 May 1946, aged 83.[28]

Franklin D. Roosevelt (1882–1945)

Roosevelt is a man with many memorials: his profile on the American dime; the strapline 'FRANKLIN D. ROOSEVELT, FOUNDER' at the foot of all NFIP posters until the late 1950s; the dedication of Erich Korngold's Symphony in F sharp (1951); and the Roosevelt Campobello International Park, established in 1964 around his 'cottage' overlooking the Bay of Fundy.

The subtitle of Naomi Rogers' book, *Polio before FDR*, implies that he was a watershed in the history of polio.[29] And so he was, given all that he did to mark polio out as an enemy for destruction and to mobilise his nation against it. Conversely, polio was a watershed for Roosevelt, transforming him from an aloof Ivy Leaguer with hereditary wealth into a man who could look anyone in the eye and say that he understood hardship and bad luck, courage and resilience. His brush with The Crippler undoubtedly built up Roosevelt's ability to charm the common man as well as those in power – the rare ability which carried him into the White House and set him on course to guide America through the Great Depression and the Second World War.

Roosevelt's story includes a twist and a footnote. The twist concerns the illness that struck him down on Campobello in July 1921. Neither of the two doctors who saw him initially thought that he had polio. The diagnosis was made two weeks later, when Robert W. Lovett was called in from Boston. Lovett was puzzled by the clinical picture, but by then it was too late to do a diagnostic lumbar puncture.

A detailed re-examination of Roosevelt's medical records has suggested that he had Guillain-Barré syndrome (GBS), not polio.[30] If true, there would be a great deal of irony in the post-FDR history of polio. However, the jury is still out. The unusual features of Roosevelt's illness – the absence of preceding minor symptoms, paralysis that ascended slowly and symmetrically, and severe pain and hypersensitivity – are certainly commoner in GBS, but can also occur in polio. A diagnosis that an expert found difficult at the bedside in 1921 cannot be any easier now. The failure to do a lumbar puncture, the only test of the day that could have distinguished between the two conditions, means that we shall never know.

The footnote relates to the Roosevelt dime (Figure 11.4), which has survived unchanged longer than any other American coin. In 2003, a Republican senator tried to have FDR's profile replaced by that of former President Ronald Reagan. The proposal was roundly defeated, thanks to the spirited intervention of Nancy *née* Davis, the film actress who had fought back against 'The Crippler' on the silver screen in 1940. She was well placed to claim that 'Ronnie wouldn't want' to have FDR removed from the dime; her married name was Nancy Reagan.[31]

Basil O'Connor (1892–1972)

O'Connor, FDR's legal partner and fellow-conspirator, continued to run the National Foundation for Infantile Paralysis (NFIP) after Roosevelt's

Figure 12.6 Albert Sabin, Jonas Salk and Basil O'Connor in the late 1950s. Reproduced by kind permission of the March of Dimes

death (Figure 12.6). O'Connor saw the NFIP safely through the years of Salk-Sabin conflict and its metamorphosis in 1958 into the March of Dimes Birth Defects Foundation.

This change was essential for the enterprise to survive. At that time, the total number of dimes raised by the NFIP would have stretched a third of the way from the White House to the Moon, but the organisation had become the victim of its own success: its funds were drying up because polio had been swept out of America.

O'Connor also continued to live life to the full, and sometimes beyond. His flirtation with cardiovascular risk factors is apparent from the photographs which show him chain-smoking, obviously stressed and with a steadily expanding waistline from those permanently reserved tables at expensive restaurants. His first heart attack, in 1951, left no obvious scars. The second, during a working trip to Phoenix, Arizona in March 1972, killed him.

A prominent fixture on the New York scene until the end, O'Connor's obituary in the *New York Times* filled three columns.[32]

Elizabeth Kenny (1886–1952)

Of all the odd-ones-out ranged against polio, Elizabeth Kenny was perhaps the oddest. In the end, she came a long way from the untrained assistant who had to make up her own rules in the Australian outback. During her American years, she was in her element, with rapturous supporters and polls, books and a movie about her life, and audiences with President Roosevelt. Doctors, scientists and the NFIP continued to snub her, but she was hard to stop – such as when she bluffed her way into the First International Poliomyelitis Conference in New York in 1948, with a forged press pass.[33]

Eventually, her luck ran out, with the onset of heart failure. In July 1952, aged 72, she had already planned her last trip home to Australia to die when she struggled to Copenhagen for the Second International Conference on Poliomyelitis. There, she met Tom Rivers, leader of the orthodox virologists and the arch-enemy who had always refused to see her. This time, Rivers brought himself to shake Kenny's hand, because he could see that she had not long to live.

Rivers later acknowledged, grudgingly, that 'on the whole, she did some good'. Coming from him, this was fulsome praise. And it would be hard to find a neater epitaph.[33]

Jonas Salk (1914–95)

Jonas Salk was left a restless soul after the excitement over his vaccine died down. Perhaps unsettled by the strain of celebrity, he drifted away from the safe haven of his laboratory towards much broader horizons: the interface between biology and the humanities. He took to wandering the United States in search of the right place for his new vision – an institute for free thinkers, liberated from the boundaries of discipline that bedevil conventional universities. O'Connor often accompanied him on these trips; that special relationship still held.

In 1960, Salk found his ideal spot, on the cliffs overlooking the Pacific at La Jolla, just north of San Diego. The path to his dream was not entirely smooth.[34] The mayor of San Diego, a polio survivor, managed to outwit local residents who opposed the project – but only just. Worse for wear from celebrating after the proposal was approved, the mayor somehow urinated on the documents which he had just signed. Happily, the plans went through, with start-up funding from the March of Dimes. The Salk Institute for Biological Studies opened its doors in 1962. With its expansive ethos of 'cell to society', the Institute soon attracted top names from science and humanities, including Francis Crick and Jacob Bronowski. Its rise has been meteoric, and it is ranked today among the world's top biomedical research centres.

In this new setting, Salk reinvented himself as a philosopher, writing books with profound titles such as *Man unfolding* and *Survival of the wisest*. These books were never bestsellers, but Salk's transformation from laboratory scientist was convincing enough for the *New York Times* to dub him 'The Father of Biophilosophy' in 1966. Salk also maintained

Figure 12.7 Françoise Gilot and Jonas Salk, 1970. Reproduced by kind permission of The Bettmann Archives/ORBIS

conventional research interests, setting off in the mid-1980s on the trail of a vaccine against AIDS. He failed to find it, but 25 years after his AIDS vaccine programme folded, nobody else has succeeded either.

Salk's personal life also saw great change, beginning in 1968 with his divorce from the long-suffering Donna. The following year, at a party in La Jolla, he met Françoise Gilot, a talented artist in her own right, but more famous as the mistress who had borne three children by Pablo Picasso.[35] They married in 1970 in France, having flown there in secret under the alias of Mr and Mrs Peterson. They made a good couple. Françoise used to accompany her husband on trips to scientific conferences, which often provided excellent sales opportunities for her paintings. And Salk, biologist, philosopher and occasional mystic, seemed content at last (Figure 12.7).

On 23 June 1995, Salk died of heart failure in La Jolla. He had seen adulation and then disappointment as his vaccine was replaced by Sabin's during the 1960s. If he had lived another four years, he would have had the satisfaction of knowing that his vaccine was back where it deserved to be, protecting the children of America against polio.

Albert Sabin (1906–93)

Like Salk, Albert Sabin became a wandering star, although without matching his rival's personal reincarnation or the stellar impact of his new vision. Secure in the knowledge that his vaccine had ousted Salk's in the United States, Sabin moved to Israel in 1969. There, he spent three years as president of the Weizmann Institute in Rehovot – world class, but already being overtaken by a certain institute set on the cliffs just north of San Diego.

In 1972, aged 66, Sabin returned to the academic circuit in the United States, now a grand old man and no longer the ruthless researcher, made dyspeptic by a belly full of fire. He certainly looked the part: hair that had turned white twenty years earlier, a Hemingwayesque beard (also white) and a benign, patricianly smile that would have looked odd on the face of the man who told Salk in 1960 that he was going to 'kill the killed vaccine'.

Some combination of personality and career pressure brought unhappiness and tragedy into Sabin's family life. His wife Sylvia, whose name Sabin had not mentioned in his terse wedding announcement telegram to Peter Olitsky, died in 1966. She killed herself with an overdose of sleeping tablets, and the date of her death – 26 August, her husband's sixtieth birthday – seems unlikely to be coincidental. Perhaps this was the event for which Sabin's two daughters 'never forgave him'.[36] A brief second marriage failed, but was soon forgotten in the move to Rehovot. The third time, after returning from Israel in 1972, Sabin was much luckier with Heloisa Dunshee de Abranches, a lively Brazilian woman later described as 'small and wonderfully seaworthy'.[37]

Heloisa tended Sabin through his declining years, which began just a decade later in his mid-70s. Degeneration of the vertebrae in Sabin's neck compressed his spinal cord and enabled him to experience at first hand some of the miseries suffered by polio victims: paralysis from the waist down, secondary pneumonia and even a respiratory arrest. Spinal surgery was only partially successful in restoring his mobility, and left him with pain so excruciating that he considered taking Sylvia's way out.

An interview for *People* magazine in July 1984 gives us an enduring snapshot of Sabin in the last ten years of his life; apart from pain and immobility, the Sabin locked in mortal combat with Salk 30 years earlier is instantly recognisable. His opinions are blunt, and his tongue is sharp; a quotation from the diplomatic Dorothy Horstmann describes Sabin as 'not enormously tolerant... but usually right'. However, like the inflexible Flexner, Sabin still has his blind spot: all allegations that his vaccine could ever cause paralysis are 'lousy and untrue'. And when Sabin's gaze falls on the photo of Sylvia, he simply says 'Tragic' and adds 'Esta vida – that's life'.[37]

Sabin died on 4 March 1993 at the age of 87. The cause of death, heart failure, was the same as Salk's a couple of years later. A fulsome obituary in *Nature* mourned the passing of a great scientist, a wise man and a friend who would be missed.[38] As he had retained the rank of lieutenant colonel in the US Army, Sabin was buried in Arlington National Cemetery. The back of his tombstone carries a simple inscription: 'SABIN. Developer of the vaccine that made possible the global eradication of poliomyelitis'.

'*The* vaccine' were the key words, guaranteed to enrage Salk. Sabin died happy in the belief that his vaccine – and only his – held the key to defeating polio. And as we now know, he was wrong.

Hilary Koprowski (1916–2013)

Given Sabin's track record in human relationships, it might have been difficult to find someone to write nice things about him after his death. The author of the obituary in *Nature* was a particularly surprising choice: Hilary Koprowski.[38] It was also Koprowski who stood beside Sabin's grave in Arlington and delivered the funeral address, with the same admiration and affection. As Koprowski wrote later, 'personally and in public, I mourned his passing'.

This display of warmth was unexpected, as there was a long history of Sabin being at Koprowski's throat and, to a lesser extent, vice versa. Sabin tried repeatedly to annihilate Koprowski's rival oral vaccine, notably with the revelation that the version recently given to hundreds of thousands of African children was contaminated with a dangerous 'vacuolating' virus. To inflict maximum damage, Sabin announced this to a full house of their peers, at the First International Oral Polio Vaccine Conference in New York in 1960. Sven Gard, the Swedish virologist on sabbatical in Koprowski's laboratory, butted in to refute Sabin's claim, but the damage was done. The stain of suspicion left over Koprowski's vaccine undoubtedly helped to tip the scales in favour of Sabin when an oral vaccine was chosen two years later to revive the flagging Salk vaccination programme in America.

However, Koprowski had good reason to be grateful to Sabin, as he revealed towards the end of his *Nature* obituary:

> In a letter to me written just a year ago, reviewing a paper speculating that AIDS started with polio vaccination in the Belgian Congo, Sabin expressed his opinion that this was 'a most irresponsible and uncritical communication'.[38]

At the time – 8 April 1993 – Tom Curtis' article had appeared in *Rolling Stone* a year earlier, but the full torrent of *The River* was yet to be unleashed.[39] The 'irresponsible and uncritical communication' was the hypothesis paper linking Koprowski's vaccine to AIDS, which Elsworth and Stricker had submitted to *Virology Research* – and which the editor had sent in confidence to Koprowski, who in turn forwarded a copy to Sabin.[40]

Why had Sabin been so uncharacteristically helpful? Perhaps, sensing that the end of his life was looming, he felt that it was time to bury the hatchet, or perhaps he saw an omen that could eventually threaten his own vaccine. In any event, Sabin's posthumous support undoubtedly helped Koprowski through the dark times that came later.

On the day that *Nature* printed Sabin's obituary, Koprowski was 77 years old and emeritus professor at the private Thomas Jefferson University in Philadelphia. The previous year, he had left the Wistar Institute after 34 years as director. In that time, he turned the place around culturally as well as scientifically, bringing in music as well as scientific rigour.[41] However, clouds had a habit of condensing over branch points in Koprowski's colourful and iconoclastic career, and his departure from the Wistar was accompanied by an undisclosed out-of-court settlement.

It is an understatement to say that Koprowski has remained active in research. By his ninetieth birthday in 2006, he had chalked up nearly 900 papers and his fiftieth grant from the ferociously competitive National Institutes of Health. At the time of writing, aged 96, he still had fingers in many scientific pies, including a rabies vaccine produced from harmless viral proteins expressed in genetically modified plants; in theory, this could induce immunity simply by being eaten. He also remained a man of culture. His piano playing was no longer up to the standard of the Warsaw Conservatoire, but November 2011 saw the première in Philadelphia of a witty musical satire which he wrote about his near-disastrous spell in hospital with a broken hip.

Koprowski was the first to make a polio vaccine that worked, only to have it beaten down in favour of two latecomers. As the last man left standing, his resilience was remarkable. But he still preferred not to talk about *The River* and how it poisoned his life.[42]

Hilary Koprowski died in his ninety-sixth year, on 11 April 2013.

Vivi Andersen (1940–71)

It would be wrong to close this section without a tribute to the people for whom this story has the greatest significance of all – the victims of polio. For this reason, we need to return to Blegdams Hospital in Copenhagen in the summer of 1952, and a 12-year-old girl whom we have already met.

Vivi Ebert was the tetraplegic patient expected to die on the morning of 27 August because no iron lung was available. Instead, her life was saved by Bjorn Ibsen's last-ditch attempt to ventilate her through a tracheostomy. Vivi was one of the 200 paralysed patients kept alive during the polio outbreak by the relays of medical students and other volunteers, taking turns to squeeze the rubber bag that pushed air into their lungs. They did this 20 times every minute for as long as it took for spontaneous breathing to return – or for the patient to die from other causes.

Vivi was 'bagged' for several days before she could breathe on her own again, and was successfully weaned off artificial ventilation. The endotracheal tube was pulled out, and the narrow window cut in the front of her throat was left to heal.

It was another two months before Vivi was well enough to go home, and months more before she could return to school. But she persevered, and in due course completed her education, trained as a secretary and became Mrs Vivi Andersen.

However, polio had not released its grip on her. Despite intensive physiotherapy at Blegdams, she regained little strength in her limbs. Vivi went home still tetraplegic and in a wheelchair, from which she could escape only with the help of others. At best, her breathing remained shallow, which predisposed her to repeated chest infections – when the extra effort of respiration quickly wore her out and sent her back into hospital.

It was during a chest infection that she was admitted as an emergency to Blegdams in September 1971. Intensive care had moved on a long way since her first admission 19 years earlier, but not far enough to save her.

Vivi Andersen, *née* Ebert, was just 31 years old when she died.[43]

Glittering prizes

At last the Dodo said, "Everybody has won and all must have prizes".
Lewis Carroll, *Alice in Wonderland*, 1865

Recognition for scientists and doctors comes in many forms, from a patient's words of gratitude to the honours reserved for true leaders in the field. Many of those drawn into the history of polio were celebrated for their work, although the correlation between reward and achievement does not always appear strong.

Both Salk and Sabin won the Lasker Award, America's equivalent of the Nobel Prize, respectively in 1956 and 1965. Sabin might have resented the Lasker awarded to O'Connor, particularly as this was seven years before his own. Salk was presented with a specially struck Congressional Gold medal, handed over on behalf of the grateful nation by President Eisenhower in January 1956. Sabin had to wait until 1970 to receive the National Medal of Science from President Nixon, but the citation rewarded his patience. In eerie anticipation of the inscription on his tombstone, it praised 'the vaccine which has eliminated poliomyelitis as a major threat to human health'.

Hilary Koprowski won none of these accolades, but could content himself with a string of honours from around the world, including the Belgian Royal Order of the Lion, the French Légion d'Honneur and, in 2007, the Albert B. Sabin Gold Medal.

The greatest reward of all for human endeavour has a curious origin, in a peculiar whitish-grey substance which we have already met. This is diatomite, used to make the hollow Berkefelt candles with which Landsteiner isolated his filterable virus. Diatomite is a highly porous siliceous clay which had found a remarkable, big-bang application some 30 years before the candles were invented. It proved to be ideal for absorbing and stabilising nitroglycerin, a powerful liquid explosive that had the unfortunate drawback of going off ahead of schedule. The resulting product, stable enough to be cast into sticks and thrown around, was dynamite (literally). The patents generated vast royalties and its inventor, wishing to be remembered for altruism rather than weapons of destruction, used the money to endow annual prizes in the great domains of human achievement. He was, of course, Alfred Nobel.

Nobel Laureates who appear in the history of polio include Karl Landsteiner (1930) for discovering blood groups, not the virus that slipped through the Berkefelt candle and transmitted polio; Macfarlane Burnet (1960) for elucidating the genetic basis of the immune response; and Francis Crick and James D. Watson (1962) for cracking the double-helical structure of DNA and 'its significance for information transfer in living organisms'. More relevant to polio itself was the Nobel Prize awarded in 1954 to John Enders, Thomas Weller and Frederick Robbins, for 'their discovery of the ability of poliomyelitis viruses to grow in cultures of various types of tissue'.[44]

The *Dictionary of Medical Eponyms* reveals that Salk 'won the Nobel Prize for developing the first effective vaccine against poliomyelitis'.[45] This is entirely in line with public expectation; it is also entirely wrong. Salk is probably the most famous non-recipient of the Nobel Prize. Like Sabin and Koprowski, he was nominated several times without success.

According to Alfred Nobel's will, the prize in each category is awarded to 'the person who, during the preceding year, shall have conferred the greatest benefit to mankind'. In 1951, the Prize for Physiology or Medicine was awarded to Max Theiler of the Rockefeller Institute, for his yellow fever vaccine. With polio disappearing fast so soon after the introduction of Salk's vaccine, surely he deserved to follow in Theiler's footsteps?

Nobel made a further stipulation, that the Prize winner must be 'the person who shall have made the most important discovery within the domain of physiology or medicine'. This criterion weighed against Salk. His recipe worked, thanks to his dedicated experiments with the ingredients, but the use of formalin to inactivate viruses – including the poliovirus – was old science. Even Maurice Brodie, 20 years before Salk, had referred back to papers from Flexner and others who had dabbled with formalin around 1910.

Predictably, Sabin was always scornful of Salk's 'kitchen chemistry', muttering that 'anyone could do what he did'. Other scientists with no obvious grudge were also dismissive, including two Nobel Laureates. Thomas Weller wrote off Salk with 'he just wasn't a very good scientist'. Renato Dulbecco explained (without fear of reprisal, as this was in Salk's obituary in *Nature*) that Salk 'did not make any innovative discoveries'.[46] And Salk tripped himself up with his most famous saying. When asked if he would patent his vaccine, he replied that it was impossible: could you

Figure 12.8 Sven Gard (1905–98), shown at left with Vilem Skrovanek and Viktor Zhdanov at the WHO Expert Committee on Poliomyelitis, in Washington, DC, June 1960. Gard, the 'father of Swedish virology' and a member of the Nobel Prize Committee, developed an inactivated polio vaccine and was a fierce critic of Salk. Viktor Zhdanov played a key role in introducing Sabin's vaccine to the USSR and in setting up the smallpox eradication campaign; later, he led the USSR's germ warfare programme and the production of 'weapons-grade' smallpox. Reproduced by kind permission of the World Health Organisation

patent the Sun?[47] To Salk's detractors, that was precisely the problem, and this was a sun with nothing new under it.

A fierce critic of Salk's research was Sven Gard, Professor of Virology at the Karolinska Institute in Stockholm (Figure 12.8). Gard had made his reputation by purifying polioviruses from thousands of mouse brains, and later followed Salk into pursuing a formalin-inactivated polio vaccine.The two men haggled repeatedly over the shape of the graph that predicted how long polioviruses had to be steeped in formalin before they could be guaranteed inactive and therefore safe to use in a vaccine. Gard insisted that Salk's famous straight-line graph was wrong; instead, the line was a shallow curve, which meant that much longer incubation was needed.[48] To Gard, the Cutter Incident was proof that Salk was wrong. Gard was later unpersuaded by the evidence that a faulty production line was to blame, and refused to accept that Salk's formulation was safe.

From Salk's viewpoint, it is unfortunate that Gard was a member of the Nobel Prize Committee, and the man who wrote the expert assessment of Salk's nominations. The Committee's deliberations of the time, recently made public after the traditional 50-year embargo, paint an unedifying picture of cronyism and personal vendettas.[48]

For whatever reason, Gard pushed hard for Enders, Weller and Robbins to be awarded the Prize in 1954, praising their discovery as 'the most important in the whole history of virology'. He bulldozed aside a majority vote in favour of another candidate, Vincent du Vigneaud, who had discovered the structure of the hormone, oxytocin. The Enders trio duly received their telegrams from Stockholm.

Gard then worked just as hard to sabotage Salk's nominations for the Prize. In 1955, Gard's evaluation of his rival's research emphasised the long history of formalin inactivation and Salk's contentious inactivation graph, and concluded 'Salk's publications on the poliovirus cannot be considered as Prize worthy'.[48]

Salk's nomination was therefore passed over. And it was the same the following year, and every year until polio vaccines were old hat and nominations for Salk eventually dried up.

Salk, Sabin and Koprowski were not the only ones to be disappointed by the decisions of the Nobel Prize Committee. Both Hideyo Noguchi and Simon Flexner, who got so much right and so much wrong, were nominated five times, but failed to progress. And even Edward C. Rosenow was nominated twice, the second time in 1948, for 'having proved' that his pleomorphic streptococcus was the cause of polio.

Off the wall

A ragged line of seventeen verdigrised bronze busts provides recognition of another sort for some of those who made 'important contributions to

Figure 12.9 Jonas Salk, Eleanor Roosevelt and Basil O'Connor at the inauguration of the Polio Hall of Fame at Warm Springs, Georgia, on 2 January 1958. Five of the seventeen bronze busts are visible: Roosevelt and O'Connor (top right), and from left to right below them, David Bodian, John Enders and Salk. Reproduced by kind permission of the March of Dimes

the knowledge and treatment of poliomyelitis'. These are displayed on the white marble 'Wall of Fame' outside the Founders' Hall at Warm Springs, now properly known as the Roosevelt Warm Springs Institute for Rehabilitation (Figure 12.9).

Of the seventeen, four are Europeans: Jacob von Heine, Oscar Medin, Ivar Wickman and Karl Landsteiner. The rest are Americans, including the obvious choices: Jonas Salk, Albert Sabin, John Enders, David Bodian, Tom Rivers and Tom Francis – and, at top right and looking back over the scientists, Franklin D. Roosevelt and Basil O'Connor. With them are the epidemiologists John Paul, the gentlemanly historian of polio, and Joseph Melnick, whose mobile fly-catching laboratory was a fixture at polio epidemics during the 1930s; and Charles Armstrong, who moved from the cotton rat to nasal sprays. So too are two basic scientists who experimented with early formalin-inactivated vaccines, Howard Howe and Isabel Morgan. Howe abandoned his vaccine after limited and inconclusive trials in man, while Morgan stopped short in monkeys.

Appropriately, Flexner is missing. But so too is Hilary Koprowski, the first of the vaccine pioneers to produce a vaccine that worked, and who took it far beyond the limits where Howe and Morgan ran out of ideas and courage.

Grand Challenge

> The WHO set a goal of global eradication by the year 2000. This was not achieved but there is reason to believe that it can occur in 5+ years.
>
> Frederick C. Robbins, in 2004[49]

To date, only one human infection – smallpox – has been completely exterminated by mankind. That triumph was announced in 1980, when Jenner's cowpox vaccine had been in use for 180 years and had culminated in an intensive 11-year global eradication programme coordinated by the World Health Organisation (WHO).[50]

Soon after the demise of smallpox, polio came to head the WHO's hit list of infections to be annihilated. It was a good prospect. Both Salk's and Sabin's vaccines were effective, and had swept entire countries free of polio with impressive speed. Sabin's OPV lent itself particularly well to mass vaccination campaigns in the developing world, because it was cheap and easy to give. The poliovirus also helped to set itself up for eradication, as humans are essentially the only reservoir of infection (great apes can catch polio, but do not pose a realistic threat). Once vaccination had eliminated polio from mankind, this would be the end of the story.

The grand notion of exterminating polio altogether, rather than just protecting children against it, was first discussed in Bethesda, Maryland, at the International Conference on the Eradication of Infectious Diseases. The meeting was held on 27–28 May 1980, just three weeks after the WHO's momentous announcement that the world was now completely free of smallpox.

It took another five years for the Pan American Health Organisation (PAHO) to declare its intention to eradicate polio from the whole of the Western Hemisphere. With the support of the United Nations Children's Fund (UNICEF) and Rotary International, and using Sabin's OPV, this goal was achieved in just six years, in August 1991. The last act of endemic polio in the Western world was to paralyse the left leg of a Peruvian toddler, Luis Fermín Tenorio Cortez – although, according to the WHO's rules, another three polio-free years had to elapse before that victory could be confirmed.[51]

By this time, an even more ambitious enterprise was under way. The Global Polio Eradication Initiative (GPEI) had been launched in 1988, to enact the World Health Assembly's resolution to eradicate polio by 2000. The GPEI brought together epidemiological and scientific firepower, vast fund-raising networks, big names and publicity machines with global presence: the WHO, UNICEF, the United Nations, the Centers for Disease Control, Rotary International and the Bill and Melinda Gates Foundation.

Back in the 1970s, the smallpox eradication campaign had to overcome extraordinary obstacles, including floods and civil war in Bangladesh, WHO apparatchiks who wanted to spend the money on other projects, and Shitala Mata, the Indian goddess of smallpox.[50] The GPEI has confronted problems of its own: terrain rendered hostile by geography and acts of man, murderous campaigns of terror to drive a wedge between vaccinators and the children at risk of polio – and the realisation that its preferred weapon, Sabin's OPV, was not up to the job where it was most needed.

Sabin's OPV worked like a dream in well-nourished, healthy children in temperate America, but often became a nightmare when put to the test in the developing world. The vaccine viruses quickly 'die' at high temperatures, and an elaborate 'cold chain', from refrigerators to iceboxes in the field, is needed to keep OPV viable in tropical heat. Even then, OPV may not provide useful immunity in communities where diarrhoeal illnesses are common in children.[52] The success rate (measured by production of protective antibody levels) is as low as 30 per cent, compared with over 90 per cent in American children. This is apparently due to interference from

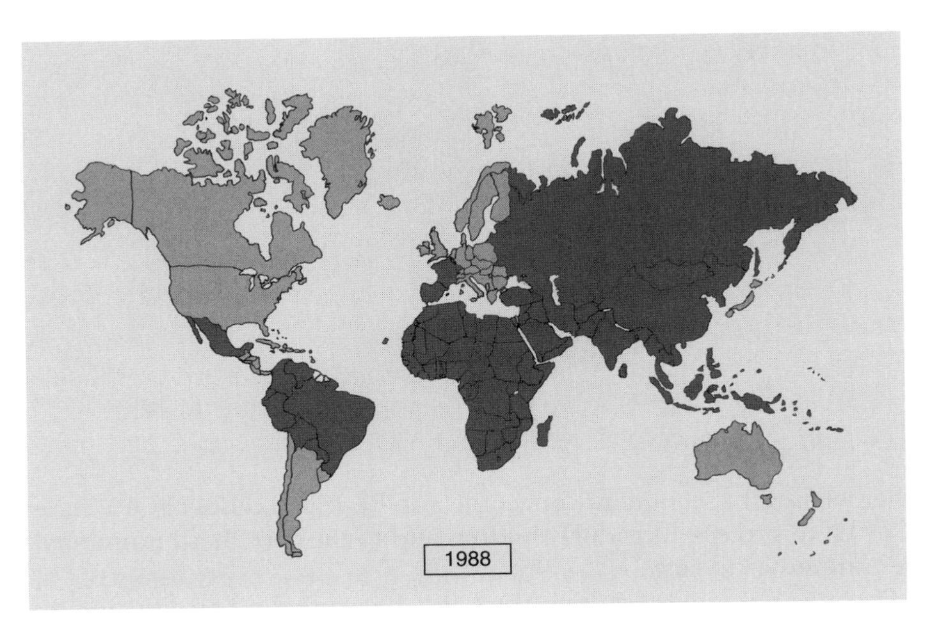

Figure 12.10 Disappearance of endemic polio (black) from the world between 1998 and 2001. Data from the World Health Organisation

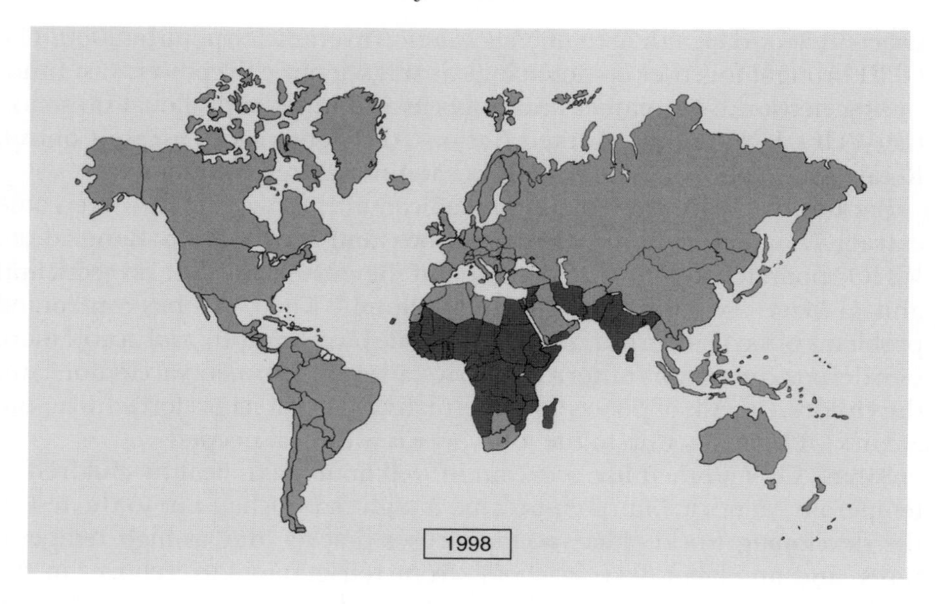

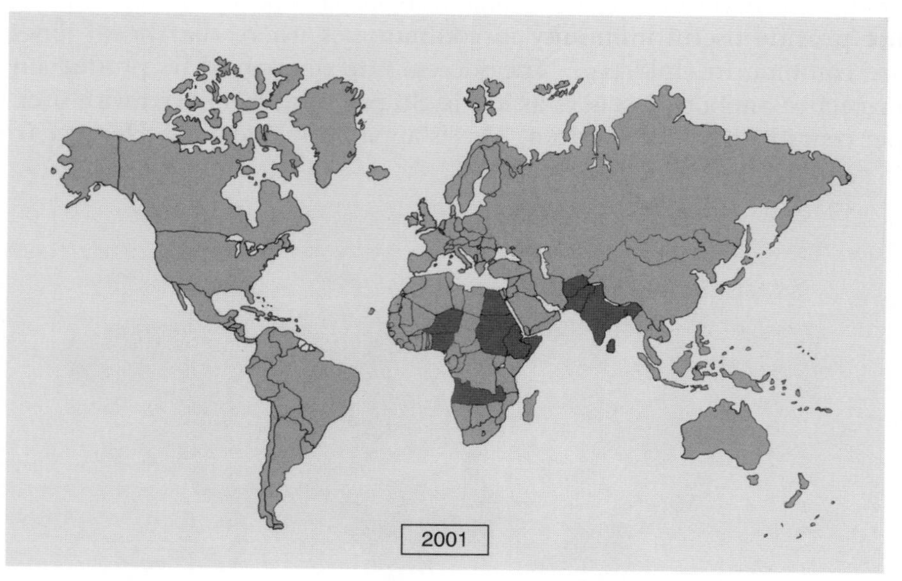

Figure 12.10 Continued

other viruses that commonly infect the gut. Repeated doses of OPV have to be given to protect individual children and to produce 'herd immunity' in the community at large.

Despite all these difficulties, the GPEI has pulled off astonishing feats, such as the biggest single vaccination offensive ever (145 million Indian children in a single sweep) and vaccinating over 80 per cent of all the under-fives in the world each year (which involves giving 2 billion doses of OPV to 575 million children in 94 countries).

The impact on polio has been dramatic. In 1988, the year of the GPEI's birth, polio was endemic in one-third of the world's countries and paralysed 350,000 people each year. Since then, the areas in the grip of endemic polio have steadily shrivelled away (Figure 12.10).

By 2009, there were just four endemic countries left, and after India was declared polio-free in January 2011, only Pakistan, Afghanistan and northern Nigeria remained. Compared with 1988, the numbers of paralytic polio cases have plummeted by over 99.9 per cent, to 6,350 in 1998 (the tenth year of the GPEI), 1625 in 2008 and only 222 in 2012 (Figure 12.11). Along the way, Type 2 polio was exterminated, in 1999.[53]

Unfortunately, this is a battle in which a 99.9 per cent victory is not enough. Extinction must be absolute, because polio will remain a live threat global health for as long as a chain of poorly vaccinated people can be formed to carry the virus out of its last boltholes to the planet at large. From current vaccination coverage, the WHO estimates that an epidemic of 200,000 cases could be unleashed if polio breaks out, setting the eradication clock back by over 20 years.[53]

This is also a campaign that the WHO and its partners in the GPEI cannot allow to fail. It has already run for 25 years, over twice as long

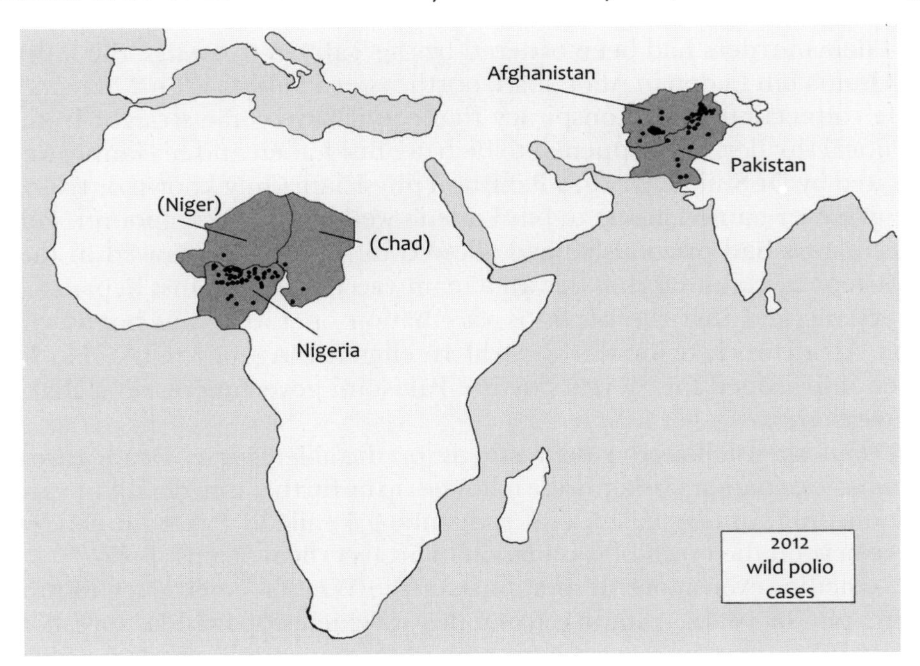

Figure 12.11 Distribution of wild-type polio cases in 2012. Endemic polio was confined to Northern Nigeria, Pakistan and Ahghanistan; a few cases were exported from Nigeria to adjacent Niger and Chad. Data from the World Health Organisation

as the smallpox eradication campaign, and has swallowed up over $10 billion. If it miscarries, it will be the most expensive public health flop in history, and could destroy the credibility of all those driving the initiative. Much more important, it would jeopardise the lives and health of hundreds of millions of people around the world.

In May 2012, the WHO announced that the extermination of polio had become a 'global public health emergency'. The eradication campaign was stepped up, and relevant governments strong-armed into increasing their support, but even these measures have failed to deliver the final blow. Indeed, the grip of vaccination on Pakistan and Afghanistan loosened, leaving rapidly increasing numbers of children in those countries unprotected.

So what went wrong?

American football

In December 2012 – which a year earlier had still seemed a feasible deadline for eradication – polio made headlines in newspapers around the world, but for the wrong reasons. Seven polio vaccinators had been shot dead on the streets of Pakistan; most were women, and the youngest was a 17-year-old schoolgirl who had been helping her older sister for the morning.[54]

Their murders had been ordered by the Taliban to avenge the killing of Osama bin Laden in Abottabad, northeastern Pakistan, on 2 May 2011. The connection was a conspiracy that could have come straight from a political thriller, but happened to be true. Bin Laden and his family were located by Dr Shikal Afridi, a Pakistani physician. Only known or trusted people ever gained access to bin Laden's well-guarded compound. Polio vaccinators had previously been allowed in; Dr Afridi followed in their footsteps by claiming that he ran a team vaccinating against hepatitis. It later emerged that the hepatitis vaccination operation was bogus, and that Afridi worked for the Central Intelligence Agency (CIA). He has been imprisoned for 33 years by the Pakistani government, on a charge of treason.[55]

Events then followed a depressingly predictable course. Death threats against vaccination workers were followed up by further murders. The vaccination programme stalled, and suddenly, 3.5 million Pakistani children were beyond the reach of vaccination to protect them against polio.[54]

Remedies even more desperate than the WHO's 'emergency footing' were put in place, ranging from pro-vaccine propaganda spread by powerful figures in government and Islam to armed guards for vaccination teams. An option that failed spectacularly in England 150 years ago was also seriously considered – legislation compelling parents to have their children vaccinated.

The stalemate will probably only be resolved when the Taliban decide that polio vaccination is too small a pawn to bother with. It is also possible that they can be convinced of the benefits of polio vaccination; this might seem a long shot, but there are successful precedents. During the smallpox eradication campaign in the Horn of Africa in 1977, a helicopter pilot brought down by Ethiopian rebels managed to vaccinate his captors while waiting to be rescued.[56] On a grander scale, Ciro de Quadros, who led the offensive to rid the Western Hemisphere of polio during the late 1980s, persuaded left-wing guerrillas from San Salvador to suspend hostilities during 'Days of Peace' to allow the vaccination campaign to continue.[57]

We can only hope that human nature will transform itself from problem to solution and find a way through the impasse. In the meantime, the immediate future looks bleak.

The world was a simpler place when Roosevelt directed operations against America's two great enemies – aggressors overseas and polio on home soil. He never faced the agonising dilemma of having to choose which enemy to go for, knowing that the choice would allow the other one to slip through his fingers. Since then, the nature of both enemies has changed beyond recognition.

There is, however, savage irony in the origin of the obstacle that may yet prevent polio vaccines – America's 'gift to the world' and the positive spin-off of an American nightmare – from achieving their final triumph.

13

Looking Forward to a Retrospective?

> However, as history is never finished, the story of poliomyelitis is not yet over ... the road will lead uphill for many years to come.

So ends John Paul's *A history of Poliomyelitis.* His words, written in 1971, still ring true today. Indeed, realists as well as pessimists are now having to admit that the prospects of exterminating polio may turn out to be 'an impossible dream'. These words were not the ranting of some uninformed and outdated Cassandra, but the expert opinion of Drs Neetu Vashisht and Jacob Puliyel, two senior paediatricians in Delhi, writing in May 2012.[1]

Today, polio still represents one of the greatest causes of disability worldwide, with an estimated 10–20 million people affected.[2] Across most of the planet, however, an entire generation has lived entirely free from polio, and it is as though the disease has already been written off.

As barometers of change, the organisations that fought so hard to keep polio in the public consciousness have had to channel their activities in other directions. In the United Kingdom, the British Polio Fellowship (formerly the Infantile Paralysis Fellowship) now concentrates on the plight of patients with post-polio syndrome – an exercise which involves re-educating doctors who learned medicine from textbooks that dismissed polio in a couple of lines. Across the Atlantic, the National Foundation for Infantile Paralysis (NFIP) slipped away from polio as long ago as 1958 in order to survive, because the threat of the disease had collapsed so dramatically. The organisation has reverted to the name from its greatest days of glory – the March of Dimes – but its focus is now on finding cures and treatments for birth defects.

Hoping not to tempt Providence, this seems an appropriate moment to look back on mankind's odd relationship with the poliovirus, and to speculate about the shape of things that might come in a polio-free world.

Fellow traveller

Polio has been part of our recorded history for just over two centuries. Compared with the millennia of suffering inflicted by pestilences such as plague, leprosy, smallpox and malaria, this is just a brief encounter. It has also been an interactive, two-way relationship. The process of civilisation has interfered with the poliovirus's way of 'life', and we have suffered the consequences of our own actions.

It was our success in putting clear water between ourselves and our excreta that transformed polio into an epidemic monster. Later, the nightmare of The Crippler was largely man-made, especially in America. And finally, it is human nature – politics, ideology and sheer bloody-mindedness – which is now preventing polio from being wiped out for all time. Overall, it could be argued that the impact of polio on mankind is more our fault than anything to do with the poliovirus.

The threats and burdens of polio have inspired great acts of courage, nobility and intellectual brilliance, ranging from the patients who cope every day with the legacies of a cruel disease, to the monumental achievements of scientists such as Karl Landsteiner, Macfarlane Burnet and John Enders. However, polio did not always bring out the best in people. Others have exploited those same threats and burdens for baser motives – and the list of villains includes those whom we would prefer to be above such things, namely doctors and scientists.

Never mind the quality, feel the width

Looking back into the history of medicine, scientists, doctors and their work are often illuminated by a harsh light. However, the lenses through which we view the past are inevitably distorted by what we know now, and whatever we choose to focus on must be calibrated against what was usual and reasonable for the time.

The Golden Age of polio research was the middle third of the twentieth century, three thrilling decades of massive turbulence across much of biomedicine, thanks to the technological revolutions of electron microscopy, tissue culture and molecular biology. When understanding charges ahead so quickly, cutting-edge science can be relegated to the mundane in just a few years.

However, even with due allowance for the advance of knowledge, the history of polio is punctuated by science that was poor by the standards of the day. Peculiar ideas survived and thrived, and in the face of strong evidence that they were fatally flawed. The most tenacious clung on for 20 years or more after they should have been given a decent burial, such as Edward Rosenow's self-miniaturising poliomyelitic streptococcus; the globoid bodies that sprouted so selectively at the Rockefeller; and the experiments contrived to prove that polio was transmitted via the nose or through insect bites. Some now appear daft but harmless, such as George Draper's claim to spot people susceptible to polio from their facial features, or Claus Jungenblut's 'diagnostic' skin test, which Albert Sabin tactfully put down. These trivia did harm indirectly, by diverting attention away from more profitable lines of enquiry; others, such as George Retan's brain washout therapy, were dangerous in their own right.

How were such poor ideas born in the first place, or not killed off in infancy? The reasons are probably much the same as they always have been, and still are today: slipshod experiments, interpreted by wishful thinking rather than statistics, and a reluctance to let contradictory findings snuff out an appealing flash of inspiration. Lurking in the background, and probably more common than we like to believe, is straightforward deceit, with everything from the gentle massage of wayward findings to the complete fabrication of data.

Why would scientists and doctors do such things? Some may be intrinsically dishonest, while others crack under the ever-present and unrelenting pressure to produce. Researchers (and their career prospects) are only as good as their last paper and their next grant; when the law of the jungle is 'publish or perish', and good journals are reluctant to print negative findings or dead hypotheses, the temptations are obvious.

We know little of what really went on behind the scenes in polio research, especially when the pressure of competition ran high. One worrying insight was revealed by Tom Rivers, who claimed that Hideo Noguchi told him never to retract errors, because 'it'll take them fifteen years to find out you were wrong'. This comment convinced Rivers that Noguchi was dishonest and that all his work was suspect.[3]

If true, this is a sad addendum to Noguchi's story, which already has a tragic ending. Others, including Peter Olitsky and Nobel Laureate Peyton Rous, contradicted Rivers' view of Noguchi and jumped to defend this 'modest, honest and immensely serious scientist'.[4] We shall never know the truth. For whatever reason, though, the scientific literature on polio remains littered with papers that should have been retracted. And Rosenow, Simon Flexner, Noguchi, Draper and Jungenblut were in plentiful company for never confessing that they had been wrong.

How could badly flawed research find its way into scientific journals? Answer: because of the failure – accidental or otherwise – of peer review. This is the mystical quality control by which scientists are supposed to evaluate other researchers' work with fairness and objectivity. The process readily breaks down in the real world, where 'peers' may be colleagues and friends, or competitors and enemies. Also, peer review stands little chance against powerful people with their own agendas – such as Flexner, who forced so many American polio researchers to look at polio with his own biased gaze.

Flexner was not the only scientist to be blinded by stubbornness and self-belief. Jonas Salk and Sabin were each convinced that his own vaccine was perfect, to the point where each refused to accept hard evidence that it could cause paralysis – a delusion that Sabin clung to until the end of his life.

Rough treatment

Most attempts to treat polio look bad from the standpoint of the twenty-first century. Today, we can accept that desperate circumstances demand desperate remedies, and can barely imagine the horror of the front line during a polio outbreak, watching children slide inexorably towards death by suffocation. Equally, it is impossible to imagine subjecting a sick, frightened child to 'treatments' such as branding the back with a red-hot iron, or draining off cerebrospinal fluid through a needle pushed up inside the skull.

At the time, however, the medical establishment and parents alike pinned their hopes on such treatments. Many were dreamed up by doctors at the top of the profession, and nobody was bold enough to suggest that their inventions might be useless or dangerous. At least something was being done; even barbaric measures might have been preferable to the agony of seeing children die, knowing that there had been other possibilities of saving them.

There was virtually no hard evidence that any polio treatments did good rather than harm. Today, new therapies are put to the test within the rigid framework of the randomised, controlled, double-blind clinical trial. Patients, selected to be as uniform as possible, are randomly allocated to receive either active treatment or a 'placebo' dummy. 'Double-blind' means that neither the patients nor those looking after them know whether they are taking active or placebo, until after the end of the study. This is a necessary deceit, as the 'placebo effect' – feeling better while taking something that could potentially work – can dramatically improve symptoms such as pain and immobility. Conversely, when the placebo effect is stripped away, the genuine therapeutic action of many supposed wonder drugs has turned out to be negligible.

Thousands of patients, carefully followed up for months or years, may be needed for a trial to have enough statistical muscle to prove that the treatment is significantly better than placebo (or is much the same). Large-scale controlled clinical trials should also reveal the side effects of the treatment under test, if we remember that patients taking placebo can sometimes develop convincing and worrying symptoms.

The vast majority of polio treatments, from Rosenow's antiserum to Sister Kenny's physical therapy, were never properly put to the test. To be fair, this would often have been a huge undertaking because of the unpredictability and variability of polio. Moreover, conducting proper clinical trials in polio would have gone against the grain of contemporary medical practice. Up to the 1950s, clever statistics and solid evidence were not needed to justify using a treatment; the say-so of an 'expert', who may (or not) have strung together a few encouraging if highly selected cases into a paper or two, was quite enough.

In 1935, the year which finally ushered out convalescent serum and the vaccines of William H. Park, Maurice Brodie and John Kolmer, medical journals were constantly buzzing with new and exciting therapies. Liver extract was used to treat asthma in the belief that this was a hormonal problem, while patients with fever were given intravenous injections of activated charcoal because this was good at mopping up toxins in the test-tube.[5,6] Both of these breakthroughs came from centres of excellence and were written up in respectable journals. Neither had the faintest shred of rationale, and although liver extract may have been innocuous, intravenous charcoal undoubtedly killed some of the subjects whose fever 'failed to respond to treatment'.

Polio therapy eventually led the way, in the massive field trials of Salk's vaccine in 1954. By then, the design of clinical trials was becoming a science in its own right, but one which Salk held in contempt. It was at Salk's insistence that Basil O'Connor sacked the clinical trials experts and bullied hundreds of American doctors into testing the vaccine according to Salk's wishes – without proper controls.[7]

Defenceless

An area that appears particularly murky now was human experimentation and especially the ease with which researchers could sidestep ethical constraints when they wanted to push ahead with their science.

In a civilised society, the laws of medical ethics should be as inviolable as the laws of physics. Clear ethical guidelines, supposedly universal, were written down for the first time in 1947. They were prompted by the litany of atrocities committed by Nazi 'medicine', exposed during the Nuremberg Medical War Crimes Trial of 1945–6. Article 1 of the Nuremberg Code states that any person taking part in an experiment 'must have legal capacity and free consent, without intervention of force, fraud or deceit'.[8] In other words, experimental subjects must understand fully what they are letting themselves in for, risks and all, and must then agree to take part of their own free will.

Until the 1950s, the doctor's decision to do an experiment was enough to turn a child into a human guinea pig. Consent, informed or otherwise, is very rarely mentioned in medical papers during the first half of the twentieth century. Expectations that the horrors revealed at Nuremberg would force in a new respect of ethical experimentation were quickly dashed, most obviously in America.

All three of the successful polio vaccine pioneers – Koprowski, Salk and Sabin – broke ethical boundaries in early trials of their vaccines.

Koprowski and Salk deliberately chose mentally backward children. This horrified a few individuals such as Tom Rivers, the editor of the *Lancet* and Sabin (temporarily), but otherwise caused no lasting ripples on the smooth surface of American medicine. The climate had changed a few years later, when Sabin applied to do his own vaccine trials in Willowbrook State School for mentally impaired youngsters. When Sabin was refused permission (by the NFIP, not Willowbrook), he resorted to using prisoners who were offered incentives to take part, including cash and the promise of early release. Although not as blatant as the trials of Koprowski and Salk, Sabin still violated Article 1 of the Nuremberg Code. Writing in 1971, John Paul took the view that none of these trials would have gone ahead if ethical guidelines of the day had been properly applied.[9]

The vaccine trials failed to cause outrage because this was par for the course for America, which had a long-standing ethical blind spot. During the 1930s, German medicine's downward spiral into Hell had been observed with detachment from across the Atlantic. News items from Germany about forced sterilisation, 'racial hygiene' and eugenics appeared regularly in leading American periodicals such as the *Journal of the American Medical Association*,[10] but failed to goad commentators or editors into saying that this was wrong. By the mid-1950s, America was yet to put its own house in order, even though memories of Nuremberg – where the United States had led the prosecution – were less than ten years old.[11]

Two places that intersect with the story of polio later shot to prominence as world-class examples of ethical abuse. Between 1956 and 1970, inmates of Willowbrook School were fed extracts of faeces from other children suffering from hepatitis, to study how the infection was transmitted. Parents could jump the long waiting list for a place at Willowbrook by signing their children into the experiment, which was run by Dr Saul Krugman of New York University School (coincidentally, a cousin of Sabin). When news of the study broke in 1971, it sparked a spirited defence by Krugman and many others, who saw nothing wrong in what they had done.[12]

At the time of Salk's vaccine trials, the US Public Health Department was halfway through its 40-year study (1932–1972) of untreated syphilis in Black subjects in Tuskegee, Alabama. This was next door to the NFIP's rehabilitation centre for those who were not white enough to have a place at Warm Springs. Penicillin, highly effective in syphilis, was withheld for 30 years after US Public Health doctors first used it to treat the disease. Many papers were published, but no questions asked; eventually, a quarter of the Tuskegee patients died of syphilis.[13] The whistle was blown in 1971, followed in 1974 by compensation for the victims and finally – '65 years late' – an apology from President Bill Clinton in May 1997.[14]

Clinton tore into those who ran the Tuskegee study for having 'diminished the stature of man by abandoning the most basic ethical precepts'.

The deafening silence from the medical and scientific community during the four decades of the trial suggests that this was a systemic disease, rather than an isolated outbreak.

Against that background, the polio vaccine trials seem almost ethical.

Nightmare

> So, first of all, let me assert my firm belief that the only thing we have to fear is fear itself – nameless, unreasoning, unjustified terror which paralyses needed efforts to convert retreat into advance.
>
> Franklin D. Roosevelt, First Inaugural Address as President of the United States of America, 4 March 1933

When novelist Alistair MacLean wrote his thriller, *The Satan Bug*, in 1962 he had to dream up a virus capable of wiping out all humanity.[15] He chose a highly virulent poliovirus, which had been turned into an unstoppable killer in a Ministry of Defence laboratory deep in the English countryside. This doomsday germ was plausible enough to blast MacLean's 'spine-chilling, throat-clutching' book straight into the best-seller list. The theme was timely and tapped neatly into fears about germ warfare as an even nastier alternative to nuclear annihilation.

It later turned out that there was substance to the paranoia, as the Russian 'Biopreparat' laboratory at Koltosovo, Novosibirsk, had been developing weapons-grade smallpox during the Cold War. This was to be delivered in bomblets that were kept nicely cool by a special refrigeration unit in the warhead of the SS–20 intercontinental ballistic missile. For added sophistication, the Soviets' favourite battle strain of smallpox, 'India', had undergone a Satan Bug-type transformation in the Biopreparat laboratory, to make it radiation-resistant – a clever touch to mop up any who might survive a thermonuclear attack.[16]

To take us neatly back to the story of polio, the smallpox germ warfare programme at Biopreparat was directed during the 1980s by Viktor Zhdanov (Figure 12.8), the politically adroit Health Minister and academician who had helped to bring Sabin's vaccine to the USSR in 1958.[16]

Wild-type polioviruses would need a lot of work to turn them into a useful biological weapon. Thanks to its high infectivity, polio would soon rampage through a community, for example after being slipped into drinking water (which does not contain enough chlorine to inactivate it), or milk or food. However, even the most virulent strains such as Mahoney would incapacitate just a few percent of its victims, compared with the 30–50 per cent death-rate with Biopreparat's supercharged 'India' smallpox strain. We can only speculate whether the ability of wild-type polioviruses to paralyse and kill could be artificially enhanced as in *The Satan Bug* – or indeed whether this has been tried. However, now that the genetic sequences that determine neurovirulence are known to the last nucleotide, the possibility is much closer to reality than when MacLean wrote his novel.

Even if it did not kill or paralyse many of its victims, polio could still be of use to an enemy as a weapon of terror. With typical English reserve, the *Postgraduate Medical Journal* of May 1949 described the social disruption caused by polio outbreaks as 'out of proportion' to the damage actually caused.[17] This was a masterpiece of understatement, especially in 1950s America where the aura of fear generated by polio provided the raw material for the NFIP's 'campaign of terror'. Since then, other viruses have brought nightmares of their own – Lassa fever, acquired immune deficiency syndrome (AIDS), severe acute respiratory syndrome (SARS), swine flu. However, none has matched The Crippler's power to paralyse with fear, setting it close behind the man-made horror of the atom bomb.[18] Not bad for a virus that was never going to rise above mid-list for its morbidity and mortality.

In 1961, the year before *The Satan Bug* was first published, MacLean wrote another novel, *Fear is the Key*.[19] As well as providing the title for Chapter 6 in this book, this would have made a neat mission statement for the NFIP and the man whose words head this section.

If and when

Two names – Rahima Bhanu and Ali Maow Maalin – and the date of 8 May 1980 have gone down in history as milestones marking the end of the 180-year journey to eradicate smallpox. The names are of the last victims of smallpox in its final two boltholes, on a small island in the mouth of the Ganges and the port of Merca on the coast of Ethiopia. The date was when the World Health Organisation (WHO), having waited two years to make sure that no outbreaks had been missed anywhere in the world, declared that 'Target Zero', the complete extermination of smallpox, had been achieved.[20]

At the time of writing, smallpox was still the only human infection ever to be cleared from the face of the planet by mankind. Rinderpest, a viral 'plague' of cattle, was completely eradicated in 2011. With luck, this will soon be followed by another human infection, the guinea worm. This is a deeply unpleasant, spaghetti-like parasite which is traditionally winched out of its burrow under the victim's skin by winding it around a stick. The extermination of the guinea worm will be welcomed by all (the 'Save the Guinea Worm' campaign, allegedly striving to defend the planet's biodiversity,[21] turns out to be a spoof).

Polio's last stand may well be in a village or refugee camp in the high mountains along the Pakistan-Afghanistan border. The name of the last victim will be recorded below the others for that year, although it will take many more months before the column can be finally ruled off. In the WHO's headquarters in Geneva, the corks will stay firmly in the champagne bottles while waiting for a decent interval – three long years – to ensure that this is not a false dawn. Only then will it be announced that polio has been exterminated.

This triumph will be the vindication of the Global Polio Eradication Initiative's (GPEI's) resolve and the fulfilment of a wild ambition that became realisable when the results of the trial of Salk's vaccine were announced on 12 April 1955.

However, most of the world's inhabitants will pay the news only fleeting attention, if at all. For those in the Western Hemisphere, polio has been extinct for a generation, and intrudes as much into their daily lives as the Black Death or scurvy. Even in most of the battlegrounds that have seen action in the last decade, polio has been demoted by the success of vaccination into a minor player among other childhood infections.

In some ways, this will be a pyrrhic victory. Polio vaccination will have to continue as usual around the world for some years, until it is absolutely certain that all polioviruses, including those derived from oral vaccine and multiplying in the intestines of chronic carriers, are extinct. Without the protection of vaccination, we would risk the catastrophe of a massive 'virgin soil' pandemic should polio somehow return. Sabin would not approve of the post-eradication vaccination strategy. To remove the risk of spreading the oral vaccine-derived polioviruses in which Sabin never believed, Salk's vaccine would be used across the world, as it is now in the United States and many other countries.[22]

The eradication of polio will have cost billions of dollars ($10 billion up to the end of 2012), largely provided by GPEI, but with contributions from some governments that are cash strapped and have other diseases to fight. Congratulations will soon be followed by pointed discussion about whether all that money would have been better spent on trying to stamp out malaria or vaccinate children against another intestinal virus, the rotavirus. It will be argued that the total extermination of polio has added little to the stalemate of near eradication, as only 200 cases worldwide slipped through in 2012. By contrast, rotavirus-induced diarrhoea and malaria each kill around half a million children every year.

Following the example of smallpox eradication, there will also be some frantic tidying up. Under the command of the WHO, research programmes using live poliovirus will be wound up, and the stocks of virus held by the tens of thousands of laboratories around the world will be destroyed. This process has already been completed in the United Kingdom, and is under way in many other countries. This will be a mammoth undertaking, especially as poliovirus may be lurking unknown in countless laboratory freezers, for example in faeces samples collected for unrelated research.

Reference samples of poliovirus will probably be retained, frozen in liquid nitrogen and locked away under conditions of the highest security. These may well end up next to the freezers that contain the two remaining stocks of smallpox virus. With true post-Cold War symmetry, one is at the Centers for Disease Control (CDC) in Atlanta, Georgia, and the other in Koltsovo, Novosibirsk – by unhappy coincidence, on the same site where weapons-grade 'India' smallpox was created by Biopreparat during the 1980s.

As with smallpox, there will be fierce debate about whether these final hostages of the vanquished enemy should be kept under lock and key (because even extinct diseases may need to be studied in the future) or killed off (because nothing is added except the risk that the virus might escape). And if the example of smallpox is followed faithfully, the discussion will bubble up every few years but fail to reach any conclusion, leaving the poliovirus dormant and safe in its −70°C sanctuary for the time being.

Even if these final relics are destroyed, polio could still return from beyond the grave. The mutations that transformed a harmless Coxsackie virus into a virulent poliovirus could be recapitulated.[23] And of course, now that the RNA sequences of strains such as Mahoney have been published, together with the step-by-step recipe for constructing it in the laboratory, a do-it-yourself synthetic poliovirus could appear at any time.[24]

In a post-eradication world with growing gaps in the defences of vaccination, the biggest polio-susceptible population in the history of the world would soon build up. In that setting, a polio pandemic would be devastating. The Crippler could once again sweep across continents, and with greater ferocity and terror than we have ever seen during all our time with polio.

I began this book in the spring of 2011 with the naive hope that I could end it with a pithy epitaph for polio. Now, it is depressingly clear that this will have to wait, possibly for many years to come. However, we should still take this opportunity to reflect on the disease which has shaped the modern world to a greater extent than we might suspect.

At first sight, polio pales into insignificance beside the apocalyptic pestilences that obviously changed the course of human history, such as the Black Death and smallpox, which wiped out entire civilisations. Yet polio tweaked numerous chains of events in more subtle ways, ranging from the scientific advances that spun out from polio research, to the formulation of rights for people with disabilities.

And because of what it did to one man, the lives of hundreds of millions of people across the world were changed forever: not just those who met polio, or who were spared that encounter, but all those who were caught up in the cataclysm of the Second World War and its aftershocks.

Who knows how the twentieth century might have turned out without the poliovirus's 'Day of Infamy' on 8 August 1921, on Campobello Island in the Bay of Fundy.

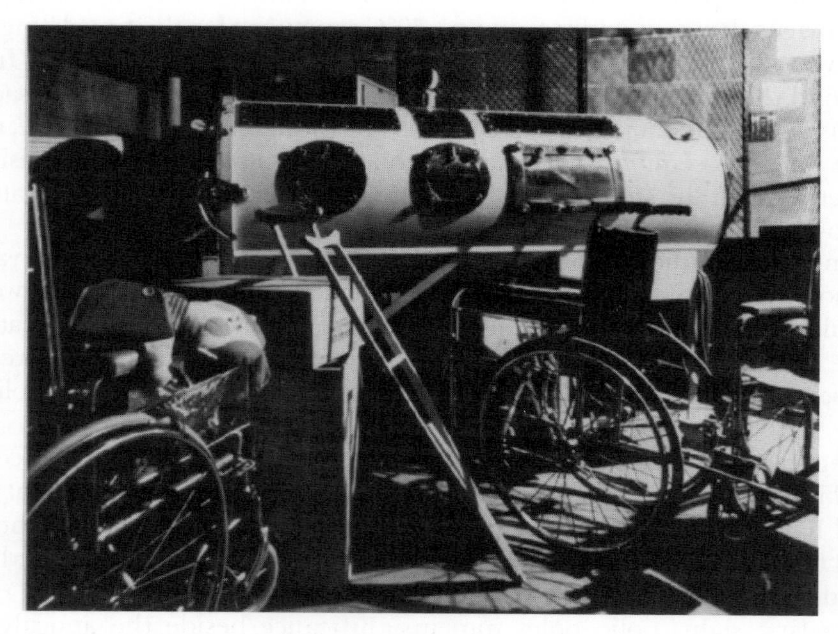

Reproduced by kind permission of the World Health Organisation

Notes

1 A Plague from Nowhere

1. Russel CK. A contribution to the study of acute poliomyelitis based on the observation of thirty-eight recent cases with two autopsies. *Montreal Medical Journal.* 1910;39:465–67.
2. Caverly CS. Preliminary report of an epidemic of paralytic disease, occuring in Vermont, in the summer of 1894. *Yale Medical Journal.* 1894;1:1–20.
3. Lavinder CH, Freeman AW, Frost WH. Epidemiologic studies of poliomyelitis in New York City and the North Eastern United States during the year 1916. *Public Health Bulletin Washington.* 1918;91.
4. Porter R. *The greatest benefit to mankind: a medical history of humanity from antiquity to the present.* London: Fontana. 1999;483.
5. Paul JR. *A history of Poliomyelitis.* New Haven and London: Yale University Press, 1971.
6. Underwood M. *A treatise on the diseases of children, with general directions for the management of infants from the birth, London.* London: Matthews; 1789.
7. Heine JC. *Beobachtungen über Lähmungzustände der unteren Extremitäten und deren Behandlung.* Stuttgart: Kohler; 1840.
8. Von Heine J. *Spinale Kinderlähmung.* Stuttgart: Cotta; 1860.
9. Vogt W. Die essentielle Lähmung der Kinder. Bern: Haller; 1858.
10. West C. On some forms of paralysis incidental to infancy and childhood. *London Med Gaz.* 1843;32:829.
11. Wickman I. *Beiträge zur Kenntnis der Heine-Medinschen Krankheit: Poliomyelitis acuta und verwandter Erkrankungen.* S. Karger; 1907.
12. Robertson RG. *Rotting face: smallpox and the American Indian.* Caldwell, Idaho: Caxton Press; 2001.
13. Pait CFC-P. Virology of poliomyelitis. *Calif Med [Internet].* 1950;73:391–93. Available from: http://www.ncbi.nlm.nih.gov/pubmed/14778005
14. Lloyd-Davies M, Lloyd-Davies TA. *The Bible: medicine and myth.* Cambridge: Silent Books; 1991.
15. Roberts C, Manchester K. *The archaeology of disease.* 3rd edition. Stroud: The History Press; 2010. pp. 181–182.
16. Hamburger O. [A case of infantile paralysis in Ancient Egypt; in Danish]. Ugeskr f Laeger 1911; 73:1565. English translation: Hansen E. A probable case of infantile paralysis in Ancient Egypt. Hosp Bull Univ Maryland 1912–1913; 8: 192–94.
17. Ruhräh J. Poliomyelitis? *American Journal of Diseases of Children.* 1932;43:195.
18. Salzmann JG. *Plurium pedis musculorum defectum.* Strasbourg: Imprimerie Jean-Henri Heitz; 1734.
19. Andry N. *L'Orthopédie.* Bruxelles: Georges Friex; 1943.
20. Maloney WJ. Michael Underwood: a surgeon practising midwifery from 1764 to 1784. *Journal of the History of Medicine and Allied Sciences.* 1950;5:289–314.
21. Underwood M. *A treatise on the diseases of children, with general directions for the management of infants from the birth.* Philadelphia: Gibson; 1793.

22. Mettler CC, Mettler FA. *History of medicine: a correlative text, arranged according to subjects*. Blakiston; 1947.
23. Monteggia GB. *Instituzione chirurgicale,* 2nd edition. Milano: Giuseppe Maspero, 1813; paras. 558–560.
24. Badham J. Paralysis in childhood: Four remarkable cases of suddenly induced paralysis in the extremities occurring in children without any apparent cerebral or cerebrospinal lesion. *London Medical Gazette*. 17:215–20.
25. Colmer G. Paralysis in teething children. *American Journal of Medical Sciences*. 1843;5:248.
26. Axelsson P. 'Do not eat those apples; they've been on the ground!': polio epidemics and preventive measures, Sweden 1880s–1940s. *Asclepio*. 2009;61:26–7.
27. Cordier S. Relation d'une épidémie de paralysie atrophique de l'enfance. *Lyon Méd*. 1888;57:5–12.
28. Jacobi MP. *Infantile spinal paralysis*. Pepper's System of Medicine, Philadelphia, Lea Brothers & Co. 1886;5:1113–64.
29. Kaufmann SHE. *Robert Koch's highs and lows in the search for a remedy for tuberculosis*. Nature Medicine Special Web Focus: Tuberculosis [Internet]. 2000; Available from: http://www.nature.com/nm/focus/tb/historical_perspective.html
30. Medin O. Ueber eine Epidemie von spinaler Kinderlähmung. *Verhandl. d. 10. Internatl. med. Kongr.* 1891;37–47.
31. Axelsson P. 'Do not eat those apples; they've been on the ground!': polio epidemics and preventive measures, Sweden 1880s–1940s. *Asclepio*. 2009;61:30.
32. Wickman OI. *Studien über Poliomyelitis acuta: Zugleich ein Beitrag zur Kenntnis der Myelitis acuta*. Berlin: S. Karger; 1905.
33. Hardy A. Poliomyelitis and the Neurologists: the view from England, 1896–1966. *Bulletin of the History of Medicine*. 1997;71:249–72.
34. Axelsson P. Ivar Wickmans akademiska motgång. *Läkartidningen*. 2003;100: 140–42.
35. Harlow JM. Passage of an iron rod through the head. *The Boston Medical and Surgical Journal*. 1848;39:389–93.
36. Thompson G. On the physiology of general paralysis of the insane and of epilepsy. *Journal of Mental Science*. 1875;20:67–74.
37. Rilliet F. De la paralysie essentielle chez les enfants. *Gaz. méd. Paris*. 1851;6: 681,704.
38. Charcot JM, Joffroy A. Cas de paralysie infantile spinale avec lésions des cornes antérieures de la substance grise de la moëlle épinière. *Arch Physiol Norm Pathol*. 1870;3:134–40.
39. Vulpian A. Cas d'atrophie musculaire graisseuse datant de l'enfance. Lésions des cornes antérieures de la substance grise de la moëlle épinière. *Arch Physiol Norm Pathol*. 1870;3:316–25.
40. Cornil V. Paralysie infantile; Cancer des seins; Autopsie; Altération de la moëlle épinière, des nerfs et des muscles; généralisation du cancer. *C.R Soc Biol (Paris)*. 1863;5:187.
41. Heidelberger M. *Karl Landsteiner 1868–1943*. Washington DC: National Academy of Sciences; 1969.
42. Wolbach SB. The filterable viruses, a summary. *J Med Research*. 1912;27:1–25.
43. Landsteiner K, Popper E. Ueberträgung der Poliomyelitis acuta auf Affen. *Z Immunitätsforsch*. 1909;2:377–90.
44. Levaditi C, Landsteiner K. La transmission de la paralysie infantile au chimpanzé. *Compt. rend. Acad. de Sci.* 1909;149:1014–16.

45. Landsteiner K, Levaditi C. La transmission de la paralysie infantile aux singes. *C.R Soc. Biol.* 1909;67:592–94.

2 The Crippler

1. Ballester R, Porras MI. La lucha europea contra la presencia epidémica de la poliomielitis: una reflexion histórica. *Dynamis.* 2012;32:273–85.
2. Martin W. Poliomyelitis in England and Wales, 1947–1950. *Brit J Soc Med.* 1951;2: 236–46.
3. Editorial. Poliomyelitis Precautions. *BMJ.* 1949:323–24.
4. Hopkins C, Dismakes W, Glick T, Warren R. Surveillance of paralytic poliomyelitis in the United States. *J Am Med Ass.* 1969;210:694–700.
5. Wright P, Kim-Farley R, De Quadros C, et al. Strategies for the global eradication of poliomyelitis by the year 2000. *N Engl J Med.* 1991;325:1774.
6. Fact sheet 114. *Poliomyelitis [Internet].* WHO. Available from: http://www.who.int/mediacentre/factsheets/fs114/en/
7. Plotkin S, Orenstein W (eds). *Vaccines.* 4th ed. Philadelphia: Saunders; 2004. p. 438; Center for Disease Control and Prevention. International notes certification of poliomyelitis eradication – The Americas 1994. From Morbidity and Mortality Weekly Report. 7 October 1994;43:270–72.
8. Kelland K, Ahmad J. *Insight: fear and suspicion in Pakistan hamper global polio fight.* Reuters. 2012;27 September.
9. Lewin P. *Infantile paralysis. Anterior poliomyelitis.* Philadelphia: WB Saunders & Co.; 1941.
10. Frauenthal H, Manning J. *A manual of infantile paralysis.* Philadelphia: FA Davis; 1914, p. 113.
11. Bayer P. The management of the acute phase of poliomyelitis. *Postgraduate Medical Journal.* 1949;25:9–12.
12. Draper G. Significant problems in acute anterior poliomyelitis. *JAMA.* 1931;97(16): 1139–41.
13. Lewin P. *Infantile paralysis.* 1941;6.
14. Frauenthal H, Manning J. *A manual of infantile paralysis.* p. 122.
15. Marie P. Hémiplégie cérébrale infantile et maladies infectieuses. *Progr Méd (Paris).* 1885;13:167–69.
16. Strümpell A. Ueber die acute Encephalitis der Kinder (Polioencephalitis acuta, cerebrale Kinderlähmung). *Allg Wien med Ztg.* 1884;29:612–20.
17. Paul H. *The control of diseases (social and communicable).* Edinburgh & London: E&S Livingstone; 1964, pp. 374–75.
18. Hobbes T. Leviathan, chapter 13. London, 1651.
19. Trojan DA, Cashman NR. Post-poliomyelitis syndrome. *Muscle & Nerve.* 2004;31:6–19.
20. Lepine R, V C. Sur un cas de paralysie générale spinale antérieure subaiguë, suivie d'autopsie. *Gaz Méd (Paris).* 1875;4:127–29.
21. Wiechers DO. Late effects of polio: historical perspectives. *Birth Defects Orig Artic Ser.* 1987;23:1–11.
22. Wiechers DO, Hubbell SL. Late changes in the motor unit after acute poliomyelitis. *Muscle & Nerve.* 1981;4:524–28.
23. Perry J, Barnes G, Gronley JK. The postpolio syndrome: an overuse phenomenon. *Clinical orthopaedics and related research.* 1988;233:145–62.
24. Modlin J, Coffey D. Poliomyelitis, polio vaccines, and the post-polio syndrome. In *Infections of the central nervous system.* Philadelphia, Pa: Lippincott-Raven Publishers; 1997, pp. 57–72.

25. Johnstone DF. Some mistaken diagnoses in the common infectious fevers. *BMJ*. 1944;2:555.
26. McAlpine D, Buxton PH, Kremer M, Cowan DJ. Acute poliomyelitis. *BMJ*. 1947;2:1019–23.
27. Frauenthal H, Manning J. *A manual of infantile paralysis* p. 216.
28. Melnick JL. Enterovirus type 71 infections: a varied clinical pattern sometimes mimicking paralytic poliomyelitis. *Rev Infect Dis*. 1984;6:S387–90.
29. Jeha LE, Sila CA, Lederman RJ, Prayson RA, Isada CM, Gordon SM. West Nile virus infection: a new acute paralytic illness. *Neurology*. 2003;61:55–9.
30. Sabin A, Wright W. Acute ascending myelitis following a monkey bite, with the isolation of a virus capable of reproducing the disease. *J Exp Med*. 1934;59:115–36.
31. Frauenthal H, Manning J. *A manual of infantile paralysis* p. 277.
32. Retan G. The treatment of acute poliomyelitis by intravenous injection of hypotonic salt solution. *J Ped*. 1937;11:647–64.
33. Guillain G, Barré J, Strohl A. Sur un syndrome de radiculonévrite avec hyperalbuminose du liquide céphalo-rachidien sans réaction cellulaire. Remarques sur les caractères cliniques et graphiques des réflexes tendineux. *Bulletins et mémoires de la Société des Médecins des Hôpitaux de Paris*. 1916;40:1462–70.
34. Landry J. Note sur la paralysie ascendante aiguë. *Gazette hebdomadaire de médecine et de chirurgie*. 1859;6:472–74, 486–88.
35. Nobuhiro Y, Hartung H-P. Guillain–Barré Syndrome. *N Engl J Med*. 2012;366: 2294–304.
36. Goldman A, Schmalstieg E, Freeman D, et al. What was the cause of Franklin Delano Roosevelt's paralytic illness? *J Med Biography*. 2003;11:232–40.
37. *Poliomyelitis laboratory manual*. Geneva: WHO, 2004.
38. Nottay B, Yang CF, Holloway BP et al. Identification of vaccine-related polioviruses by hybridization with specific RNA probes. *J Clin Microbiol*. 1995;33.
39. Bates P, Pellow J. *Horizontal man: the story of Paul Bates*. London: Longmans, Green and Co.; 1964.
40. Offit PA. *The cutter incident: how America's first polio vaccine led to the growing vaccine crisis*. New Haven: Yale University Press; 2005.
41. Porter R. *The greatest benefit to mankind. A medical history of humanity*. London: Fontana Press; 1997. p. 403.
42. Thomas FP, Beres A, Shevell MI. A Cold Wind Coming: Heinrich Gross and child euthanasia in Vienna. *J Child Neurol*. 2006;21:3234.
43. Frauenthal H, Manning J. *A manual of infantile paralysis*. Philadelphia: FA Davis; 1914, p. 112.
44. Adamson JD, Moody JP, Peart AFW, Smillie RA, Wilt JC, Wood WJ. Poliomyelitis in the Arctic. *Can Med Assoc Journal*, 1949;61:339–48.
45. Smallman-Raynor MR, Cliff AD. *Poliomyelitis: a world geography: emergence to eradication*. Oxford University Press, USA; 2006.
46. Peart AF, Rhodes AJ. An outbreak of poliomyelitis in Canadian Eskimos in wintertime. *Can Journal Public Health*. 1949;40:405.
47. Mitchell C. 57 years of silence. *Winnipeg Free Press*, 26 July 2009.
48. Lassen H. The epidemic of poliomyelitis in Copenhagen, 1952. *Proc R Soc Med*. 1954;47:67–71.
49. Pincock S. *Bjørn Aage Ibsen. The Lancet [Internet]*. 2007;370:1538. Available from: http://linkinghub.elsevier.com/retrieve/pii/S014067360761650X
50. Interview with GW, 14 February 2013.
51. Interview with GW, 2 November 2012.
52. Interviews with Anna Tóth and Magda Pribojszki, January 2013.

3 The Virus That Never Was

1. Quincy J. *The American Medical Lexicon: on the plan of Quincy's Lexicon Physico-medicum, with many Retrenchments, additions, and improvements; comprising an explanation of the Etymology and signification of the terms used in anatomy, physiology, surgery, materia me.* New York: T&J Swords; 1811.
2. Webster N. *A brief history of epidemic and pestilential diseases: with the principal phenomena of the physical world, which precede and accompany them, and observations deduced from the facts stated: in two volumes.* Hartford: Hudson & Goodwin; 1799.
3. Halliday S. Death and miasma in Victorian London: an obstinate belief. *BMJ.* 2001;323(7327):1469–71.
4. Osler W. *The principles and practice of medicine.* New York: D Appleton and Co; 1892.
5. Medin O. Ueber eine Epidemie von spinaler Kinderlähmung. *Verhandl. d. 10. Internatl. med. Kongr.* 1891;37–47.
6. Axelsson P. 'Do not eat those apples; they've been on the ground!': polio epidemics and preventive measures, Sweden 1880s-1940s. *Asclepio.* 2009;61: 26–27.
7. Caverly CS. Preliminary report of an epidemic of paralytic disease, occuring in Vermont, in the summer of 1894. *Yale Medical Journal.* 1894;1:1–20.
8. Lovett RW. The occurrence of infantile paralysis in Massachusetts in 1907. *The Boston Medical and Surgical Journal.* 1908;159:131–39.
9. Sandler B. *Diet prevents polio.* Milwaukee, WI: The Lee Foundation for Nutritional Research; 1951.
10. Scobey RR. Food poisoning as the etiological factor in poliomyelitis. *Arch Pediatr.* 1946;63:322–54.
11. Scobey RR. Is human poliomyelitis caused by an exogenous virus? *Science.* 1954; 51:117.
12. Toomey JA., Takacs W, Tischer. MD. Attempts to recover poliomyelitis virus from fruit, well water, chicken cords and dog stools. *J. Pediat.* 1943;23:168–71.
13. McBean E. *The Poisoned Needle.* Mokelumne Hill, California: Mokelumne Hill Press; 1993.
14. Wickman OI. *Studien über Poliomyelitis acuta: Zugleich ein Beitrag zur Kenntnis der Myelitis acuta.* Berlin: S. Karger; 1905.
15. Australian Medical Gazette. 24 August 1897.
16. Cone TE. Milk sickness (tremetol poisoning). In Kiple KF, Graham RR, Frey D, Browne A, (eds) *The Cambridge world history of human disease.* Cambridge: Cambridge University Press; 1993.
17. Osler W. *The principles and practice of medicine,* New York: D Appleton and Co; 1892, p. 266.
18. Pitzman M. The cause and prevention of infantile paralysis (polio): which is subtitled 'A fundamentally different theory as to the cause and (if true) simple positive preventative against polio'. 1947.
19. Halstead BW, Schantz EJ. *Paralytic shellfish poisoning.* Geneva: WHO; 1984.
20. De Lisle H. Poisoning from the rough skinned newt. *Herpetology.* 2010;13:7–12.
21. Centers for Disease Control and Prevention. Epidemiologic notes and reports: scombroid poisoning – Illinois, South Carolina. *MMWR.* 1989;38:140–2, 147.
22. Herter CA. Report of a case of lead paralysis with special reference to cytological changes in the nervous system and the distribution of lead. *NY Med J.* 1895;61: 665–67.
23. Wharton J. *The arsenic century. How Victorian Britain was poisoned at home, work and play.* Oxford: OUP; 2010.

24. Septimus Reynolds E. An account of the epidemic outbreak of arsenical poisoning occurring in beer-drinkers in the north of England and the midland counties in 1900. *Lancet*. 1901;1:166–70.
25. Barltrop D. Lead poisoning. *Archives of Disease in Childhood*. 1971;46:233–35.
26. McKhann C, Vogt E. Lead poisoning in children. *JAMA*. 1933;101:1134–35.
27. Osler W. *The principles and practice of medicine*, p. 1011.
28. Kidd JG, Langworthy OR. Jake paralysis. Paralysis following the ingestion of Jamaica ginger extract adulterated with triortho-chesyl phosphate. *Bulletin of the Johns Hopkins Hospital*. 1933;52:39.
29. Morgan JP. The Jamaica ginger paralysis. *JAMA*. 1982;248:1864–67.
30. Parascandola J. The Public Health Service and Jamaica Ginger paralysis in the 1930s. *Public Health Reports*. 1995;110:361–63.
31. Hayes WJ, Laws ER. *Handbook of pesticide toxicology*. San Diego: Academic Press. 1991, p. 769.
32. Biskind MS, Morton S. Public health aspects of the new insecticides. *American Journal of Digestive Diseases*. 1953;20:331–41.
33. Eskenazi B, Chevrier J, Rosas LG, Anderson HA, Bornman MS, Bouwman H, et al. The Pine River statement: human health consequences of DDT use. *Environmental Health Perspectives*. 2009;117:1359–67.
34. Carson R. *Silent Spring*. Boston: Houghton Mifflin; 1962.
35. Hill KR. A fatal case of DDT poisoning in a child, with an account of two accidental deaths in dogs. *British Medical Journal*. 1945;2:845–47.
36. Case RAM. Toxic Effects of DDT in man. *Brit Med J*. 1945;2:842–45.
37. Lazar T. DDT pancakes. *Brit Med J*. 1946;1:932.
38. Biskind MS. *Statement on clinical intoxication from DDT and other new insecticides. Presented before the Select Committee to investigate the use of chemicals in food products*. United States House of Representatives, Westport, CT, 12 December 1950.
39. Scobey R. *The poison cause of poliomyelitis*. Statement to the US House of Representatives, Washington DC, April 1952.
40. Morgan JP, Tulloss TC. The Jake Walk Blues: a toxicological tragedy mirrored in popular music. *JEMF Quarterly*, 1977;122–26.
41. Segalla S. The 1959 Moroccan oil poisoning and US Cold War disaster diplomacy. *J North African Studies*. 2012;17:315–36.
42. Dennis DT. Jake Walk in Vietnam. *Ann Int Med*. 1977;86:665.
43. Tosi L, Righetti C, Adami L, Zanette G. October 1942: a strange epidemic paralysis in Saval, Verona, Italy. Revision and diagnosis 50 years later of tri-ortho-cresyl phosphate poisoning. *J Neurol Neurosurg & Psychiatr*. 1994;57:810–13.

4 Germs of Ideas

1. Landsteiner K, Popper E. Uebertragung der Poliomyelitis acuta auf Affen. *Z Immunitätsforsch*. 1909;2:377–90.
2. Dochez A, Peabody F, Draper G. *A clinical study of acute poliomyelitis*. New York: Monographs of the Rockefeller Institute for Medical Research; 1912.
3. Frauenthal H, Manning J. *A manual of infantile paralysis*. Philadelphia: FA Davis; 1914, pp. 80–2.
4. Rosenow EC. Relation of a streptococcus to epidemic poliomyelitis – studies in etiology, diagnosis and specific treatment. *California Med*. 1952;76:396–401.
5. Fennel EA. Streptothrix Interproximalis. N. SP. an obligate microaerophile from the human mouth. *J Infect Dis*. 1918;22:567–72.

6. Jackson L. Demonstration of micrococci in the bones in rickets and scurvy. *J Infect Dis.* 1918;457–61.
7. Harris W. The experimental production of pellagra in the monkey. *J Am Med Ass.* 1913;60:1948–55.
8. Osler W. *The principles and practice of medicine.* New York: D Appleton and Co; 1892, p. 780.
9. Sachs B. The present-day conception of acute anterior poliomyelitis. *J Bone Joint Surg Am.* 1908;S2–6:173–83.
10. Thomsen O. So-called filterable virus of tuberculosis. *Ungeskrift for Laeger, Copenhagen.* 1929;91:729–32.
11. Cannon P. *Ludvig Hektoen, 1863–1951. A biographical memoir.* Washington DC: National Academy of Sciences; 1954.
12. Hektoen L, Mathers G, Jackson L. Microscopic demonstration of cocci in the central nervous system in epidemic poliomyelitis. *J Inf Dis.* 1918;22:87–94.
13. Rosenow EC, Wheeler GW. The etiology of epidemic poliomyelitis. *J Inf Dis.* 1918;22:281–312.
14. Rosenow E, Towne E, Hess C. The elective localization of streptococci from epidemic poliomyelitis. *J Inf Dis.* 1918;22:313–44.
15. Rosenow EC. Observations with the Rife microscope on filter-passing forms of microorganisms. *Science.* 1932;76:192–93.
16. Rosenow E. The production of an antipoliomyelitis serum in horses by inoculation of the pleomorphic streptococcus from poliomyelitis. *JAMA.* 1917;69:261–66.
17. Rosenow EC. Report on the treatment of fifty-eight cases of epidemic poliomyelitis with immune horse serum. *J Inf Dis.* 1918;22:379–426.
18. Long P, Olitsky P, Stewart F. The role of streptococci in experimental poliomyelitis of the monkey. *J Exp Med.* 1928;48:431–48.
19. Amoss HL, Eberson F. Therapeutic experiments with Rosenow's antipoliomyelitic serum. *J Exp Med.* 1918;27:309–17.
20. Howitt B. Poliomyelitis. A review of the literature. *California and Western Med.* 1930;33:596–601.
21. Commentary. Etiology of poliomyelitis. *Am J Publ Health.* 1933;23:300.
22. Dr. Rosenow's obsession. *Time.* 23 July 1944.
23. Editorial on poliomyelitis. *Postgrad Med J.* 1949;25:8.
24. Smith T. Hideyo Noguchi, 1876–1928. *Bull NY Acad Sci.* 1929;5:877–85.
25. Noguchi H. *Snake venoms: an investigation of venomous snakes with special reference to the phenomena of their venoms.* Washington, D.C.: Carnegie Institute; 1909.
26. Noguchi H, Moore J. A demonstration of *Treponema pallidum* in the brain in cases of general paralysis. *J Exp Med* 1913;27:232–38.
27. Noguchi H. *Leptospira icteroides* and yellow fever. *Proc Natl Acad Sci.* 1920;6:110–11.
28. Flexner S, Lewis P. The nature of the virus of epidemic poliomyelitis. *J Am Med Ass.* 1909;53:592–94.
29. Flexner S, Noguchi H. Experiments on the cultivation of the microörganism causing epidemic poliomyelitis. *J Exp Med.* 1913;27:461–85.
30. Amoss HL. The cultivation and immunological reactions of the globoid bodies in poliomyelitis. *J Exp Med.* 1917;25:545–55.
31. Tsen E. The etiology of epidemic poliomyelitis. *J Exp Med.* 1918;26: 269–87.
32. Heist J, Solis-Cohen M, Kolmer J. Studies on epidemic poliomyelitis. I. The isolation and cultivation of the globoid bodies. *J Inf Dis.* 1918;22: 169–74.
33. Wilson G. Faults and fallacies in microbiology. *J Gen Microbiol* 1959;21:1–15.
34. Logrippo G. Concerning the nature of the globoid bodies. *J Bacteriol.* 1936;31:245–53.

35. Faber H. Flexner and Noguchi's globoid red herrings: a footnote to the history of poliomyelitis. *Journal Inf Dis.* 1971;124:231–34.
36. Burnet F, Macnamara J. Immunological differences between strains of poliomyelitic virus. *Brit J Exp Path.* 1931;12:57–61.
37. Plotkin SA, Carp RI, Graham AF. The polioviruses of man. *Annals of the New York Academy of Sciences.* 1960;101:357–89.
38. Crick F, Watson J. Structure of small viruses. *Nature.* 1956;177:473–76.
39. Finch J, Klug A. Structure of poliomyelitis virus. *Nature.* 1959; 183:1709–14.
40. Schaffer F, Schwerdt C. Crystallisation of purified MEF-1 poliomyelitis virus particles. *Proc Natl Acad Sci USA.* 1955;41:1020–23.
41. Reagan RL, Schenck DM, Brueckneir AL. Morphological observations by electron microscopy of the Brunhilde strain of poliomyelitis virus. *J Inf Dis,* 1950;86:295–96.
42. Palacios G, Oberste M. Enteroviruses as agents of emerging infectious diseases. *J Neuro Virol.* 2005;11:424–33.
43. Goldberger J. *Goldberger on pellagra.* New Orleans: Louisiana State Uni Press; 1964.
44. Nobel Prize Nomination and selection of Medicine Laureates [Internet]. [accessed 22 January 2013]. Available from: http://www.nobelprize.org/nobel_prizes/medicine/nomination
45. Bull C. The pathologic effects of streptococci from cases of poliomyelitis and other sources. *J Exp Med.* 1917;25:557–80.
46. Rosenow EC, Rosenow OF. Influence of streptococcal infections on the compulsive behavior of criminals. *Postgrad Med.* 1951;10:423–32.
47. Lynes B. *The cancer cure that worked! Fifty years of suppression.* S. Lake Tahoe, CA: BioMed Publishing Group; 1987.
48. Brown N. The Rife microscope in the Science Museum collection. London: Science Museum; 1996; Bracegirdle B. Rife and his microscopes. *Queckett J Microscopy.* 2003; 39:459–73.
49. Ogawa M. *Robert Koch's 74 days in Japan. Kleine Reihe 27.* Berlin: Mori-Ogai Gedenkstätte der Humboldt-Universität zu Berlin; 2003.

5 Lost in Transmission

1. Braverman I, Wexler D, Oren M. A novel mode of infection with hepatitis B: penetrating bone fragments due to the explosion of a suicide bomber. *Isr Med Assoc J.* 2002;4:528–29.
2. Melnick J. Enteroviruses: polioviruses, coxsackieviruses, echoviruses and newer enteroviruses. In Fields B, Knipe D, Howley P et al (eds) *Fields virology.* Philadelphia: Lippincott-Raven; 1996, pp. 655–712.
3. Peart AF, Rhodes AJ. An outbreak of poliomyelitis in Canadian Eskimos in wintertime. *Canadian J Publ Health.* 1949;40:405.
4. Bates C. Now wash your hands: they are much dirtier than you think. *Daily Mail,* 15 October 2012.
5. Burnet M, White D. *Natural history of infectious disease.* 4th ed. Cambridge: Cambridge University Press; 1972, p. 94.
6. Plotkin SA, Carp RI, Graham AF. The polioviruses of man. *Ann N Y Acad Sci.* 1962. 1962;101:357–89.
7. Downie A, Dumbell K. Survival of variola virus in dried exudate and crusts from smallpox patients. *Lancet.* 1947;1:550–53.

8. Dick G, Dane D, Fisher O, et al. 2. A trial of SM Type 1 attentuated poliomyelitis virus vaccine. *Brit Med J*. 1957;1:65–70.
9. Minor P. Poliovirus. In Nathanson N, Ahmed R, (eds) *Viral pathogenesis*. Philadelphia: Lippincott-Raven; 1996, pp. 555–74.
10. Dunnebacke T, Levinthal J, Williams R. Entry and release of poliovirus as observed by electron microscopy of cultured cells. *J Virol*. 1969;4:505–13.
11. Rissler J. Zur Kenntniss der Veränderungen des Nervensystems bei Poliomyelitis anterior acuta. *Nordiskt Medicinskt Arkiv*. 1888;29:1–63.
12. Kling CA, Wernstedt WE, Pettersson A. Recherches sur le mode de propagation de la paralysie infantile épidémique (maladie de Heine-Medin). *Zeit Immunitätforsch*. 1912;12:316–23.
13. Sawyer W. An epidemiological study of polio. *Am J Trop Dis Prev Med*. 3:164–75.
14. Trask J, Vignec A, Paul J. Poliomyelitis virus in human stools. *J Am Med Ass*. 1938;111:6–11.
15. Melnick JL. Poliomyelitis virus in urban sewage in epidemic and in nonepidemic times. *American Journal of Epidemiology*. 1947;45:240–53.
16. Bodian D. A reconsideration of the pathogenesis of poliomyelitis. *Am J Hyg*. 1952;55:414.
17. Howe HA, Bodian D, Morgan IM. Subclinical poliomyelitis in the chimpanzee and its relation to alimentary reinfection. *American J Hyg*. 1950;51:85–108.
18. Bodian D, Paffenbarger R. Poliomyelitis infection in households: frequency of viremia and specific antibody response. *Am J Hyg*. 1954;60:83–98.
19. Horstmann D, McCollum R, Mascola A. Viremia in human poliomyelitis. *J Exp Med*. 1954;99:355–69.
20. Lambert SM. A yaws campaign and an epidemic of poliomyelitis in Western Samoa. *J Trop Med Hyg*. 1936;39:41–6.
21. Hill AB, Knowelden J. Inoculation and poliomyelitis: a statistical investigation in England and Wales in 1949. *British Medical Journal*. 1950;2:1–6.
22. Nathanson N, Langmuir A. The Cutter incident. poliomyelitis following formaldehyde-inactivated poliovirus vaccination in the United States during the spring of 1955. II. Relationship of Poliomyelitis to Cutter vaccine. *Am J Hyg*. 1963;78:29–60.
23. Nathanson N, Bodian D. Experimental poliomyelitis following intramuscular virus injection. *Bull Johns Hopkins Hospital*. 1961;103:308–15.
24. Aycock W. Tonsillectomy and poliomyelitis. I. Epidemiological considerations. *Medicine (Baltimore)*. 1942;21:65–94.
25. Caverly CS. Preliminary report of an epidemic of paralytic disease, occurring in Vermont, in the summer of 1894. *Yale Med J*. 1894;1:1–20.
26. Russell WR. Poliomyelitis. The pre-paralytic state, and the effect of physical activity on the severity of paralysis. *Brit Med J*. 1947;2:1023–28.
27. Brown G, Francis T, Pearson H. Rapid development of carrier state and detection of poliomyelitis virus in stool nineteen days before onset of paralytic disease. *J Am Med Ass*. 1945;129:121–23.
28. Lewin P. *Infantile paralysis. Anterior poliomyelitis*. Philadelphia: WB Saunders & Co.; 1941. p. 38.
29. Farkas A. Rest and poliomyelitis. *Arch Ped*. 1952;69:6–23.
30. Jungeblut CW, Meyer K, Engle ET. Inactivation of poliomyelitis virus and of diphtheria toxin by various endocrine principles. *J Immunol*. 1934;27:43–56.
31. Lewin P. *Infantile paralysis*. 1941;36.
32. Toomey JA, Takacs WS, Pirone PP. Accelerated production of poliomyelitis. *Arch Ped*. 1949;78:1.

33. Sandler BPC-P. The production of neuronal injury and necrosis with the virus of poliomyelitis in rabbits during insulin hypoglycemia. *Am J Pathol.* 1941;17: 69–84.
34. Sandler B. *Diet prevents polio.* Milwaukee, WI: The Lee Foundation for Nutritional Research; 1951.
35. Kempf JE, Pierce ME, Soule MH. Failure to produce neuronal injury and necrosis with poliomyelitis virus in rabbits during insulin hypoglycemia. *Proc Soc Exp Biol Med.* 1941; 48:187–88.
36. Draper G. *Infantile paralysis.* 2nd ed. New York: Appleton-Century; 1935, pp. 52–73.
37. Draper G. The nature of the human factor in infantile paralysis. *Am J Med Sci.* 1932;184:111–16.
38. Levine M, Neal J, Park W. Relation of physical characteristics to susceptibility to anterior poliomyelitis. *J Am Med Ass.* 1933;100:160–62.
39. Flexner S, Lewis PA. Experimental epidemic poliomyelitis in monkeys. *J Exp Med.* 1910;12:227–32.
40. Flexner S, Clark PF. A note on the mode of infection in epidemic poliomyelitis. *Proc Soc Exp Biol Med.* 1912; 10: 1–10.
41. Schultz EW, Gebhardt LP. Olfactory tract and poliomyelitis. *Proc Soc Exp Biol Med.* 1934; 31:728–33.
42. Sabin A, Olitsky P, Cox H. Protective action of certain chemicals against infection of monkeys with nasally instilled poliomyelitis virus. *J Exp Med.* 1936;63:877–92.
43. Rutty CJ. *Do something! Do anything! Poliomyelitis in Canada, 1927–1967.* PhD thesis, Toronto: University of Toronto; 1996. pp. 112–96.
44. Armstrong C, Harrison WT. Prevention of intranasally-inoculated poliomyelitis of monkeys by instillation of alum into the nostrils. *Public Health Reports (1896–1970).* 1935;725–30.
45. Smith B. The Victorian poliomyelitis epidemic 1937–1938. In: Callwell J, (ed.) *What we know about health transition: the cultural, social and behavioural determinants of health: the proceedings of an International Workshop, Canberra, May 1989.* Canberra: Australian National University; 1990. p. 868, 874.
46. Paul J. *The nose as a portal of entry? A history of poliomyelitis.* New Haven and London: Yale University Press; 1971, pp. 240–51.
47. Schultz E, Gebhardt L. Zinc sulphate prophylaxis in poliomyelitis. *J Am Med Ass.* 1937;108:2182–87.
48. Tisdall FF, Brown A, Defries RD, Ross MA, Sellers AH. Zinc-sulphate nasal spray in the prophylaxis of poliomyelitis. *Can. Public Health J.* 1937;28:523–43.
49. De Kruif P. *Men against the maiming disease. Part 2, The fight for life.* London: Scientific Book Club; 1940. p. 185.
50. Editorial. Zinc sulphate spray for prevention of poliomyelitis. *Brit Med J.* 1938;1: 953–54.
51. Sabin A. The olfactory bulbs in human poliomyelitis. *Am J Dis Child.* 1940;60: 1313–18.
52. Howe H, Bodian D. Poliomyelitis in the chimpanzee: a clinical pathological study. *Bull Johns Hopk Hosp.* 1941;69:149–81.
53. Paul JR. *A history of Poliomyelitis,* p. 251.
54. Frauenthal H, Manning J. *A manual of infantile paralysis.* Philadelphia: FA Davis; 1914, pp. 7–15.
55. Ten Broeck C. *Experiments to determine if paralyzed domestic animals and those associated with cases of infantile paralysis may transmit this disease.* 45th Report of Mass State Board of Health. Boston: Wright & Potter; 1914.

56. Freeman A. Paralysis in dogs at Little Rock, Arkansas. *Annual Report of the Surgeon General of the US Public Health Service for 1917.* Washington, DC: US Government, 1917, p. 30.
57. Rogers N. Dirt, flies, and immigrants: explaining the epidemiology of poliomyelitis, 1900–1916. *Journal of the History of Medicine and Allied Sciences.* 1989;44: 486–505.
58. Frauenthal H, Manning J. *A manual of infantile paralysis*, pp. 42–8.
59. Rosenau M, Brues C. Some experimental observations concerning the transmission of poliomyelitis through the agency of *Stomoxys calcitrans. Mass State Board Hlth Month Bull.* 1912;7:314–16.
60. Power ME, Melnick JL, Bishop MB. A study of the 1942 fly population of New Haven. *Yale J Biology & Med.* 1943;15:693–705.
61. Howard C, Clark P. Experiments in insect transmission of the virus of poliomyelitis. *Proc Soc Exper Biol Med.* 1940;43:718–23.
62. Toomey JA, Pirone P. Can Drosophila flies carry poliomyelitis virus? *J Infect Dis.* 1947;81(1):135–38.
63. Frauenthal H, Manning J. *A manual of infantile paralysis*, pp. 48–63.
64. Tuttle W. *Daddy's gone to War: the Second World War in the lives of America's children.* Oxford: OUP; 1993, p. 195.
65. Gear J. Epidemiology of poliomyelitis in Africa. *Annales de la Société Belge de Médecine Tropicale.* 1958;39:927–37.
66. Frauenthal H, Manning J. *A manual of infantile paralysis*, p. 35.
67. Paul J. The Los Angeles epidemic of 1934. *A history of Poliomyelitis*, pp. 212–24.
68. Bodian D, Morgan I, Howe H. Differentiation of types of poliomyelitis viruses. III. The grouping of fourteen strains into three basic immunological types. *Am J Hyg.* 1949;49:234–40.
69. Oshinsky D. *Polio. An American story.* Oxford: OUP; 2006, p. 119.
70. Le Cornu A, Rowan A. Trends in the use of non-human primates in biomedical research programmes. *Lab Animals.* 1978;12:235–42.
71. Sabin A. Pathogenisis of poliomyelitis. *Science.* 1956;123:1151.
72. Brodie M, Elvidge A. The portal of entry and the transmission of the virus of poliomyelitis. *Science.* 1934;79:235–36.
73. Kling C, Levaditi C, Lepine P. La pénétration du virus poliomyélitique à travers la muqueuse du tube digestif chez le singe et sa conservation dans l'eau. *Bull. Acad. de Méd.* 1929;102:158–65.
74. Sabin AB, Ward R. The natural history of human poliomyelitis I. Distribution of virus in nervous and non-nervous tissues. *J Exp Med.* 1941;73:771–93.
75. Armstrong C. The experimental transmission of poliomyelitis to the eastern cotton rat. *Publ Health Rep.* 1939;54:1719–23.
76. Niewisk S, Prince G. Diversifying animal models: the use of hispid cotton rats (*Sigmodon hispidus*) in infectious diseases. *Lab Animals.* 2002;36:357–72.
77. Armstrong C. Successful transfer of the Lansing strain of poliomyelitis virus from the cotton rat to the white mouse. *Publ Health Rep.* 1939;54:2302–05.
78. Li C, Schaeffer M. Adaptation of type I poliomyelitis virus to mice. *Proc Soc Exp Biol Med.* 1953;82:477–52.
79. Levaditi C. Le virus poliomyélitique. In Lépine P, (ed.) *Les ultravirus des maladies humaines.* Paris: Maloine; 1938, p. 572.
80. Sabin AB, Olitsky PK. Cultivation of poliomyelitis virus in vitro in human embryonic nervous tissue. *Proc Soc Exp Biol Med.* 1936; 34: 357–59.
81. Sabin A. Non-cytopathic variants of poliomyelitis viruses and resistance to superinfection in tissue culture. *Science.* 1954;120:357–59.

82. Enders J, Weller T, Robbins F. Cultivation of the Lansing strain of poliomyelitis virus in cultures of various human embryonic tissue. *Science*. 1949;109:85–57.
83. Norrby E, Prusiner S. Polio and Nobel Prizes: looking back 50 years. *Ann Neurol*. 2007;61:391.
84. Axelsson P. 'Do not eat those apples; they've been on the ground!': polio epidemics and preventive measures, Sweden 1880s-1940s. *Asclepio*. 2009;61:23–37.
85. Lewin P. *Infantile paralysis*, p. 5.
86. Bernier R. Some observations on poliomyelitis lameness surveys. *Rev Infect Dis*. 1984;6:S371–75.
87. Nathanson N, Martin JR. The epidemiology of poliomyelitis: enigmas surrounding its appearance, epidemicity, and disappearance. *Am J Epidemiol*. 1979;110:672–92.

6 Fear Is the Key

1. Weekly infectious disease reports. *Brit Med J*. 1949;2:1023.
2. Ibid., quoting Dr. Leonard Scheele, American Surgeon-General.
3. Axelsson P. 'Do not eat those apples; they've been on the ground!': polio epidemics and preventive measures, Sweden 1880s–1940s. *Asclepio*. 2009;61:23–37.
4. Bayer P. Management of the acute phase of poliomyelitis. *Post Grad Med J*. 1949;25:9–12.
5. Lee J. *Poliomyelitis in the Lone Star State: a brief examination in rural and urban communities*. MA thesis, Texas State University. San Marcos, pp. 41–4.
6. Ibid. pp. 47–52.
7. Ibid. pp. 54–6.
8. Ibid. p. 52, 56, 57, 60, 61.
9. Ibid. p. 41.
10. Peart AF, Rhodes AJ. An outbreak of poliomyelitis in Canadian Eskimos in wintertime. *Can J Publ Health*. 1949;40:405–19.
11. Adamson JD, Moody JP, Peart AFW, Smillie RA, Wilt JC, Wood WJ. Poliomyelitis in the Arctic. *Can Med Ass J*. 1949;61:339–48.
12. Editorial. 'Polio hysteria'. *J Ped*. 1954;45:123–24.
13. Offit P. *The cutter Incident*. New Haven and London: Yale University Press; 2005, p. 32.
14. Hardy A. Poliomyelitis and the Neurologists: the view from England, 1896–1966. *Bulletin of the History of Medicine*. 1997;71:249–72.
15. Editorial. *Postgrad Med J*. 1949;25:1.
16. *Hull Daily Mail*, 25 August 1949.
17. *Bath Chronicle Weekly Gazette*, 13 August 1949.
18. *Gloucestershire Echo*, 9 August 1949.
19. *Derby Daily Telegraph*, 27 August 1949.
20. Martin W. Poliomyelitis in England and Wales, 1947–1950. *Brit J Soc Med*. 1951;5:136–46.
21. Axelsson P. *Do not eat those apples*, pp. 23–4.
22. Bayer P. Management of the acute phase of poliomyelitis, 9.
23. *CDC Cases & Deaths [Internet]*. Available from: http://www.cdc.gov/vaccines/pubs/pinkbook/downloads/appendices/G/cases&deaths
24. Duncan C, Duncan S, Scott S. The effects of population density and malnutriton on the dynamics of whooping cough. *Epidemiol Infect*. 1998;121:325–34.
25. Paul H. *The control of diseases (social and communicable)*. 2nd ed. Edinburgh & London: E&S Livingstone Ltd; 1964. p. 93; pp. 223–26.

26. Paul J. *The epidemic of 1916: a history of poliomyelitis*. New Haven and London: Yale University Press; 1971, pp. 148–60.
27. Oshinsky D. *Polio. An American story*. Oxford: OUP; 2005, pp. 20–1.
28. Rogers N. *Dirt and disease. Polio before FDR*. New Brunswick, NJ: Rutgers University Press; 1990. pp. 30–49.
29. *Annual report of the surgeon general of the public health service of the United States, for the fiscal year 1917*. Washington DC; 1917, pp. 30–1, 188–203.
30. Kluger J. *Conquering polio. Smithsonian magazine [Internet]*. 2005; Available from: http://www.smithsonianmag.com/science-nature/polio.html
31. *New York Times*, 26 July 1916.
32. Ditunno J Jr, Herbison G. Franklin D. Roosevelt: diagnosis, clinical course, and rehabilitation from poliomyelitis. *Am J Phys Med Rehabil*. 2002;81:557–66.
33. Goldman A, Schmalstieg E, Freeman D et al. What was the cause of Franklin Delano Roosevelt's paralytic illness? *J Med Biogr*, 2003;11:232–40.
34. Takaro T. The man in the middle. *Dartmouth Medicine*. 2004;52–7.
35. Smith J. *Patenting the sun*. New York: William Morrow; 1990. p. 207.
36. Offit P. *The cutter Incident*. pp. 19–20.
37. Oshinsky D. *Polio. An American story*. pp. 48–52.
38. Ibid.. p. 52.
39. Ibid., pp. 54–5.
40. Lee J. *Poliomyelitis in the Lone Star State*, p. 40.
41. Smith J. *Patenting the sun*, p. 82.
42. Lee J. *Poliomyelitis in the Lone Star State*, p. 74.
43. *100 Years of US Consumer Spending: 1950 [Internet]*. [cited 14 February 2013]. Available from: http://www.bls.gov/opub/uscs/1950.pdf
44. Smith J. *Patenting the sun*, p. 161.
45. Ibid., p. 85.
46. Jones J. *Bad blood: the Tuskegee syphilis experiment*. New York: Free Press; 1981.
47. Benison S. *Tom Rivers: reflections on a life in medicine and science*. Cambridge, Mass: The MIT Press; 1967, p. 310.
48. Poliomyelitis. *Papers and discussion presented at the Third International Poliomyelitis Conference*. Philadelphia: JB Lippincott; 1955.
49. Wilson J. *Margin of safety: the story of poliomyelitis vaccine*. London: Collins; 1963, p. 64, 69.
50. De Kruif P. *Microbe Hunters*. London: Jonathan Cape; 1927.
51. Benison S. *Tom Rivers*, pp. 181–82.
52. Lewis S. *Arrowsmith. With an afterword by Mark Schorer*. New York: New American Library; 1961.
53. Brodie M, Park WH. Active immunization against poliomyelitis. *Am J Publ Health*. 1936;26:119–25.
54. Wilson J. *Margin of safety*, p. 65.
55. Smith J. *Patenting the sun*, p. 71.
56. Offit P. *The cutter Incident*, p. 21.
57. Smith J. *Patenting the sun*, p. 257.
58. Ibid., p. 201.
59. Clemmesen S. *Closing Session*, 12 July 1957. Poliomyelitis. *Papers and discussion presented at the Fourth International Poliomyelitis Conference*. Philadelphia: JB Lippincott; 1959.
60. Matysiak A. *Albert B. Sabin: the development of an oral vaccine against poliomyelitis*. PhD thesis, Cincinnati: University of Cincinatti; 2005, p. 143.
61. Oshinsky D. *Polio. An American story*, pp. 53–5.
62. Ibid., p. 153.

63. 'Slap at Sabin for his opposition to Salk program'. *Cincinatti Post,* 24 June 1955.
64. Smith J. *Patenting the sun*, pp. 65–6.
65. Benison S. *Tom Rivers*, p. 280.
66. Smith J. *Patenting the sun*. p. 115.
67. Benison S. *Tom Rivers*. p. 281.
68. O'Connor B. *Opening address*. Poliomyelitis. *Papers and discussion presented at the Second International Poliomyelitis Conference*. Philadelphia: JB Lippincott; 1952.
69. Pope Pius XII, address in the Campidoglio, Rome. 1. Poliomyelitis. *Papers and discussion presented at the Third International Poliomyelitis Conference*. Philadelphia: JB Lippincott; 1955.
70. Offit P. *The cutter Incident*, p. 28.
71. Wilson J. *Margin of safety*, p. 66.
72. Ibid., p. 67.
73. Poliomyelitis. *Papers and discussion presented at the First International Poliomyelitis Conference*. Philadelphia: JB Lippincott; 1949, p. 338.
74. Ibid., p. 335.
75. North B. *Something to lean on. The first sixty years of the British Polio Fellowship, 1939–1999*. South Ruislip, Middlesex: British Polio Fellowship; 1999, pp. 1–3.
76. Ibid., p. 3, 13, 19.
77. Ibid., pp. 20–1.
78. Ibid., p. 11.
79. Ibid., p. 7, 9, 36.
80. Ibid., pp. 90–5.
81. Ibid., p. 21.
82. Smith J. *Patenting the sun*. p. 211.
83. Oshinsky D. *Polio. An American Story*. p. 90.
84. Smith J. *Patenting the sun*. p. 83.
85. *The 1945 Film Daily Year Book of Motion Pictures*. New York: Wid's Films and Film Folk, Inc.; 1945. p. 344.
86. Smith J. *Patenting the sun*. p. 170.
87. Oshinsky D. *Polio. An American story*. p. 153.
88. Smith J. *Patenting the sun*. p. 200.
89. Gallagher H. *FDR's splendid deception*. St Petersburg, Florida: Vandamere Press; 1999.
90. 'Franklin Roosevelt will swim to health'. *Atlanta Journal Sunday Magazine*. 26 October 1924, p. 7.
91. Smith J. *Patenting the sun*. p. 50.

7 First Do No Harm

1. Cure for infantile paralysis. *New York Times,* 9 March 1911.
2. Porter R. *The greatest benefit to mankind: a medical history of humanity from antiquity to the present*. London: Fontana. 1997;483.
3. Bunker H, Kirby G. Treatment of general paralysis of the insane by inoculation with malaria. *J Am Med Ass*. 1925;84:563–68.
4. Bureau of investigation. The Cunningham 'Tank treatment'. *J Am Med Ass*. 1928;90: 1494–95.
5. Fever therapy cabinet, Burdick Corp. *J Am Med Ass* 1938;111:423–24.
6. Jones J. *Bad blood: the Tuskegee syphilis experiment*. New York: Free Press; 1981.

7. Boëns H. *La variole, la vaccine et les vaccinides en 1884*. Reprinted from the Bulletin de l'Académie Royal Médicale de Belgique, vol 18 no. 1. Bruxelles: H Manceaux; 1884, p. 107–08.
8. Moed L, Shwayder TA, Chang MW. Cantharidin revisited: a blistering defense of an ancient medicine. *Arch Dermatol* 2001;137:1357–60.
9. Jacobi MP. *Infantile spinal paralysis*. Pepper's System of Medicine, Philadelphia, Lea Brothers & Co. 1886;5:1113–64.
10. Underwood M. *A treatise on the diseases of children, with general directions for the management of infants from the birth*. Philadelphia: Gibson; 1793.
11. Woodbury F. Report on lecture by Prof. S.D. Gross, Infantile paralysis. *Phil Med Times*. 1872;2:408–09.
12. Osler W. *The principles and practice of medicine*. New York: D Appleton and Co; 1892, pp. 833–34.
13. Frauenthal H, Manning J. *A manual of infantile paralysis, with modern methods of treatment*. Philadelphia: FA Davis; 1914, pp. 270–75.
14. Ibid., p. 277.
15. Ibid., p. 130.
16. Farkas A. Rest and poliomyelitis. *Arch Ped*. 1952;69:6–23.
17. Lovett R. *The treatment of infantile paralysis*. Philadelphia: Blakeston's Sons; 1916.
18. Paul J. *A history of poliomyelitis*. New Haven: Yale University Press; 1971, p. 336–9.
19. Frauenthal H, Manning J. *A manual of infantile paralysis*, p. 308.
20. Ibid., p. 309.
21. Ibid., p. 283.
22. Ibid., pp. 309–10.
23. Jones RW (Sir Robert), quoted in Frauenthal HM, Manning JVV. *A manual of infantile paralysis*, pp. 341–42.
24. Ibid., p. 348.
25. Porter R. *The greatest benefit to mankind*, p. 383.
26. Walter A. Observations of tenotomia and myotomia, for the cure of deformed members; anatomically, physiologically and therapeutically considered. With seventy-four cases. *Select Med Library and Eclectic J Med*. 1840;4:385–421.
27. Frauenthal H, Manning J. *A manual of infantile paralysis*, p. 352.
28. Ibid., p. 357.
29. Ibid., p. 361.
30. Jones RW (Sir Robert), quoted in Frauenthal HM, Manning JVV. *A manual of infantile paralysis*, p. 298.
31. Lewin P. *Infantile paralysis. Anterior poliomyelitis*. Philadelphia: WB Saunders & Co.; 1941.
32. Hardy A. Poliomyelitis and the neurologists: the view from England, 1896–1966. *Bull Hist Med*. 1997;71:252.
33. Paul H. *The control of diseases (social and communicable)*. Edinburgh & London: E&S Livingstone; 1964.
34. Walker A. *Sister Elizabeth Kenny: maverick heroine of the polio treatment controversy*. Rockhampton: Central Queensland Univ Press; 2003.
35. Kenny E. *My battle and victory: history of the discovery of poliomyelitis as a systemic disease*. London: Robert Hale; 1955.
36. Cohn V. *Sister Kenny: the woman who challenged the doctors*. Minneapolis: University of Minnesota Press; 1973.
37. Report of the Queensland Royal Commission on Treatment of infantile paralysis by Sister Kenny's method. *BMJ*. 1938;1:350–51.

38. Paul J. *A history of poliomyelitis*, pp. 341–43.
39. Benison S. *Tom Rivers: reflections on a life in medicine and science*. Cambridge, Mass: MIT Press; 1967, pp. 282–83.
40. Pohl J, Kenny E. *The Kenny concept of infantile paralysis and its treatment*. Minneapolis: Bruce; 1943.
41. Wade M. *Straws in the wind: early epidemics of poliomyelitis in Johannesburg, 1918–1945*. Pretoria: University of South Africa; 2006, p. 152.
42. Lewin P. *Infantile paralysis*, p. 134.
43. Frauenthal H, Manning J. *A manual of infantile paralysis*, p. 279.
44. Henderson Y. The return of the pulmotor as a 'resuscitator': a back step towards the death of thousands. *Science*. 1943;98:547–51.
45. West J. *Yandell Henderson, 1873–1944. A biographical memoir*. Washington DC: National Academy of Sciences; 1998.
46. Sherwood R. Obituary. Philip Drinker, 1894–1972. *Ann Occup Med*. 1973;16:93–4.
47. Paul J. Therapeutic methods: the iron lung. *In A History of Poliomyelitis*, pp. 324–34.
48. Drinker P, Shaw L. Apparatus for prolonged administration of artificial respiration: I. A design for adults and children. *J Clin Invest*. 1929;7:229–47.
49. Drinker P, McKhann CF. The use of a new apparatus for the prolonged administration of artifical respiration. I. A fatal case of poliomyelitis. *Journal of the American Medical Association*. 1929;92:1658–65.
50. Woollam CHM. The development of apparatus for intermittent negative pressure respiration. *Anaesthesia*. 1976;31:666–85.
51. Branson RD. A tribute to John H. Emerson. Jack Emerson: notes on his life and contributions to respiratory care. *Respir Care*. 1998;43:567–71.
52. Dalziel J. On sleep and apparatus for promoting artificial respiration. In Murray J, (ed.) *Report of the British Association for the Advancement of Science*. 1839, pp. 127–28.
53. Trubuhovich R V. Notable Australian contributions to the management of ventilatory failure of acute poliomyelitis. *Crit Care*. 2006;8:383–93.
54. Menzies F. Mechanical respirators [letter]. *BMJ*. 1939;1:35.
55. Lawrence G. The Smith-Clarke respirator. *Lancet*. 2002;359:716.
56. Dr. Marshall Barr, Berkshire Medical Heritage Centre, personal communication.
57. Bayer P. The management of the acute phase of poliomyelitis. *Postgraduate Medical Journal*. 1949;25:9–12.
58. Bates P, Pellow J. *Horizontal man: the story of Paul Bates*. London: Longmans, Green and Co.; 1964.
59. Murray A. An improvised iron lung. *BMJ*. 1956;2:1361.
60. Lewin P. *Infantile paralysis*, p. 132.
61. Berry M. *Personal communication*, 28 September 2012.
62. Bates P, Pellow J. *Horizontal man*. p. 45.
63. Gallagher H. *Black bird fly away: disabled in an able-bodied world*. St. Petersburg, Florida: Vandamere Press; 1998, p. 60.
64. Solomon P. Sensory deprivation: a review. *Am J Psychiatr*. 1957;114:357–62.
65. Smith J. *Patenting the sun*, p. 41.
66. North B. *Something to lean on. The first sixty years of the Bristish Polio Fellowship, 1939–1999*. South Ruislip, Middlesex: British Polio Fellowship; 1999, pp. 34–5.
67. Iron lung patient dies in power cut. Sydney Morning Herald. 29 May 2008.
68. Lassen H. *The management of respiratory and bulbar paralysis in poliomyelitis*. Geneva: WHO; 1955, pp. 157–211.
69. Frauenthal H, Manning J. *A manual of infantile paralysis*, pp. 285–87.
70. Specter of paralysis stalks Carolina. *Literary Digest*. 1935;17, 39.

71. Retan GM. The development of the therapeutic use of forced perivascular (spinal) drainage. *Journal of the American Medical Association.* 1935;105:1333–40.
72. Brain washing therapy to treat brain inflammation. *Collier's* 23 July 1938.
73. Retan G. The treatment of acute poliomyelitis by intravenous injection of hypotonic salt solution. *J Ped.* 1937;11:647–64.
74. Kramer SD, Geer HA, Himes AT. Use of continuous intravenous administration of hypotonic sodium chloride (Retan Treatment) in acute experimental poliomyelitis in monkeys. *J. Immunol* 1942;44:175–94.
75. Retan G. The Retan technic of spinal drainage. *JAMA.* 1943;121:71.
76. Hemilä H. Vitamin C and infectious diseases. In: Packer L, Fuchs J, (eds) *Vitamin C in health and disease.* New York: Marcel Dekker; 1997, pp. 471–503.
77. Jungeblut CW. Further observations of the poliocidal property of pregnant mare serum. *Exp Biol Med.* 1935;33:137–41.
78. Jungeblut CW. Further observations on vitamin C therapy in experimental poliomyelitis. *J Exp Med.* 1937;66:459–77.
79. Sabin AB. Vitamin C in relation to experimental poliomyelitis with incidental observations on certain manifestations in Macacus rhesus monkeys on a scorbutic diet. *J Exp Med.* 1939;69:507–15.
80. Klenner FR. The treatment of poliomyelitis and other virus diseases with vitamin C. *Southern Med Surg.* 1947;111:209–14.
81. McCormack WJ. Ascorbic acid as a chemotherapeutic agent. *Archives of Pediatrics.* 1952;69:151–55.
82. Saul A. Hidden in plain sight. The pioneering work of Frederick Robert Klenner. *J Orthomolecular Med.* 2007;22:31–8.
83. Peters R, Stocken L, Thompson R. British anti-Lewisite (BAL). *Nature.* 1945;156:616–19.
84. Eskwith IS. Empirical administration of BAL in one case of poliomyelitis. *Arch Ped.* 1951;81:684–86.
85. Scobey RR. The poison cause of poliomyelitis and obstructions to its investigation. *Arch Ped.* 1952;69:172–93.
86. Reid PF. Alpha-cobratoxin as a possible therapy for multiple sclerosis: A review of the literature leading to its development for this application. *Critical Rev Immunol.* 2007;27:291–302.
87. Paul J. Convalescent serum therapy. In *A history of poliomyelitis.* pp. 190–99.
88. Hammon W, Coriell L, Wehrle P. Evaluation of Red Cross gamma globulin as a prophylactic agent for poliomyelitis. IV. Final report of results based on clinical diagnosis. *JAMA.* 1953;151:1272–85.
89. *National Advisory Committee for the Evaluation of Gamma Globulin in the Prophylaxis of Poliomyelitis: An Evaluation of the Efficacy of Gamma Globulin in the Prophylaxis of Paralytic Poliomyelitis as Used in the United States 1953.* Washington DC: U.S. Public Health Service; 1954.
90. Benison S. *Tom Rivers.* p. 487.
91. Johnson M. Poliomyelitis vaccination. *BMJ.* 1949;2:757.

8 Dead or Alive

1. Cohen L. Anterior poliomyelitis with reference to the occurrence of two attacks in the same individual (with report of two cases). *New Engl J Med.* 1935;213: 601–04.

2. Bazin H. Pasteur and the birth of vaccines made in the laboratory. In Plotkin SA, (ed.) *History of vaccine development*. New York: Springer; 2011, pp. 33–45.
3. Williams G. The disinterested divulger of a salutary blessing. In *Angel of Death: the story of smallpox*. Basingstoke: Palgrave Macmillan; 2010, pp. 175–208.
4. Jenner E. *An inquiry into the causes and effects of the Variolae Vaccinae: a disease discovered in some of the Western Counties of England, particularly Gloucestershire, and known by the name of the Cow Pox*. London: Sampson Low; 1798.
5. Williams G. First steps in the right direction. In *Angel of Death*, pp. 52–70.
6. Gheorgiu M. Antituberculous BCG vaccine: lessons from the past. In Plotkin S, (ed.) *History of vaccine development*, New York: Springer; 2011, pp. 47–50.
7. Flexner S, Lewis P. Experimental poliomyelitis in monkeys: seventh and eighth notes. *J Am Med Assoc*. 1910; 54:1789–95; and 1910; 55:662–70. *J Am Med Ass*. 1910;54, 55:1789–95, 662–70.
8. Römer P. *Die epidemische Kinderlähmung (Heine-Medinsche Krankheit)*. Berlin: Springer; p. 1911.
9. Flexner S, Amoss H. Survival of the poliomyelitis virus for six years in glycerol. *J Exp Med*. 1917;25:539–43.
10. Rhoads C. Immunity following the injection of monkeys with mixtures of poliomyelitis virus and convalescent human serum. *J Exp Med*. 1931;53:115–21.
11. Frozen monkeys. Current Comment, 17 August. *J Am Med Ass*. 1935;105:517.
12. Boy Scout Jamboree canceled because of poliomyelitis. Medical News, 17 August. *J Am Med Ass*. 1935;105:521.
13. Maurice Brodie. Obituaries. *Can Med Ass J*. June 1939, p. 632.
14. Brodie M, Elridge A. The portal of entry and the transmission of the virus of poliomyelitis. *Science*. 1934;79:235–37.
15. Department of Health. *Directive for dealing with outbreaks of smallpox*. London: HM Stationery Office; 1962. p. 24.
16. Brodie M. Active immunization against poliomyelitis. *J Exp Med*. 1932;56:493–505.
17. Brodie M. Active immunization in monkeys against poliomyelitis with germicidally inactivated virus. *Science*. 1934;79:594–95.
18. Paul J. Ill-fated vaccine trials of 1935. In *A history of poliomyelitis*. New Haven: Yale University Press; 1971. p. 256.
19. Schaeffer M. William H. Park (1863–1939): his laboratory and his legacy. *Am J Publ Health*. 1985;75:1296–302.
20. Benison S. *Tom Rivers: reflections on a life in medicine and science*. Cambridge, Mass: MIT Press; 1967, pp. 183–84.
21. Smith J. *Patenting the sun*. New York: William Morrow; 1990, pp. 71–2.
22. Brodie M. Active immunization of children against poliomyelitis with formalin-inactivated virus suspension. *Proc Soc Exp Biol Med*. 1934;32:300–02.
23. Brodie M, Park W. Active immunization against poliomyelitis. *NY State J Med*. 1935;35:815–18.
24. Brodie M, Park W. Active immunization against poliomyelitis. *J Am Med Ass*. 1935;105:1089–93.
25. Specter of Paralysis stalks Carolina. *Literary Digest*. 1935;17.
26. Paul J. *A history of poliomyelitis*, pp. 256–57.
27. Gilbert R. *The tormented President: Calvin Coolidge, death, and clinical depression*. Connecticut: Westport; 2003, pp. 151–63.
28. Kolmer J. Susceptibility and immunity in relation to vaccination in acute anterior poliomyelitis. *J Am Med Ass*. 1935;105:1956–62.
29. Kolmer J. Vaccination against acute anterior poliomyelitis. *Am J Publ Health*. 1936;26:126–35.

30. Benison S. *Tom Rivers*, p. 188.
31. Benison S. *Tom Rivers*, p. 189.
32. Leake J. Poliomyelitis following vaccination against this disease. *J Am Med Soc.* 1935;105:2152.
33. Benison S. *Tom Rivers*, p. 190.
34. Brodie M, Park W. Active immunization against poliomyelitis. *Am J Publ Health.* 1936;26:119–25.
35. Marks H. The 1954 Salk poliomyelitis vaccine field trial. *Clin Trials.* 2011;8:224.
36. Schneider J. Polio vaccination. *JAMA.* 1962;179:988.
37. Burnet F, Macnamara J. Immunological differences between strains of poliomyelitic virus. *Brit J Exp Path.* 1931;12:57–61.
38. Paul J. *A history of poliomyelitis*, pp. 233–37.
39. Ibid., p. 256.
40. Koprowski H. After Pasteur: history of new rabies vaccines. In: Plotkin SA, (ed.) *History of vaccine development,* pp. 103–08.
41. Programme of the American Medical Assocation 86th Annual Meeting, Atlantic City, New Jersey, June 1935. *JAMA.* 1935;104:1739–41.
42. New Remedies. *JAMA.* 1935;104:1605.
43. Ibid., p. 1712, 1728, 1753.
44. Ibid., p. 1728.
45. Flexner S. Concerning active immunization in poliomyelitis. *Science.* 1935;82:420–21.
46. Paul J. *A history of poliomyelitis,* pp. 261–62.
47. Morgan I. Immunization of monkeys with formalin-inactivated poliomyelitis viruses. *Am J Epidemiol.* 1948;48:394–406.
48. Paul J. *A history of poliomyelitis,* p. 261.
49. Jackson R. Migration of gray squirrels. *Science.* 1935;82:549–50.

9 Front Runner

1. Smith J. *Patenting the sun. Polio and the Salk vaccine.* New York: Morrow; 1990, pp. 101–05.
2. Wilson JR. Margin of safety. The story of poliomyelitis vaccine. London: Collins; 1963, p. 75.
3. Smith J. *Patenting the sun,* pp.103–05.
4. Smith J. *Patenting the sun,* pp. 106–07.
5. Bodian D, Morgan I, Howe H. Differentiation of types of poliomyelitis viruses. III. The grouping of fourteen strains into three basic immunological types. *Am J Hyg.* 1949;49:234–40.
6. Wilson J. *Margin of safety.* p. 75.
7. Enders J, Weller T, Robbins F. Cultivation of the Lansing strain of poliomyelitis virus in cultures of various human embryonic tissue. *Science.* 1949;109:85–57.
8. Youngner J. Monolayer tissue cultures I: preparation and standardization of suspensions of trypsin-dispersed monkey kidney cells. *Proc Soc Exp Biol Med.* 1954;85:202–05.
9. Salk JE, Youngner J. Use of color change of phenol red as the indicator in titrating poliomyelitis virus or its antibody in a tissue culture system. *Am J Hyg.* 1954;60:214–21.
10. Kane F. The Second International Poliomyelitis Conference. A report. *Ulster Med J.* 1952;21:49–60.
11. Smith J. *Patenting the sun,* pp. 171–73.

12. Horstmann D, Paul J. The incubation period in human poliomyelitis and its implications. *J Am Med Ass.* 1947;135:11–4.
13. Salk J, Krech U, Youngner J, Al. E. Formaldehyde treatment and safety testing of experimental poliomyelitis vaccines. *Am J Publ Health.* 1954;44:563–70.
14. Committee on Typing of the National Foundation for Infantile Paralysis. Immunological classification of poliomyelitis viruses: a cooperative program for the typing of one hundred strains. *Am J Hyg.* 1951;54:191–274.
15. Beecher H. Ethics and clinical research. *New Eng J Med.* 1966;274:1354.
16. Smith J. Patenting the sun, pp. 136–42.
17. Offit P. *The cutter Incident.* New Haven and London: Yale University Press; 2005, p. 39–40.
18. Salk J, Bennett BL, Lewis LJ, Ward EN, Youngner JS. Studies in human subjects on active immunization against poliomyelitis. I. A preliminary report of experiments in progress. *JAMA.* 1953;151:1081–98.
19. Smith J. *Patenting the sun*, pp. 144–46.
20. The scientist speaks for himself, *CBS broadcast*, 26 March. 1953.
21. The letters of Albert Sabin [Internet]. Albert B. *Sabin Digitization Project*, University of Cincinatti. [cited 20 Febraury 2013]. Available from: http://www.libraries.uc.edu/liblog/2011/10/05/the-albert-b-sabin-digitization-project-sabin-and-salk/
22. Hellman H. *Greatest feuds in medicine: ten of the liveliest disputes ever.* New York: Wiley; 2001, pp. 136–41.
23. Milzer A, Leveson S, Shaughnessy H, et al. Immunogenicity studies in human subjects of trivalent tissue culture poliomyelitis vaccine inactivated by ultraviolet light. *Am J Publ Hlth.* 1954;44:26–33.
24. Offit P. *The cutter Incident*, p. 43.
25. Albert Sabin to Aimes McGuinness, 15 December 1953. [Internet]. Jonas Salk Papers. MSS 1. University Library, Mandeville Department of Special Collections, University of California, San Diego. 1953 [cited 20 Febraury 2013]. Available from: http://drc.libraries.uc.edu/bitstream/handle/2374.UC/672775/general_1985-89_039.pdf
26. Smith J. *Patenting the sun*, p. 88.
27. Meldrum M. 'A calculated risk': the Salk polio vaccine field trials of 1954. *BMJ.* 1998;317:1233–36.
28. Wilson J. *Margin of safety*, pp. 87–8.
29. Oshinsky D. *Polio. An American story.* Oxford: OUP; 2006, p. 160.
30. Marks H. The 1954 Salk poliomyelitis vaccine field trial. In: Goodman S, Marks H, Robinson K, (eds) *100 landmark clinical trials.* John Wiley and Sons; 2011, p. 224.
31. Smith J. *Patenting the sun*, p. 231.
32. Smith J. *Patenting the sun*, p.221–23.
33. Offit P. *The cutter Incident*, pp. 45–8.
34. Oshinsky D. *Polio: an American story.* p. 180.
35. Carter R. *Breakthrough. The saga of Jonas Salk.* Wichell W, (ed.) New York: Trident Press; 1966.
36. Smith J. *Patenting the sun*, p. 253.
37. Ibid., pp. 70–1.
38. Meeting of the Advisory Committee on Technical Aspects of the Poliomyelitis Field Trials, 30–31 January, 1954. Alabama Department of Public Health, Polio

Correspondence, SG014427, Alabama Department of Archives and History. Montgomery, Alabama; 1954.
39. Smith J. *Patenting the sun*, pp.266–67.
40. Carter R. *Breakthrough*, pp. 229–30.
41. Francis Jr TJ. Papers, Bentley Library, University of Michigan, Ann Arbor.
42. Smith J. *Patenting the sun*, pp. 280–85.
43. Ibid., p. 318.
44. Wilson J. *Margin of safety*, pp. 96–8.
45. Smith J. *Patenting the sun*, pp. 313–20.
46. Offit P. *The cutter Incident*, p. 62.
47. Ibid., p. 48.
48. Smith J. *Patenting the sun*, p. 322–27.
49. Fábregas L, Bails J. Youngner proud to be a part of history, still angered by Salk's slight [Internet]. 2005 [cited 14 February 2013]. Available from: http://triblive.com/x/pittsburghtrib/news/specialreports/s_319390.html#axzz2LRG0yO5m
50. Smith J. *Patenting the sun*, pp. 345–47.
51. Wilson J. *Margin of safety*, p. 100.
52. Smith J. *Patenting the sun*, p. 347.
53. Wilson J. *Margin of safety*. p. 86, pp. 115–16.
54. Offit P. *The cutter Incident*, pp. 59–61.
55. Ibid., p. 62.
56. Ibid., pp. 72–5.
57. Ibid., pp. 75.
58. Ibid., pp.63–5.
59. Ibid., p.83.
60. Nathanson N, Langmuir A. The Cutter incident. Poliomyelitis following formaldehyde-inactivated poliovirus vaccination in the United States during the spring of 1955. I. Background. *Am J Hyg*. 1963;78:16.
61. Nathanson N, Langmuir A. The Cutter incident. Poliomyelitis following formaldehyde-inactivated poliovirus vaccination in the United States during the spring of 1955. II. Relationship of poliomyelitis to Cutter vaccine. *Am J Hyg*. 1963;78:29–60.
62. Nathanson N, Langmuir A. The Cutter incident. Poliomyelitis following formaldehyde-inactivated poliovirus vaccination in the United States during the spring of 1955. III. Comparison of the clinical character of vaccinated and contact cases occurring after use of high-rate lots of. *Am J Hyg*. 1963;78:61.
63. Gard S. Discussion on Induction of long-term immunity to paralytic poliomyelitis by use of non-infectious vaccine. *Papers and discussions presented at the Third international Poliomyelitis Conference, Rome, 6–11 September 1954*. Philadelphia: JB Lippincott; 1955.
64. Wilson J. *Margin of safety*, p. 108.
65. Ibid., p. 116.
66. Wilson J. *Margin of safety*, p. 116.
67. Ibid., p. 113–15.
68. Offit P. *The cutter Incident*, p. 119.
69. Wilson J. *Margin of safety*, pp. 117–19.
70. Offit P. *The cutter Incident*, p. 133–43.
71. Ibid., pp. 113–15.
72. CDC Cases & Deaths [Internet]. Available from: http://www.cdc.gov/vaccines/pubs/pinkbook/downloads/appendices/G/cases&deaths

10 Poles Apart

1. Williams G. More fatal than smallpox. In *Angel of Death.* Basingstoke: Palgrave Macmillan; 2010, pp. 256–82.
2. Jenner E. *The origin of the vaccine inoculation.* Massachusetts, London: DN Shury; 1801.
3. Sabin A. Properties and behaviour of orally administered attenuated poliovirus vaccine. *J Am Med Ass* 1957;164:1216–23.
4. Berkovich S, Pickering J, Kibrick S. Paralytic poliomyelitis in Massachussetts, 1959: a study of the disease in a well-vaccinated population. *New Engl J Med.* 1961;264:1323–27.
5. Vaughn R. *Listen to the music: the life of Hilary Koprowski.* Berlin: Springer; 2000.
6. Theiler M, Smith H. The use of yellow fever virus modified by in vitro cultivation for human immunization. *J Exp Med.* 1937;65:787–800.
7. McDade JE. Historical Aspects of Q Fever. In Marrie TJ, (ed.) *Q Fever,* Volume I: The Disease. CRC Press; 1990, p. 8.
8. Koprowski H, Cox HR. Studies on chick embryo adapted rabies vaccine. I. Culture characteristics and pathogenicity. *J Immunol.* 1948;60:533–54.
9. Interview with GW, Ardmore, Pennsylvania, 7 November 2011.
10. Armstrong C. Successful transfer of the Lansing strain of poliomyelitis virus from the cotton rat to the white mouse. *Publ Health Rep.* 1939;54:2302–05.
11. Koprowski H, Plotkin S. History of Koprowski vaccine against poliomyelitis. In Plotkin S, (ed.) *History of vaccine development.* New York: Springer; 2011, pp. 155–66.
12. Maxcy K. *Proceedings of a Round-Table Conference on Immunization in Poliomyelitis, Hershey, Pennsylvania, March 15–17 1951.* New York: National Foundation for Infantile Paralysis; 1951, p. 1.
13. Carter R. *Breakthrough. The saga of Jonas Salk.* Wichell W, (ed.) New York: Trident Press; 1966, p. 171.
14. Benison S. *Tom Rivers: reflections on a life in medicine and science.* Cambridge, Mass: MIT Press; 1967, p. 465–67.
15. Koprowski H, Jervis GA, Norton TW. Immune responses in human volunteers upon oral administration of a rodent-adapted strain of poliomyelitis virus. *Am J Hyg.* 1952; 55:109–116.
16. Poliomyelitis: a new approach. *Lancet.* 1953; 259:552.
17. Plotkin SA. CHAT oral polio vaccine was not the source of human immunodeficiency virus type 1 group M for humans. *Clin Infect Dis.* 2001;32:1068–84.
18. Koprowski H, Norton TW, Hummeler K et al. Immunization of infants with living attenuated poliomyelitis virus. *JAMA.* 1956;162:1281–88.
19. Tyrrell D. *George Williamson Auchinvole Dick. Munk's Roll 1997.* London: Royal College of Physicians; 1997, p. 107.
20. Personal communication. Alan Trudgett, Belfast, 27 May 2011.
21. Personal communication. Joan Williams, Perth, 17 March 2012.
22. Dick G, Dane D. Vaccination against poliomyelitis with live virus vaccines. 3. The evaluation of TN and SM virus vaccines. *BMJ.* 1957;1:70–4.
23. Dane DS, Dick GWA, Connolly JH et al. Vaccination against poliomyelitis with live virus vaccines. 1. A trial of TN type II vaccine. *BMJ.* 1957;1: 59–65.
24. Dick GWA, Dane DS, Fisher OD et al. Vaccination against poliomyelitis with live virus vaccines. 2. A trial of SM type I attenuated poliomyelitis virus vaccine. *BMJ.* 1957;1:65–70.

25. Dane DS, Dick GWA, Briggs M et al. Vaccination against poliomyelitis with live virus vaccines. 8. Changes in Sabin type I oral vaccine virus after multiplication in the intestinal tract. *BMJ.* 1961;2:269–71.
26. Plotkin SA. Recent results of mass immunization against poliomyelitis with Koprowski strains of attenuated live poliovirus. *Am J Publ Health.* 1962; 52: 946–60.
27. Courtois G, Flack A, Jervis GA et al. Preliminary report on mass vaccination of man with live attenuated poliomyelitis virus in the Belgian Congo and Ruanda-Urundi. *BMJ.* 1958;2:187–90.
28. Plotkin S. History of rubella vaccines and the recent history of cell culture. In Plotkin S, (ed.) *History of vaccine development*, p. 226.
29. Don Francis, interview with GW, Berkeley, Gloucs., 20 May 2011.
30. Oshinsky D. *Polio. An American story.* Oxford: OUP; 2005, p. 138.
31. Skeptics Forum [Internet]. Available from: http://tech.dir.groups.yahoo.com/group/skeptics-forum/message/1006
32. Matysiak A. *Albert B. Sabin: the development of an oral vaccine against poliomyelitis.* University of Cincinatti; 2005, pp. 22–33.
33. Benison S. *Tom Rivers*, p. 182.
34. Sabin AB. 'Stained slide' microscopic agglutination test. Application to (1) rapid typing of pneumococci; (2) determination of antibody. *Am J Publ Health* 1929;26: 492–96.
35. Sabin AB, Park WH, Jungeblut CW. Nature of skin reactions produced by heat-inactivated poliomyelitis virus: reaction of persons convalescing from poliomyelitis and of normal persons to intracutaneous injections of heat-inactivated virus. *Arch Int Med.* 1933;51:878–889.
36. Sabin A, Wright W. Acute ascending myelitis following a monkey bite, with the isolation of a virus capable of reproducing the disease. *J Exp Med.* 1934;59(2):115–36.
37. Smith J. *Patenting the sun. Polio and the Salk vaccine.* New York: Morrow; 1990. p. 147–8.
38. Olitsky P, Sabin AB. Comparative effectiveness of various chemical sprays in protecting monkeys against nasally-instilled poliomyelitis virus. *Proc Soc Exper Biol Med.* 1937;36:532–35.
39. Sabin AB to Olitsky PK. Telegram, 12 September 1935, Box OL3, Sabin File 7, Peter Olitsky papers, American Philosophical Society, Philadelphia.
40. Sabin AB, Olitsky PK. *Cultivation of poliomyelitis virus in vitro in human embryonic nervous tissue.* Proceedings of the Society for Experimental Biology and Medicine. Society for Experimental Biology and Medicine (New York, NY). Royal Society of Medicine; 1936, pp. 357–59.
41. Sabin AB, Olitsky PK. Influence of pathway of infection on pathology of olfactory bulbs in experimental poliomyelitis. *Proc Soc Exper Biol Med.* 1936; 35: 300–01.
42. Matysiak A. *Albert B. Sabin*, p. 88.
43. Sabin A. The olfactory bulbs in human poliomyelitis. *Am J Dis Child.* 1940; 60:1313–18.
44. Matysiak A. *Albert B. Sabin.* p. 100.
45. Flexner S. Concerning active immunization in poliomyelitis. *Science.* 1935; 82:420–21.
46. Matysiak A. *Albert B. Sabin*, pp. 101–03.
47. Ibid., p. 110.
48. Oshinsky D. *Polio. An American story*, pp. 142–43.
49. Benison S. *Tom Rivers*, p. 465.

50. Offit P. *The cutter Incident*. New Haven and London: Yale University Press; 2005, p. 47.
51. Sabin A. Present position of immunization against poliomyelitis with live virus vaccines. *BMJ*. 1959;1:663–82.
52. Sabin AB, Boulger LR. History of Sabin attenuated poliovirus oral live vaccine strains. *J. Biol. Stand.* 1973;1:115–18.
53. Matysiak A. *Albert B. Sabin*, p. 138.
54. Rothman D, Rothman S. *The Willowbrook wars. A decade of struggle for social justice.* New York: Harper and Row; 1984.
55. Matysiak A. *Albert B. Sabin*. p. 166.
56. Sabin AB, Hennessen WA, Winsser J. Studies on variants of poliomyelitis virus: experimental segregation and properties of avirulent variants of three immunologic types. *J Exp Med.* 1954;99:551–76.
57. Smith J. *Patenting the sun*. p. 123.
58. Matysiak A. *Albert B. Sabin*, pp. 156–59.
59. Horstmann DH, Melnick JL. Poliomyelitis in chimpanzees. Studies in homologous and heterologous immunity following inapparent infection. *J Exp Med.* 1950; 91:573–97.
60. Matysiak A, Albert B. *Sabin*, pp. 192–200.
61. Chumakov M, Sarmanova E, Bychkova M, Al E. Identification of Kemerovo tick-borne fever virus and its antigenic independence. *Fed Proc Trans Suppl.* 1964;23:852–54.
62. One-dose oral vaccine against polio revealed. *Washington Pos*t, 7 October 1956.
63. Sabin A. Oral poliovirus vaccine: history of its development and use and current challenge to eliminate poliomyelitis from the world. *J Infect Dis.* 1985;151, 423.
64. Sabin AB Role of my cooperation with Soviet scientists in the elimination of polio: possible lessons for relations between the U.S.A. and the USSR. *Perspect Biol Med.* 1987;31:57–64.
65. World Health Organisation, Pan-American Sanitary Bureau. Live Poliovirus Vaccines: papers and discussions held at the First International WHO Conference on live poliovirus vaccines, Washington DC, 12–17 July 1959. Geneva: WHO; 1960.
66. Hooper E. *The River: a journey back to the source of HIV and AIDS*. Boston, Mass: Little, Brown & Co; 1999, p. 212.
67. Horstmann D. The Sabin live poliovirus vaccination trials in the USSR, 1959. *Yale J of Biology and Medicine*. 1991;64:499–512.
68. Horstmann D. Report on a visit to the USSR, Poland and Czechoslovakia to review work on a live poliovirus vaccine, August–October 1959. Unpublished.
69. US Department of Health, Education and Welfare. CDC Poliomyelitis Surveillance Report no. 218. Atlanta, GA: US Dept Health Educ Welfare; 1960.
70. Sabin AB, Pelon W, Spigland I et al. Community-wide use of oral polioivirus vaccine. *Am J Dis Child*. 1961;101:46–55.
71. Matysiak A. *Albert B. Sabin*, pp. 255–56.
72. Pan-American Health Organisation. Scientific publication no. 50. Washington DC: PAHO; 1960.
73. US Congress. House of Representatives Committee on manufacture of live virus polio vaccine and results of utilization of killed virus polio vaccine, 16–17 March 1961. Washington DC: US Government; 1961, pp. 13–52.
74. Albrecht R, Bigwood D, Levy W, Al. E. Oral poliovirus vaccination program in Central New York State, 1961. *Publ Health Rep.* 1963;78:403–12.
75. Matysiak A. *Albert B. Sabin*, pp. 265–66.
76. Ibid., p. 261, 268.

77. Hoffert WR, Schneider NJ, Sigel MM et al. Serological aspects of live polio vaccine evaluation in Dade County Fla. *Am J Publ Health* 1062; 52:961–69.
78. Matysiak A. *Albert B. Sabin*, p. 268.
79. Matysiak A. *Albert B. Sabin*, p. 2.

11 In the Opposite Corner

1. Jenner E. *The origin of the vaccine inoculation*. London: DN Shury; 1801.
2. Williams G. The most beautiful discovery or a disastrous illusion? In *Angel of Death: the story of smallpox*. Basingstoke: Palgrave Macmillan; 2010, pp. 283–305.
3. Ruijs WLM, Hautvast JLA, Van Ijzendoorn G, Van Ansem WJG, Van der Velden K, Hulscher M. How orthodox protestant parents decide on the vaccination of their children: a qualitative study. *BMC Public Health.* 2012;12:408.
4. Ruijs WLM, Hautvast JLA, Van der Velden K, De Vos S, Knippenberg H, Hulscher M. Religious subgroups influencing vaccination coverage in the Dutch Bible belt: an ecological study. *BMC Public Health.* 2011;11:102.
5. Patriarca PA, Sutter RW, Oostvogel PM. Outbreaks of paralytic poliomyelitis, 1976–1995. *J Infect Dis.* 1997;175:S165–72.
6. White FMM, Lacey BA, Constance PDA. An outbreak of poliovirus infection in Alberta, 1978. *Can J Public Health.* 1981;329–44.
7. Oostvogel PM, Van Wijngaarden JK, Van der Avoort HG, Mulders MN, Conyn-Van Spaendonck MA, Rumke HC. *Poliomyelitis outbreak in an unvaccinated community in The Netherlands, 1992–93.* Lancet 1994, pp. 665–70.
8. Isolation of wild poliovirus type 3 among members of a religious community objecting to vaccination – Alberta, Canada, 1993. *MMWR* (Morb Mortal Wkly Rep) 1993;42:337–39.
9. Six members of Apostolic sect in Zimbabwe given suspended prison sentences for medically neglecting children. *Zimbabwe News,* 10 October 2010.
10. Mudzwiti M. Dad jailed for measles deaths. *Sunday Times of Zimbabwe,* 17 August 2010.
11. Zimbabwe sect member murders wife after she tried to vaccinate children against deadly measles outbreak. *The Herald,* Zimbabwe, 24 September 2010.
12. Williams G. An affront to the rights of man. *In Angel of Death*, pp. 235–56.
13. Little L. *Crimes of the cowpox ring. Some moving pictures thrown on the dead wall of official silence*. Minneapolis: The Liberator Pubishing Co.; 1906.
14. Williams G. *More fatal than smallpox. Angel of Death*, pp. 256–82.
15. McBean E. *The poisoned needle.* Mokelumne Hill, California: Mokelumne Hill Press; 1993.
16. The hidden dangers in polio vaccines. Ibid.
17. Palmer D, Palmer B. *The science of chiropractic: its principles and adjustments.* Davenport, Iowa: Palmer School of Chiropractic; 1906.
18. Lee J. *Poliomyelitis in the Lone Star State: a brief examination in rural and urban communities*. MA Thesis, Texas State University; 2005, p. 61.
19. Shepherd D. *Homoeopathy in epidemic diseases*. Saffron Walden: CW Daniel Company; 1967, pp. 74–49.
20. Henderson DA, White J, Morris L, Langmuir A. Paralytic disease associated with oral polio vaccines. *JAMA*. 1964;190:41–8.
21. Sabin A. Commentary on report on oral poliomyelitis vaccines. *JAMA*. 1964;190:52–5.
22. Sweet B, Hilleman M. The vacuolating virus, SV40. *Proc Soc Exp Biol Med.* 1960;105:420–27.

23. Bookchin D, Schumacher J. *The virus and the vaccine*. New York: St Martin's Press; 2004, pp. 82–3.
24. Smith CE, Simpson DI, Bowen ET, Zlotnic I. Fatal human disease from vervet monkeys. *Lancet*. 1967;2:1119–21.
25. Sharp P, Li W. Understanding the origins of AIDS viruses. *Nature*. 1988;336:315.
26. Bowen-Jones E, Pendry S. The threats to primates and other mammals from the bushmeat trade in Africa and how this could be diminished. *Oryx*. 1999;33:233–47.
27. LeBrun A, Cerf J, Gelfand H, et al. Vaccination with the CHAT strain of type 1 attenuated poliomyelities virus in Leopoldville, Belgian Congo 1. Description of the city, its history of poliomyelitis, and the plan of the vaccination campaign. *Bull World Health Organ*. 1960;22:203–13.
28. Elswood B, Stricker R. Polio vaccines and the origin of AIDS (letter). *Res Virol*. 1993;144:175–77.
29. Curtis T. The origin of AIDS. *Rolling Stone*, 19 March 1992, pp. 54–60.
30. Koprowski H. AIDS and the polio vaccine. *Science*. 1993;257:1024.
31. Hooper E. *The River: a journey back to the source of HIV and AIDS*. Boston, Mass: Little, Brown & Co; 1999. p. 253.
32. Ibid., p. 198.
33. Ibid., p. 411; pp. 457–88.
34. Martin B. The politics of a scientific meeting: the Origin-of-AIDS debate at the Royal Society. *Politics & the Life Sciences*. 2011, pp. 119–30.
35. Poinar H, Kuch M, Pääbo S. Molecular analyses of oral polio vaccine samples. *Science*. 2001;292:743–44.
36. Hahn BH, Shaw GM, De Cock KM, Sharp PM. AIDS as a zoonosis: scientifc and public health implications. *Science*. 2000;287:607–14.
37. Plotkin SA. CHAT oral polio vaccine was not the source of human immunodeficiency virus type 1 group M for humans. *Clinical Infectious Diseases*. 2001;32:1068–84.
38. AIDS origins [Internet]. Available from: http://www.aidsorigins.com
39. Roberts J. Polio: the virus and the vaccine. *The Ecologist*. 2004;35–45.
40. West J. *Pesticides and polio: a critique of the scientific literature [Internet]*. The Weston A. Price Foundation; 2002. Available from: http://www.westonaprice.org /environmental-toxins/pesticides-and-polio
41. Sharp P, Bailes E, Chaudhuri R, Rodenburg C, Santiago M, Hahn B. The origins of acquired immune deficiency syndrome viruses: where and when? *Philos Trans R Soc Lond B Biol Sci*. 2001;356:867–76.
42. Worobey M, Santiago M, Keele B, et al. Origin of AIDS: contaminated polio vaccine theory refuted. *Nature*. 2004;428:820.
43. Sandler B. *Diet prevents polio*. Milwaukee, WI: The Lee Foundation for Nutritional Research; 1951.
44. Kapp C. Surge in polio spreads alarm in northern Nigeria. Rumours about vaccine safety in Muslim-run states threaten WHO's eradication programme. *Lancet*. 2003;362:1631–32.
45. Jegede AS. What led to the Nigerian boycott of the polio vaccination campaign? *PLoS Med*. 2007;4:e73.
46. Clements CJ, Greenough P, Shull D. How vaccine safety can become political – The example of polio in Nigeria. *Current Drug Safety*. 2006;1:117–19.
47. Warraich H. Religious opposition to polio vaccination. *Emerg Infect Dis*. 2009;15:978.
48. Health workers boycott polio vaccination in Bajaur Agency. *Daily Times of Pakistan.*, 20 February 2007, http://www.dailytimes.com.pk/default.asp?page=2007\02\20\story_20-2-2007_pg7_29; 2007.

49. Peal Harbor: Mother of All Conspiracies [Internet]. Available from: http://whatreallyhappened.com/WRHARTICLES/pearl/www.geocities.com/Pentagon/6315/pearl.html
50. Images Of Poliomyelitis: Franklin D. Roosevelt At Campobello [Internet]. Available from: http://www.whale.to/vaccine/west2.html
51. Yanchunas D. The Roosevelt Dime at 60. *COINage Magazine*, February 2006.

12 Loose Ends and a Gordian Knot

1. Racaniello VR, Baltimore. D. Molecular cloning of poliovirus cDNA and determination of the complete nucleotide sequence of the viral genome. *Proc. Natl. Acad. Sci. USA*. 1981;78:4887–91.
2. Mueller S, Wimmer E, Cello J. Polio and poliomyelitis: a tale of guts, brains and an accidental event. *Virus Res*. 2005;111:175–93.
3. Mendelsohn CL, Wimmer E, Racaniello. VR. Cellular receptor for poliovirus: molecular cloning, nucleotide sequence, and expression of a new member of the immunoglobulin superfamily. *Cell*. 1989;56:855–65.
4. Koike S, Taya C, Kurata T, Al. E. Transgenic mice susceptible to poliovirus. *Proc. Natl. Acad. Sci. USA*. 1991;88:951–55.
5. Toyada H, Kohara M, Kataoka Y, Al E. Complete nucleotide sequences of all three poliovirus serotype genomes. Implication for genetic relationship, gene function and antigenic determinants. *J. Mol. Biol*. 1984;174:561–85.
6. Nomoto A, Omata T, Toyoda H, Al. E. Complete nucleotide sequence of the attenuated poliovirus Sabin 1 strain genome. *Proc. Natl. Acad. Sci. USA*. 1982;79: 5793–97.
7. Cello J, Paul A, Wimmer E. Chemical synthesis of poliovirus cDNA; generation of infectious virus in the absence of natural template. *Science*. 2002;297:1016–18.
8. Jiang P, Faase J, Toyoda H, Al. E. Evidence for emergence of diverse polioviruses from C-cluster coxsackie A viruses and implications for global poliovirus eradication. *Proc Natl Acad Sci USA*. 2007;104:9457–62.
9. Albert Sabin to Basil. O'Connor, 1 August 1955, File 12, Box 4, Correspondence, National Foundation for Infantile Paralysis Series, CHMC.
10. Sabin AB, Boulger LR. History of Sabin attenuated poliovirus oral live vaccine strains. *J. Biol. Stand*. 1973;1:115–18.
11. Dr. Salk promotes polio vaccine in UK. From BBC News [Internet]. BBC News. 1955 [cited 25 February 2013]. Available from: http://news.bbc.co.uk/onthisday/hi/dates/stories/may/5/newsid_2510000/2510495.stm
12. Más Lago P. Eradication of poliomyelitis in Cuba: a historical perspective. *Bull WHO*. 1999;77:681–87.
13. CDC Cases & Deaths [Internet]. Available from: http://www.cdc.gov/vaccines/pubs/pinkbook/downloads/appendices/G/cases&deaths
14. Henderson DA, White J, Morris L, Langmuir A. Paralytic disease associated with oral polio vaccines. *JAMA*. 1964;190:41–8.
15. Sabin A. Commentary on report on oral poliomyelitis vaccines. *JAMA*. 1964;190:52–5.
16. Vaccines and biologicals. Report of the interim meeting of the technical consultative group on the global eradication of poliomyelitis, Geneva 9–11 November 2002. Geneva: WHO; 2003.
17. Sutter R, Kew O, Cochi S. Poliovirus vaccine – live. In: Plotkin S, Orenstein W, (eds) *Vaccines*. 4th ed. Philadelphia: Saunders; 2004, pp. 692–95.

18. Sabin AB. Oral poliovirus vaccine: history of its development and use and current challenges to eliminate poliomyelitis from the world. *J Infect Dis* 1985; 151:420–36.
19. Halsey NA, Pinto J, Espinosa-Rosales F, Al. E. Search for poliovirus carriers among people with primary immune deficiency diseases in the United States, Mexico, Brazil, and the United Kingdom. *Bull World Health Organ*. 2004;82:3–8.
20. Guillot S, Caro V, Cuervo N et al. Natural genetic exchange between vaccine and wild poliovirus strains in humans. *J Virol*. 2000;74:8434–43.
21. Kew O, Morris-Glasgow V, Landaverde M, et al. Outbreak of poliomyelitis in Hispaniola associated with circulating type 1 vaccine-derived poliovirus. *Science*. 2002;296:356–59.
22. Robertson SE, Traverso HP, Drucker JA, al. et. Clinical efficacy of a new, enhanced-potency, inactivated poliovirus vaccine. *Lancet*. 1988;1:897–99.
23. RW Lovett MD. Obituary. *BMJ*. 1924;2:85.
24. Rous P. Simon Flexner, 1863–1946. Obituary notices of the Fellows of the Royal Society. 1949;6:409–45.
25. Ibid., p. 409.
26. Professor Simon Flexner at Cambridge. Lecture on epidemic poliomyelitis. *BMJ*. 1938;1:468–69.
27. Rous P. *Simon Flexner*, p. 427.
28. Paul J. *A history of poliomyelitis*. New Haven: Yale University Press; 1971, p. 125.
29. Rogers N. *Dirt and disease. Polio before FDR*. New Brunswick, NJ: Rutgers University Press; 1990, pp. 30–49.
30. Goldman A, Schmalstieg E, Freeman D, et al. What was the cause of Franklin Delano Roosevelt's paralytic illness? *J Med Bio*. 2003;11:232–40.
31. Yanchunas D. The Roosevelt Dime at 60. *COINage Magazine*, February 2006.
32. Whitman A. Basil O'Connor, polio crusader, dies. New York Times. 1972 Mar 10;
33. Benison S. *Tom Rivers: reflections on a life in medicine and science*. Cambridge, Mass: MIT Press; 1967, pp. 282–83.
34. Smith J. *Patenting the sun. Polio and the Salk vaccine*. New York: Morrow; 1990, p. 373–74.
35. Kalter S. At 57, Françoise Gilot recalls life with Picasso but enjoys it with scientist Jonas Salk. People [Internet]. 1979 Jul; Available from: http://www.people.com/people/archive/article/0,,20074227,00.html
36. Matysiak A, *Albert B. Sabin: the development of an oral vaccine against poliomyelitis*. PhD thesis, Cincinnati: University of Cincinnati; 2005, p. 143.
37. Bowen E. Albert Sabin. *People [Internet]*. 1984 July; Available from: http://www.people.com/people/archive/article/0,,20088192,00.html
38. Koprowski H. Obituary: Albert B. Sabin (1906–1993). *Nature*. 1993;362:499.
39. Hooper E. *The River: a journey back to the source of HIV and AIDS*. Boston, Mass: Little, Brown & Co; 1999, p. 253.
40. Elswood B, Stricker R. Polio vaccines and the origin of AIDS (letter). *Res Virol*. 1993;144:175–177.
41. Vaughn R. *Listen to the music: the life of Hilary Koprowski*. Berlin: Springer; 2000.
42. Interview with GW, Ardmore, Pennsylvania, 7 November 2011.
43. Personal communication, Dr. Preben G. Berthelsen, Copenhagen, 15 January 2013.
44. Enders J, Weller T, Robbins F. Cultivation of the Lansing strain of poliomyelitis virus in cultures of various human embryonic tissue. *Science*. 1949;109:85–57.
45. Firkin B, Whitworth J. *Dictionary of medical eponyms*. New York & London: Parthenon Publishing; 1996, p. 469.
46. Offit P. *The cutter Incident*, New Haven and London: Yale University Press; 2005, p. 129–30.

47. Interview with Edward R. Murrow, CBS Television, 12 April 1955.
48. Norrby E, Prusiner S. Polio and Nobel Prizes: looking back 50 years. *Ann Neurol.* 2007;61:391.
49. Robbins F. History of poliomyelitis vaccines. In Plotkin S, Orenstein W, (eds) *Vaccines*, 4th ed. Philadelphia: Saunders; 2004, p. 26.
50. Fenner F, Henderson D, Arita I, et al. *Smallpox and its eradication.* Geneva: WHO; 1988, pp. 842–52.
51. International notes. Certification of poliomyelitis eradication – the Americas. *MMWR.* 1994;43:720–27.
52. Patriarca P, Wright P, John T. Factors affecting the immunogenicity of oral poliovirus vaccine in developing countried: review. *Rev Infect Dis.* 1991;13:926–39.
53. Fact sheet 114: Poliomyelitis [Internet]. WHO. Available from: http://www.who.int/mediacentre/factsheets/fs114/en/
54. Riaz H, Rehman A. Polio vaccination workers gunned down in Pakistan. *Lancet Infect Dis.* 2013;13:120.
55. Boone J. Doctor who helped US in search for Osama Bin Laden jailed for 33 years. Guardian. 2012 May 23;
56. Fenner F, et al. *Smallpox and its eradication.* 1988, p. 224.
57. Plotkin SA, Vidor E. Poliovirus – inactivated. In Plotkin SA, Orenstein W, (eds) *Vaccines.* 4th ed. Philadelphia: Saunders; 2004, p. 438.

13 Looking Forward to a Retrospective?

1. Vashisht N, Puliyel J. Polio programme: let us declare victory and move on. *Indian J Med Ethics.* 2012;9:114–17.
2. Does polio still exist? Is it curable? [Internet]. 2012 [cited 23 February 2013]. Available from: http://www.who.int/features/qa/07/en/index.html
3. Benison S. *Tom Rivers: reflections on a life in medicine and science.* Cambridge, Mass: MIT Press; 1967, p. 95.
4. Benison S. *Tom Rivers*, footnotes on, pp. 95, 96.
5. Liver therapy in asthma. *Bull Méd Paris.* 1936;50:815–32.
6. Intravenous injections of charcoal. *Lancet.* 1936;2:1266.
7. Marks H. The 1954 Salk poliomyelitis vaccine field trial. *Clin Trials.* 2011;8:224.
8. Schuster E. Fifty years later: the significance of the Nuremberg Code. *New Engl J Med.* 1997;337:1436–40.
9. Paul J. *A history of poliomyelitis.* New Haven: Yale University Press; 1971, pp. 408–09.
10. See Foreign Letters in J Am Med Ass, e.g. 'Sterilization to improve the race', 1933;101:866–67; and 'From our regular correspondent in Berlin', 1934;102:630.
11. Beecher H. Ethics and clinical research. *New Engl J Med.* 1966;274:1354–60.
12. Rothman D, Rothman S. *The Willowbrook wars. A decade of struggle for social justice.* New York: Harper and Row; 1984.
13. Jones J. *Bad blood: the Tuskegee syphilis experiment.* New York: The Free Press; 1993.
14. Presidential Apology [Internet]. CDC. 1997 [cited 23 Febraury 2013]. Available from: http://www.cdc.gov/tuskegee/clintonp.htm
15. MacLean A. *The Satan Bug.* London: Fontana Books; 1964.
16. Williams G. *Angel of Death: the story of smallpox.* Basingstoke: Palgrave Macmillan; 2010. p. 361–5.
17. Editorial. Poliomyelitis. *Postgrad Med J.* 1949;25:1.
18. Offit P. *The cutter Incident.* New Haven and London: Yale University Press; 2005. p. 32.

19. MacLean A. *Fear is the Key*. London: Fontana Books; 1961.
20. Fenner F, Henderson D, Arita I, et al. *Smallpox and its eradication*. Geneva: WHO; 1988, pp. 842–52.
21. Save the Guinea Worm Foundation [Internet]. http://www.deadlysins.com/guineaworm/truth.htm, accessed 21 February 2013.
22. Sutter R, Kew O, Cochi S. Poliovirus vaccine – live. In: Plotkin S, Orenstein W, (eds) *Vaccines*. 4th ed. Philadelphia: Saunders; 2004, pp. 692–95.
23. Jiang P, Faase J, Toyoda H, Al. E. Evidence for emergence of diverse polioviruses from C-cluster coxsackie A viruses and implications for global poliovirus eradication. *Proc Natl Acad Sci USA*. 2007;104:9457–62.
24. Cello J, Paul A, Wimmer E. Chemical synthesis of poliovirus cDNA; generation of infectious virus in the absence of natural template. *Science*. 2002;297:1016–18.

Bibliography

Adamson JD, Moody JP, Peart AFW, Smillie RA, Wilt JC, Wood WJ. Poliomyelitis in the Arctic. *Canadian Medical Association Journal*. 1949;61:339–48.

Annual Report of the Surgeon General of the Public Health Service of the United States, for the fiscal year 1917. Washington DC; 1917. p. 30–1, 188–203.

Armstrong C, Harrison WT. Prevention of intranasally-inoculated poliomyelitis of monkeys by instillation of alum into the nostrils. *Public Health Reports (1896–1970)*. 1935;725–30.

Armstrong C. Successful transfer of the Lansing strain of poliomyelitis virus from the cotton rat to the white mouse. *Publ Health Rep*. 1939;54:2302–05.

Armstrong C. The experimental transmission of poliomyelitis to the eastern cotton rat. *Publ Health Rep*. 1939;54:1719–23.

Axelsson P. 'Do not eat those apples; they've been on the ground!': polio epidemics and preventive measures, Sweden 1880s–1940s. *Asclepio*. 2009;61:26–7.

Badham J. Paralysis in childhood: Four remarkable cases of suddenly induced paralysis in the extremities occurring in children without any apparent cerebral or cerebrospinal lesion. *London Medical Gazette*. 17:215.

Bates P, Pellow J. *Horizontal man: the story of Paul Bates*. London: Longmans, Green and Co.; 1964.

Bayer P. The management of the acute phase of poliomyelitis. *Postgraduate Medical Journal*. 1949;25:9–12.

Bazin H. Pasteur and the birth of vaccines made in the laboratory. In: Plotkin SA, (ed.) *History of vaccine development*. New York: Springer; 2011.

Beecher H. Ethics and clinical research. *NEJM*. 1966;274:1354.

Benison S. *Tom Rivers: reflections on a life in medicine and science*. Cambridge, Mass: The MIT Press; 1967.

Bernier R. Some observations on poliomyelitis lameness surveys. *Rev Infect Dis*. 1984;6:S371–75.

Biskind MS. Statement On Clinical Intoxication From DDT And Other New Insecticides Presented before the Select Committee to Investigate the Use of Chemicals in Food Products to the United States House of Representatives, Washington DC, Westport, CT, 12 December 1950.

Bodian D, Morgan I, Howe H. Differentiation of types of poliomyelitis viruses. III. The grouping of fourteen strain into three basic immunological types. *Am J Hyg*. 1949;49:234–40.

Bodian D, Paffenbarger R. Poliomyelitis infection in households: frequency of viremia and specific antibody response. *Am J Hyg*. 1954;60:83–98.

Bodian D. A reconsideration of the pathogenesis of poliomyelitis. *Am J Hyg*. 1952; 55:414.

Brodie M, Elridge A. The portal of entry and the transmission of the virus of poliomyelitis. *Science*. 1934;79:235–37.

Brodie M, Park W. Active immunization against poliomyelitis. *J Am Med Ass*. 1935; 105:1089–93.

Brodie M, Park WH. Active immunization against poliomyelitis. *Am J Publ Health.* 1936;26:119–25.

Brodie M. Active immunization against poliomyelitis. *J Exp Med.* 1932;56:493–505.

Brodie M. Active immunization in monkeys against poliomyelitis with germicidally inactivated virus. *Science.* 1934;79:594–95.

Burnet F, Macnamara J. Immunological differences between strains of poliomyelitic virus. *Brit J Exp Path.* 1931;12:57–61.

Caverly CS. Preliminary report of an epidemic of paralytic disease, occurring in Vermont, in the Summer of 1894. *Yale Medical Journal.* 1894;1:1–20.

Cello J, Paul A, Wimmer E. Chemical synthesis of poliovirus cDNA; generation of infectious virus in the absence of natural template. *Science.* 2002;297:1016–18.

Charcot JM, Joffroy A. Cas de paralysie infantile spinale avec lésions des cornes antérieures de la substance grise de la moëlle épinière. *Arch Physiol norm Pathol.* 1870;3:134–40.

Cohn V. *Sister Kenny: the woman who challenged the doctors.* Minneapolis: University of Minnesota Press; 1973.

Committee on Typing of the National Foundation for Infantile Paralysis. Immunological classification of poliomyelitis viruses: a cooperative program for the typing of one hundred strains. *Am J Hyg.* 1951;54:191–274.

Courtois G, Flack A, Jervis GA et al. Preliminary report on mass vaccination of man with live attenuated poliomyelitis virus in the Belgian Congo and Ruanda-Urundi. *BMJ.* 1958;2:187–90.

De Kruif, P. *Microbe Hunters.* London: Jonathan Cape; 1927.

Dick GWA, Dane DS. Vaccination against poliomyelitis with live virus vaccines. 3. The evaluation of TN and SM virus vaccines. *BMJ.* 1957;1:70–4.

Dochez A, Peabody F, Draper G. *A clinical study of acute poliomyelitis.* New York: Monographs of the Rockefeller Institute for Medical Research; 1912.

Draper G. *Infantile paralysis.* 2nd ed. New York: Appleton-Century; 1935, pp. 52–73.

Draper G. The nature of the human factor in infantile paralysis. *Am J Med Sci* 1932;184:111–16.

Drinker P, Shaw LA. Apparatus for prolonged administration of artificial respiration: design for adults and children. *J Clin Invest.* 1929;7:229–34.

Editorial on poliomyelitis. *Postgrad Med J.* 1949;25:8.

Enders J, Weller T, Robbins F. Cultivation of the Lansing strain of poliomyelitis virus in cultures of various human embryonic tissue. *Science.* 1949;109:85–57.

Faber H. Flexner and Noguchi's globoid red ferrings: A footnote to the history of poliomyelitis. *Journal of Infectious Diseases.* 1971;124:231–34.

Fenner F, Henderson D, Arita I, et al. *Smallpox and its eradication.* Geneva: WHO; 1988, pp. 842–52.

Finch J, Klug A. Structure of poliomyelitis virus. *Nature* 1959; 183:1709–14. *Nature* 1959;183:1709–14.

Flexner S, Lewis P. Experimental poliomyelitis in monkeys: seventh and eighth notes. *J Am Med Assoc.* 1910;54:1789–95; and 1910;55:662–70. *JAMA.* 1910;54, 55:1789–95, 662–70.

Flexner S, Lewis P. The nature of the virus of epidemic poliomyelitis. *JAMA.* 1909;53:592–4.

Flexner S, Lewis PA. Experimental epidemic poliomyelitis in monkeys. *The Journal of Experimental Medicine.* 1910;12:227–32.

Flexner S, Noguchi H. Experiments on the cultivation of the microörganism causing epidemic poliomyelitis. *J Exp Med.* 1913;27:461–85.

Flexner S. Concerning active immunization in poliomyelitis. *Science.* 1935;82:420–21.

Frauenthal H, Manning J. *A manual of infantile paralysis*. Philadelphia: FA Davis; 1914.

Gallagher H. *FDR's splendid deception*. St Petersburg, Florida: Vandamere Press; 1999.

Gard S. Discussion on Induction of long-term immunity to paralytic poliomyelitis by use of non-infectious vaccine. *Papers and discussions presented at the Third international Poliomyelitis Conference, Rome, 6–11 September 1954*. Philadelphia: JB Lippincott; 1955.

Goldman A, Schmalstieg E, Freeman D, et al. What was the cause of Franklin Delano Roosevelt's paralytic illness? *J Med Biogr*. 2003;11:232–40.

Guillot S, Caro V, Cuervo N et al. Natural genetic exchange between vaccine and wild poliovirus strains in humans. *J Virol*. 2000;74:8434–43.

Hammon WM, Coriell LL, Wehrle PF: Evaluation of Red Cross gamma globulin as a prophylactic agent for poliomyelitis. IV. Final report of results based on clinical diagnosis. *JAMA*. 1953;151:1272–85.

Hardy A. Poliomyelitis and the neurologists: The view from England, 1896–1966. *Bulletin of the History of Medicine*. 1997;71:249–72.

Heine JC. *Beobachtungen über Lähmungszustande der unteren Extremitäten und deren Behandlung*. Stuttgart: Kohler; 1840.

Hellman H. *Greatest Feuds in Medicine: ten of the liveliest disputes ever*. New York: Wiley; 2001, pp. 136–41.

Henderson D, White J, Morris L, Langmuir A. Paralytic disease associated with oral polio vaccines. *JAMA*. 1964;190:41–8.

Hooper E. *The River: a journey back to the source of HIV and AIDS*. Boston, Mass: Little, Brown & Co; 1999.

Horstmann D, McCollum R, Mascola A. Viremia in human poliomyelitis. *J Exp Med*. 1954;99:355–69.

Horstmann D, Paul J. The incubation period in human poliomyelitis and its implications. *JAMA*. 1947;135:11–4.

Horstmann D. The Sabin live poliovirus vaccination trials in the USSR, 1959. *Yale J of Biology and Medicine*. 1991;64:499–512.

Horstmann DH, Melnick JL. Poliomyelitis in chimpanzees. Studies in homologous and heterologous immunity following inapparent infection. *J Exp Med*. 1950; 91:573–97.

Howe H, Bodian D. Poliomyelitis in the chimpanzee: a clinical pathological study. *Bull Johns Hopk Hosp*. 1941;69:149–81.

Howe HA, Bodian D, Morgan IM. Subclinical poliomyelitis in the chimpanzee and its relation to alimentary reinfection. *American J Hyg*. 1950;51:85–108.

Jacobi MP. *Infantile spinal paralysis*. Pepper's System of Medicine, Philadelphia, Lea Brothers & Co. 1886;5:1113–64.

Jegede AS. What led to the Nigerian boycott of the polio vaccination campaign? *PLoS Med*. 2007;4:e73.

Jenner E. *An inquiry into the causes and effects of the Variolae Vaccinae: a disease discovered in some of the Western Counties of England, particularly Gloucestershire, and known by the nme of the Cow Pox*. London: Sampson Low; 1798.

Jenner E. *The origin of the vaccine inoculation*. London: DN Shury, 1801.

Jiang P, Faasew JA, Toyoda H et al. Evidence for emergence of diverse polioviruses from C-cluster coxsackie A viruses and implications for global poliovirus eradication. *Proc Natl Acad Sci USA*. 2007;104:9457–9462.

Jones J. *Bad blood: the Tuskegee syphilis experiment*. New York: Free Press; 1981.

Kenny E. *My battle and victory: history of the discovery of poliomyelitis as a systemic disease*. London: Robert Hale, 1955.

Kling C, Levaditi C, Lepine P. La *pénétration* du virus *poliomyélitique à travers* la muqueuse du tube digestif chez le singe et sa conservation dans l'eau. *Bull Acad de Méd*. 1929;102:158–65.

Kling CA, Wernstedt WE, Pettersson A. Recherches sur le mode de propagation de la paralysie infantile épidémique (maladie de Heine-Medin). *Zeit Immunitatforsch.* 1912;12:316–23.

Kolmer J. Vaccination against acute anterior poliomyelitis. *Am J Publ Health.* 1936; 26:126–35.

Koprowski H, Jervis GA, Norton TW. Immune responses in human volunteers upon oral administration of a rodent-adapted strain of poliomyelitis virus. *Amer J Hyg.* 1952;55:109–116.

Koprowski H, Norton TW, Hummeler K et al. Immunization of infants with living attenuated poliomyelitis virus. *JAMA.* 1956;162:1281–88.

Koprowski H. Obituary: Albert B. Sabin (1906–1993). *Nature.* 1993;362:499.

Kramer SD, Geer HA, Himes AT. The use of continuous intravenous administration of hypotonic sodium chloride (Retan treatment) in acute experimental poliomyelitis in monkeys. *J Immunol.* 1942;44:175–94.

Landsteiner K, Popper E. Uebertragung der Poliomyelitis acuta auf Affen. *Z Immunitätsforsch.* 1909;2:377–90.

Lassen H. The epidemic of poliomyelitis in Copenhagen, 1952. *Proc R Soc Med.* 1954;47:67–72.

Leake J. Poliomyelitis following vaccination against this disease. *J Am Med Soc.* 1935;105:2152.

Lee J. *Poliomyelitis in the Lone Star State: a brief examination in rural and urban communities.* MA thesis, San Marcos: Texas State University; 2005, pp. 41–4.

Levaditi C. Le virus poliomyélitique. In Lépine P, (ed.) *Les ultravirus des maladies humaines.* Paris: Maloine; 1938, p. 572.

Levine M, Neal J, Park W. Relation of physical characteristics to susceptibility to anterior poliomyelitis. *JAMA.* 1933;100:160–62.

Lewin P. *Infantile paralysis.* Anterior poliomyelitis. Philadelphia: WB Saunders & Co.; 1941.

Little L. *Crimes of the Cowpox Ring. Some moving pictures thrown on the dead wall of official silence.* Minneapolis: The Liberator Publishing Co; 1906.

Logrippo G. Concerning the nature of the globoid bodies. *J Bacteriol.* 1936;31:245–53.

Lovett RW. The occurrence of infantile paralysis in Massachusetts in 1907. *The Boston Medical and Surgical Journal.* 1908;159:131–39.

MacLean A. *The Satan Bug.* London: Fontana Books; 1964.

Marks H. The 1954 Salk poliomyelitis vaccine field trial. *Clin Trials.* 2011;8:224.

Marks H. The 1954 Salk poliomyelitis vaccine field trial. In: Goodman S, Marks H, Robinson K, (eds) *100 landmark clinical trials.* John Wiley and Sons; 2011, p. 224.

Martin W. Poliomyelitis in England and Wales, 1947–1950. *Brit J Soc Med.* 1951;5: 136–46.

Matysiak A. *Albert B. Sabin: the development of an oral vaccine against poliomyelitis.* PhD thesis, Cincinnati: University of Cincinnati; 2005.

McBean E. *The Poisoned Needle.* Mokulemne Hill, California: Mokulemne Hill Press, 1993.

Medin O. Ueber eine Epidemie von spinaler Kinderlähmung. *Verhandl. d. 10. Internatl. med. Kongr.* 1891;37–47.

Mendelsohn, C. L., Wimmer E, and V. R. Racaniello. Cellular receptor for poliovirus: molecular cloning, nucleotide sequence, and expression of a new member of the immunoglobulin superfamily. *Cell.* 1989. 56:855–65.

Morgan I. Immunization of monkeys with formalin-inactivated poliomyelitis viruses. *Am J Epidemiol.* 1948;48:394–406.

Morgan JP. The Jamaica ginger paralysis. *JAMA.* 1982;248:1864–67.

Mueller S, Wimmer E, Cello J. Polio and poliomyelitis: a tale of guts, brains and an accidental event. *Virus Res.* 2005;111:175–93.

Nathanson N, Bodian D. Experimental poliomyelitis following intramuscular virus injection. *Bulletin of the Johns Hopkins Hospital.* 1961;103:308.

Nathanson N, Langmuir A. The Cutter incident. Poliomyelitis following formaldehyde-inactivated poliovirus vaccination in the United States during the spring of 1955. I. Background. *Am J Hyg.* 1963;78:16.

Nathanson N, Langmuir A. The Cutter incident. Poliomyelitis following formaldehyde-inactivated poliovirus vaccination in the United States during the spring of 1955. II. Relationship of poliomyelitis to Cutter vaccine. *Am J Hyg.* 1963;78:29–60.

Nathanson N, Langmuir A. The Cutter incident. Poliomyelitis following formaldehyde-inactivated poliovirus vaccination in the United States during the spring of 1955. III. Comparison of the clinical character of vaccinated and contact cases occurring after use of high-rate lots of. *Am J Hyg.* 1963;78:61.

Nathanson N, Martin JR. The epidemiology of poliomyelitis: enigmas surrounding its appearance, epidemicity, and disappearance. *American Journal of Epidemiology.* 1979;110:672–92.

Nobel Prize Nomination and Selection of Medicine Laureates [Internet]. [accessed 22 January 2013]. Available from: http://www.nobelprize.org/nobel_prizes/medicine/nomination.

Nomoto, A., T. Omata, H. Toyoda et al. Complete nucleotide sequence of the attenuated poliovirus Sabin 1 strain genome. *Proc Natl Acad Sci USA.* 1982;79:5793–5797.

Norrby E, Prusiner S. Polio and Nobel Prizes: looking back 50 years. *Ann Neurol.* 2007;61:391.

North B. *Something to lean on. The first sixty years of the British Polio Fellowship, 1939–1999.* South Ruislip, Middlesex: British Polio Fellowship, 1999.

Offit PA. *The Cutter Incident: how America's first polio vaccine led to the growing vaccine crisis.* New Haven, CT and London: Yale University Press, 2005.

Oostvogel PM, Van Wijngaarden JK, Van der Avoort HG, Mulders MN, Conyn-Van Spaendonck MA, Rumke HC. *Poliomyelitis outbreak in an unvaccinated community in The Netherlands, 1992–1993.* Lancet, 1994;344:665–70.

Oshinsky D. *Polio. An American story.* Oxford: OUP; 2006.

Osler W. *The principles and practice of medicine.* New York: D Appleton and Co; 1892.

Patriarca PA, Sutter RW, Oostvogel PM. Outbreaks of paralytic poliomyelitis, 1976–1995. *J Infect Dis.* 1997;175:S165–72.

Paul H. *The control of diseases (social and communicable).* Edinburgh & London: E&S Livingstone; 1964.

Paul JR. *A history of poliomyelitis.* New Haven and London: Yale University Press, 1971.

Peart AF, Rhodes AJ. An outbreak of poliomyelitis in Canadian Eskimos in wintertime. *Canadian journal of public health.* 1949;40:405–19.

Plotkin SA, Carp RI, Graham AF. The polioviruses of man. *Ann NY Acad Sci.* 1960;101:357–89.

Plotkin SA. CHAT oral polio vaccine was not the source of human immunodeficiency virus type 1 group M for humans. *Clinical Infectious Diseases* 2001;32:1068–84.

Poliomyelitis. *Papers and discussion presented at the First International Poliomyelitis Conference.* Philadelphia: JB Lippincott; 1949.

Poliomyelitis. *Papers and discussion presented at the Third International Poliomyelitis Conference.* Philadelphia: JB Lippincott; 1955.

Porter R. *The greatest benefit to mankind: a medical history of humanity from antiquity to the present.* London: Fontana Press; 1997.

Racaniello, V. R., and D. Baltimore. Molecular cloning of poliovirus cDNA and determination of the complete nucleotide sequence of the viral genome. *Proc Natl Acad Sci. USA.* 1981: 78:4887–91.

Reagan RL, Schenck DM, Brueckneir AL. Morphological observations by electron microscopy of the Brunhilde strain of poliomyelitis virus. *Journal of Infectious Diseases*. 1950;86:295–96.

Report of Queensland Commission. Treatment of infantile paralysis by Sister Kenny's method. *BMJ*. 1938;1:350–51.

Retan G. The development of the therapeutic use of forced perivascular (spinal) drainage. *J Am Med Ass*. 1935;105:1333–40.

Retan G. The treatment of acute poliomyelitis by intravenous injection of hypotonic salt solution. *J Ped*. 1937;11:647–64.

Rissler J. Zur Kenntniss der Veränderungen des Nervensystems bei Poliomyelitis anterior acuta. *Nordiskt Medicinskt Arkiv*. 1888;29:1–63.

Rogers N. *Dirt and disease: polio before FDR*. Brunswick, NJ: Rutgers University Press, 1992.

Rogers N. Dirt, flies, and immigrants: explaining the epidemiology of poliomyelitis, 1900–1916. *J Hist Med All Sci*. 1989;44:486–505.

Rous P. Simon Flexner, 1863–1946. *Obituary notices of the Fellows of the Royal Society*. 1949;6:409–45.

Ruijs WLM, Hautvast JLA, Van der Velden K, De Vos S, Knippenberg H, Hulscher M. Religious subgroups influencing vaccination coverage in the Dutch Bible belt: an ecological study. *BMC Public Health*. 2011;11:102.

Ruijs WLM, Hautvast JLA, Van Ijzendoorn G, Van Ansem WJG, Van der Velden K, Hulscher M. How Orthodox Protestant parents decide on the vaccination of their children: a qualitative study. *BMC Public Health*. 2012;12:408.

Sabin A. Commentary on report on oral poliomyelitis vaccines. *JAMA*. 1964;190:52–5.

Sabin A. Pathogenesis of poliomyelitis. *Science*. 1956;123:1151.

Sabin A. The olfactory bulbs in human poliomyelitis. *Am J Dis Child*. 1940;60:1313–18.

Sabin AB, Hennessen WA, Winsser J. Studies on variants of poliomyelitis virus: experimental segregation and properties of avirulent variants of three immunologic types. *J Exp Med*. 1954;99:551–76.

Sabin AB, Olitsky PK. Cultivation of poliomyelitis virus in vitro in human embryonic nervous tissue. *Proc Soc Exp Biol Med*. 1936;34:357–59.

Sabin AB, Ward R. The natural history of human poliomyelitis I. Distribution of virus in nervous and non-nervous tissues. *J Exp Med*. 1941;73:771–93.

Sabin AB. Oral poliovirus vaccine: history of its development and use and current challenges to eliminate poliomyelitis from the world. *J Infect Dis*. 1985; 151:420–36.

Sabin AB. Present position of immunization against poliomyelitis with live virus vaccines. *BMJ*. 1959;1:667–68.

Sabin AS. Oral poliovirus vaccine: histotry of its development and use and current challenge to eliminate poliomyelitis from the world. *J Infect Dis*. 1985;151:423.

Sabin AB: Properties and behaviour of orally administered attenuated poliovirus vaccine. *JAMA*. 1957;164:1216–23.

Sabin, AB, and Boulger LR. History of Sabin attenuated poliovirus oral live vaccine strains. *J Biol Stand*. 1973;1:115–18.

Sachs B. The present-day conception of acute anterior poliomyelitis. *J Bone Joint Surg Am*. 1908;S2–6:173–83.

Salk J, Bennett BL, Lewis LJ, Ward EN, Youngner JS. Studies in human subjects on active immunization against poliomyelitis. I. A preliminary report of experiments in progress. *JAMA*. 1953;151:1081–98.

Salk JE, Youngner J. Use of color change of phenol red as the indicator in titrating poliomyelitis virus or its antibody in a tissue culture system. *Am J Hyg*. 1954;60:214–21.

Sandler B. *Diet prevents polio.* Milwaukee, WI: The Lee Foundation for Nutritional Research; 1951.

Schultz E, Gebhardt L. Zinc sulphate prophylaxis in poliomyelitis. *JAMA.* 1937;108: 2182–87.

Schuster E. Fifty years later: the significance of the Nuremberg Code. *New Engl J Med.* 1997;337:1436–40.

Scobey R. *The poison cause of poliomyelitis.* Statement to the US House of Representatives, Washington DC, April 1952.

Scobey RR. The poison cause of poliomyelitis and obstructions to its investigation. *Arch Pediatr.* 1952;69:172–93.

Scobey RR. Food poisoning as the etiological factor in poliomyelitis. *Arch Pediatr.* 1946;63:322–54.

Scobey RR. Is human poliomyelitis caused by an exogenous virus? *Science.* 1954;51:117.

Sharp P, Bailes E, Chaudhuri R, Rodenburg C, Santiago M, Hahn B. The origins of acquired immune deficiency syndrome viruses: where and when? *Philos Trans R Soc Lond B Biol Sci.* 2001;356:867–76.

Smallman-Raynor MR, Cliff AD. *Poliomyelitis: A World Geography: emergence to eradication.* Oxford University Press, USA; 2006.

Smith J. *Patenting the sun.* New York: William Morrow; 1990.

Sutter R, Kew O, Cochi S. Poliovirus vaccine – live. In: Plotkin S, Orenstein W, (eds) *Vaccines.* 4th ed. Philadelphia: Saunders; 2004, pp. 692–95.

Sweet B, Hilleman M. The vacuolating virus, SV40. *Proc Soc Exp Biol Med.* 1960;105:420–27.

Toyada, H., M. Kohara, Y. Kataoka et al. Complete nucleotide sequences of all three poliovirus serotype genomes. Implication for genetic relationship, gene function and antigenic determinants. *J Mol.Biol.* 1984. 174:561–85.

Underwood M. *A treatise on the diseases of children, with general directions for the management of infants from the birth, London.* Matthews; 1789.

Underwood M. *A treatise on the diseases of children, with general directions for the management of infants from the birth.* Philadelphia: Gibson; 1793.

Vaughn R. *Listen to the music: the life of Hilary Koprowski.* Berlin: Springer, 2000.

Von Heine J. *Spinale Kinderlähmung.* Stuttgart: Cotta; 1860.

Vulpian A. Cas d'atrophie musculaire graisseuse datant de l'enfance. Lésions des cornes antérieures de la substance grise de la moëlle épinière. *Arch Physiol Norm Pathol.* 1870;3:316–25.

Wickman I. *Beiträge zur Kenntnis der Heine-Medinschen Krankheit: Poliomyelitis acuta und verwandter Erkrankungen.* Berlin: S. Karger; 1907.

Wickman OI. *Studien über Poliomyelitis acuta: Zugleich ein Beitrag zur Kenntnis der Myelitis acuta.* Berlin: S. Karger; 1905.

Williams G. *Angel of Death: the story of smallpox.* Basingstoke: Palgrave Macmillan; 2010.

Wilson G. Faults and Fallacies in Microbiology. *J Gen Microbiol.* 1959;21:1–15.

Wilson JR.*Margin of safety: the story of poliomyelitis vaccine.* London: Collins; 1963.

Youngner JR. Monolayer tissue cultures I: preparation and standardization of suspensions of trypsin-dispersed monkey kidney cells. *Proc Soc Exp Biol Med.* 1954;85:202–05.

Index

Printed and bound by
CPI Group (UK) Ltd, Croydon, CR0 4YY